Lecture Notes in Computer Science   6480

Commenced Publication in 1973
Founding and Former Series Editors:
Gerhard Goos, Juris Hartmanis, and Jan van Leeuwen

W0192938

Marina L. Gavrilova   C.J. Kenneth Tan
Edward David Moreno (Eds.)

Transactions on Computational Science XI

Special Issue on Security in Computing, Part II

 Springer

Editors-in-Chief

Marina L. Gavrilova
University of Calgary, Department of Computer Science
2500 University Drive N.W., Calgary, AB, T2N 1N4, Canada
E-mail: mgavrilo@ucalgary.ca

C.J. Kenneth Tan
Exascala Ltd.
Unit 9, 97 Rickman Drive, Birmingham B15 2AL, UK
E-mail: cjtan@exascala.com

Guest Editor

Edward David Moreno
DCOMP/UFS - Federal University of Sergipe
Aracaju/SE, Brazil
E-mail: edwdavid@gmail.com

Library of Congress Control Number: 2010939851

CR Subject Classification (1998): C.2, K.6.5, D.4.6, E.3, K.4.4, G.2

ISSN	0302-9743 (Lecture Notes in Computer Science)
ISSN	1866-4733 (Transaction on Computational Science)
ISBN-10	3-642-17696-8 Springer Berlin Heidelberg New York
ISBN-13	978-3-642-17696-8 Springer Berlin Heidelberg New York

springer.com

© Springer-Verlag Berlin Heidelberg 2010
Printed in Germany

Typesetting: Camera-ready by author, data conversion by Scientific Publishing Services, Chennai, India
Printed on acid-free paper   06/3180

LNCS Transactions on Computational Science

Computational science, an emerging and increasingly vital field, is now widely recognized as an integral part of scientific and technical investigations, affecting researchers and practitioners in areas ranging from aerospace and automotive research to biochemistry, electronics, geosciences, mathematics, and physics. Computer systems research and the exploitation of applied research naturally complement each other. The increased complexity of many challenges in computational science demands the use of supercomputing, parallel processing, sophisticated algorithms, and advanced system software and architecture. It is therefore invaluable to have input by systems research experts in applied computational science research.

Transactions on Computational Science focuses on original high-quality research in the realm of computational science in parallel and distributed environments, also encompassing the underlying theoretical foundations and the applications of large-scale computation. The journal offers practitioners and researchers the opportunity to share computational techniques and solutions in this area, to identify new issues, and to shape future directions for research, and it enables industrial users to apply leading-edge, large-scale, high-performance computational methods.

In addition to addressing various research and application issues, the journal aims to present material that is validated – crucial to the application and advancement of the research conducted in academic and industrial settings. In this spirit, the journal focuses on publications that present results and computational techniques that are verifiable.

Scope

The scope of the journal includes, but is not limited to, the following computational methods and applications:

- Aeronautics and Aerospace
- Astrophysics
- Bioinformatics
- Climate and Weather Modeling
- Communication and Data Networks
- Compilers and Operating Systems
- Computer Graphics
- Computational Biology
- Computational Chemistry
- Computational Finance and Econometrics
- Computational Fluid Dynamics

- Computational Geometry
- Computational Number Theory
- Computational Physics
- Data Storage and Information Retrieval
- Data Mining and Data Warehousing
- Grid Computing
- Hardware/Software Co-design
- High-Energy Physics
- High-Performance Computing
- Numerical and Scientific Computing
- Parallel and Distributed Computing
- Reconfigurable Hardware
- Scientific Visualization
- Supercomputing
- System-on-Chip Design and Engineering

Editorial

The Transactions on Computational Science journal is part of the Springer series *Lecture Notes in Computer Science*, and is devoted to the gamut of computational science issues, from theoretical aspects to application-dependent studies and the validation of emerging technologies.

The journal focuses on original high-quality research in the realm of computational science in parallel and distributed environments, encompassing the facilitating theoretical foundations and the applications of large-scale computations and massive data processing. Practitioners and researchers share computational techniques and solutions in the area, identify new issues, and shape future directions for research, as well as enable industrial users to apply the techniques presented.

The current volume is devoted to Security in Computing (Part 2), and is edited by Edward David Moreno. It is comprised of 14 selected papers that represent the diverse applications and designs being addressed today by the security and cryptographic research community. This special issue is devoted to state-of-the-art research on security in computing and includes a broad spectrum of applications such as new architectures, novel hardware implementations, cryptographic algorithms, and security protocols.

We would like to extend our sincere appreciation to Special Issue Guest Editor Edward David Moreno for his dedication and insights in preparing this high-quality special issue. We also would like to thank all authors for submitting their papers to the special issue, and to all associate editors and referees for their valuable work. We would like to express our gratitude to the LNCS editorial staff of Springer, in particular Alfred Hofmann, Ursula Barth, and Anna Kramer, who supported us at every stage of the project.

It is our hope that the fine collection of papers presented in this special issue will be a valuable resource for Transactions on Computational Science readers and will stimulate further research into the vibrant area of computational science applications.

October 2010

Marina L. Gavrilova
C.J. Kenneth Tan

Security in Computing:
Research and Perspectives, Part II
Special Issue Guest Editor's Preface

In an increasingly connected world, security has become an essential component of modern information systems. Our ever-increasing dependence on information implies that the importance of information security is growing. Several examples of security applications are present in everyday life such as mobile phone communication, internet banking, secure e-mail, data encryption, etc.

The thrust of embedded computing has both diversified and intensified in recent years as the focus on mobile computing, ubiquitous computing, and traditional embedded applications has begun to converge. A side effect of this intensity is the desire to support sophisticated applications such as speech recognition, visual feature recognition, and secure wireless networking in a mobile, battery-powered platform. Unfortunately these applications are currently intractable for the embedded space.

Another consideration is related to mobile computing, and, especially, security in these environments. The first step in developing new architectures and systems that can adequately support these applications is to obtain a precise understanding of the techniques and methods that come close to meeting the needs of security, performance, and energy requirements; with an emphasis on security aspects.

This special issue brings together high-quality and state-of-the-art contributions on security in computing. The papers included in this issue deal with some hot topics in the security research sphere: new architectures, novel hardware implementations, cryptographic algorithms and security protocols, and new tools and applications. Concretely, the special issue contains 14 selected papers that represent the diverse applications and designs being addressed today by the security and cryptographic research community.

As a whole, this special issue provides a vision on research and new perspectives in security research. With authors from around the world, these articles bring us an international sample of significant work.

The title of the first paper is "SEAODV: A Security Enhanced AODV Routing Protocol for Wireless Mesh Networks", by Celia Li, Zhuang Wang, and Cungang Yang. In this paper, the authors propose SEAODV, which is a security enhanced version of AODV (the Ad hoc On Demand Distance Vector). The AODV routing algorithm is a routing protocol designed for ad hoc mobile networks. The authors use Blom's key pre-distribution scheme to establish keys to ensure that every two nodes in the network uniquely share the pairwise keys. So, SEAODV adds secure AODV extensions to the original AODV routing messages, and it is lightweight and computationally efficient, since only symmetric cryptographic operations are involved. Finally, the authors carry out several tests and conclude that SEAODV offers superior performance in terms of computation cost and route acquisition latency as compares with two other secure routing protocols, ARAN and SAODV.

In the second contribution, which is entitled "Auto-Generation of Least Privileges Access Control Policies for Applications Supported by User Input Recognition", Sven Lachmund and Gregor Hengst present means to auto-generate least privileges access control policies for applications. The authors introduce and discuss two approaches: extending a static analysis approach by user input recognition, and introducing a new runtime approach on user input recognition that is based on information tracking and aspect-oriented programming. They show a third solution, combining the other two contributions with some of the existing work. A prototype in Java is implemented, and it is shown that the total number of aspects is kept within a manageable range, proving feasibility and scalability.

In the third contribution, which is entitled "Impossibility Results for RFID Privacy Notions", Frederik Armknecht, Ahmad-Reza Sadeghi, Alessandra Scafuro, Ivan Visconti, and Christian Wachsmann focus on the security and privacy model proposed by Paise and Vaudenay (PV-model) and investigate some subtle issues such as tag corruption aspects. The PV-model is one of the most comprehensive RFID security and privacy models up to date since it captures many aspects of real world RFID systems and aims at abstracting most previous works in a single concise framework. The authors point out subtle weaknesses and deficiencies in the PV-model.

In the fourth contribution, which is entitled "Implementation of Multivariate Quadratic Quasigroups for Wireless Sensor Networks", authored by Ricardo José Menezes Maia, Paulo Sérgio Licciardi Messeder Barreto, and Bruno Trevizan de Oliveira, a new approach to solving the problem of providing PKCs (public key cryptosystems) in WSNs (wireless sensor networks) is proposed. The authors use nesC and focus on modules for the encryption and decryption of a 160-bit MQQ (Multivariate Quadratic Quasigroup) algorithm that have been implemented on platforms TelosB and MICAz sensors.

In the fifth contribution, which is entitled "Hardware Architectures for Elliptic Curve Cryptoprocessors Using Polynomial and Gaussian Normal Basis Over GF(2^233)", by Vladimir Tujillo-Olaya and Jaime Velasco-Medina, the authors present two elliptic curve cryptoprocessors suitable for the computation of point multiplication over GF(2m) using Gaussian Normal Basis (GNB) and polynomial basis (PB). In this case, efficient hardware architectures are designed for finite field multiplication, in order to select the best implementation for the cryptoprocessor design. These multiplier architectures incorporate bit-serial and digit-serial algorithms. The authors designed cryptoprocessors using the same tools, FPGA, finite field m size and hardware description language, and show that the GNB cryptoprocessor presents a higher performance than the PB cryptoprocessor (but the scalability is an advantage of polynomial basis). So, they conclude that the designed cryptoprocessors present a high performance, use a small area, and provide a good time-area trade-off.

In the sixth paper "GPU Accelerated Cryptography as an OS Service", by Owen Harrison and John Waldron, the authors provide a standard method of access to the latest GPU crypto acceleration work to all components within an operating system, with minimal loss of performance. For this process, the authors have seen that the GPU can be effectively integrated into the OCF with careful design of a driver consisting of a kernelspace OCF driver and a userspace daemon. The results obtained show that there is an average overhead of 3.4% when using the OCF for AES over a standalone implementation. In the context of RSA-1024 we see that there is a very low 0.3% average overhead when compared with a standalone version.

In the seventh paper, which is entitled "From a Generic Framework for Expressing Integrity Properties to a Dynamic MAC Enforcement for Operating Systems", Patrice Clemente, Jonathan Rouzaud-Cornabas, and Christian Toinard propose a novel framework for expressing integrity requirements associated with direct or indirect activities, mostly in terms of information flows. The paper presents formalization for the major integrity security properties of the literature. The framework enables the user to formalize the major integrity security properties. The authors use a MAC enforcement mechanism implementing that algorithm to effectively and efficiently control those system calls.

In the eighth paper, which is entitled "Performance Issues on Integration of Security Services", Fábio Dacêncio Pereira and Edward David Moreno project and develop a SSIL (Security Services Integrated Layer) for allowing the integration of security services. They investigate the efficiency and impact of behavioral models used in SSIL specialized for detecting anomalies and conclude that there are advantages in having a set of security services in a single integrated system, since the possible fragility of a service can be compensated by others.

In the ninth paper "Statistical Model Applied to NetFlow for Network Intrusion Detection", André Proto, Leandro A. Alexandre, Maira L. Batista, Isabela L. Oliveira and Adriano M. Cansian present a proposal for event detection in computer networks using statistical methods and the analysis of NetFlow data flows. The aim is to use this proposal to monitor a computer network perimeter, detecting attacks in the shortest time possible through anomalies identification in traffic and alerting the administrator when necessary. The authors carry out a test for monitoring the system to four services widely used by users on the Internet: FTP, SSH, SMTP, and HTTP. Finally, the authors conclude that this methodology can be used for events detection in large-scale networks.

The paper "J-PAKE: Authenticated Key Exchange Without PKI", authored by Feng Hao and Peter Ryan, proposes a protocol called J-PAKE, which authenticates a password with zero-knowledge and then subsequently creates a strong session key if the password is correct. The authors show that the protocol fulfills some properties, and show how to effectively integrate the ZKP (Zero-Knowledge Proof) into the protocol design and achieve good efficiency. The authors have compared their approach with de facto internet standard SSL/TLS, and demonstrate that J-PAKE has comparable computational efficiency to the EKE and SPEKE schemes with clear advantages on security. For this reason it is more lightweight in password authentication.

The paper "Distance Based Transmission Power Control scheme for Indoor Wireless Sensor Networks", by P.T.V. Bhuvaneswari, V. Vaidehi, and M. Agnes Saranya, proposes a new scheme that is an energy efficient RSS (Received Signal Strength) based distributed localization algorithm and Distance Based Transmission Power Control (DBTPC). The proposed localization algorithm consists of two stages, namely, distance estimation and coordinates estimation, and with this it improves the accuracy in relative coordinate estimation and minimizes the energy cost incurred for transmitting information between nodes.

The paper "A Novel Feature Vectors Construction Approach for Face Recognition", by Paul Nicholl, Afandi Ahmad, and Abbes Amira, discusses a novel feature vectors construction approach for face recognition using DWT (Discrete Wavelet Transform). The authors evaluate the method using different classes of tests. The first set of experiments performed focused on the choice of DWT features. It is revealed that, where direct coefficient values were used for recognition, the LL quadrant provided the best results. The second set of tests were designed to identify which wavelet filters were the most effective at extracting features for face recognition with the specified database. Finally, the authors investigated two approaches, PMA and ORA, for the feature threshold, and their results show that the PMA is an ineffective approach, with recognition accuracy decreasing by an average of 0.025% from the results obtained without DWT coefficient selection.

The paper "An Extended Proof-Carrying Code Framework for Security Enforcement", authored by Heidar Pirzadeh, Danny Dubé, and Abdelwahab Hamou-Lhadj, proposes a novel approach to solving the proof size problem while avoiding a significant increase of the TCB. The authors present an extension to a traditional proof-carrying code framework in which proofs tend to be too large to transmit. For this, their approach is based on the innovative idea of sending a program that generates the proof instead of the proof itself. Finally, they developed a virtual machine called the VEP (Virtual Machine for Extended PCC - Proof-Carrying Code) that runs on the consumer's side and that is responsible for running the proof generator program.

The last paper in this special issue, "NPT Based Video Watermarking with Non-overlapping Block Matching" by S.S. Bedi, Shekhar Verma, and Geetam S. Tomar, presents a NTP (Naturalness Preserving Transform) that is based on collusion and compression resistant watermarking techniques for video. Their experimental results confirm several theoretical findings and demonstrate the resistance of the technique to temporal frame averaging, additive noise, and JPEG based compression.

Finally, we sincerely hope that this special issue stimulates your interest in the many subjects surrounding the area of security. The topics covered in the papers are timely and important, and the authors have done an excellent job of presenting their different approaches. Regarding the reviewing process, our referees (integrated by recognized researchers from the international community) made a great effort to evaluate the papers. We would like to acknowledge their effort in providing us the excellent feedback at the right time. So, we wish to thank all the authors and reviewers. To conclude, we would also like to express our gratitude to the Editor-in-Chief of TCS, Dr. Marina L. Gavrilova, for her advice, vision, and support.

September 2010   Edward David Moreno

LNCS Transactions on Computational Science – Editorial Board

Table of Contents – Part II

Table of Contents – Part I

SEAODV: A Security Enhanced AODV Routing Protocol for Wireless Mesh Networks

Celia Li[1], Zhuang Wang[2], and Cungang Yang[2]

[1] Department of Computer Science and
Engineering, York University
[2] Department of Electrical and Computer
Engineering, Ryerson University

Abstract. In this paper, we propose a Security Enhanced AODV routing proto-
col (SEAODV) for wireless mesh networks (WMN). SEAODV employs
Blom's key pre-distribution scheme to compute the pairwise transient key
(PTK) through the flooding of enhanced HELLO message and subsequently
uses the established PTK to distribute the group transient key (GTK). PTK and
GTK authenticate unicast and broadcast routing messages respectively. In
WMN, a unique PTK is shared by each pair of nodes, while GTK is shared secretly between the node and all its one-hop neighbours. A message authentication code (MAC) is attached as the extension to the original AODV routing
message to guarantee the message's authenticity and integrity in a hop-by-hop
fashion. Security analysis and performance evaluation show that SEAODV is
more effective in preventing identified routing attacks and outperforms ARAN
and SAODV in terms of computation cost and route acquisition latency.

Keywords: AODV, Wireless Mesh Networks, MAC, Routing, Hop-by-Hop
Authentication.

1 Introduction

Wireless Mesh Network (WMN) [1][2][3] is composed of an infrastructure component (infrastructure mesh) as well as many ad hoc (client mesh) networks. The routing
and security requirements of these separate components may differ substantially due
to different characteristics of the separate mesh components. Routing protocols in ad
hoc networks generally fall into two categories: proactive and on-command protocols.
Proactive protocols are table-driven. Nodes store routing information about neighbours and periodically broadcast routing information to keep the routing tables up-to-
date. On-demand protocols involve a sender node establishing a route on-demand
only when data is needed to be sent. Although a profusion of routing protocols have
been proposed for ad hoc networks, the difference in characteristics between infrastructure and ad hoc nodes is evidence that routing protocols designed for ad hoc
networks cannot be directly applied for wireless mesh networks. Hybrid routing
[4][5][6] (e.g., HWMP) seems to be one of the promising answers to the question
of what is the trend in WMNs' routing. In hybrid routing, proactive routing is

M.L. Gavrilova et al. (Eds.): Trans. on Comput. Sci. XI, LNCS 6480, pp. 1–16, 2010.

specifically used for traffics flow to the mesh portal and on-demand routing is selected for intra-mesh traffic. Most existing routing protocols including HWMP have been designed with performance as a priority and have neglected to incorporate a significant amount of security issues of routing protocols in WMN. The aim of this paper is therefore to design a secure version of the on-demand part of HWMP and used it to securely discover a route between any pair of mesh routers in the network.

We propose SEAODV, a security enhanced version of AODV. Choosing AODV as our protocol's footstone is due to its simplicity, maturity, popularity and availability in the research over the past few years. As our main contributions, we utilize PTK and GTK to protect the unicast and broadcast routing message respectively to ensure that the route discovery process between any two nodes in WMNs is secure. We apply BLOM's key pre-distribution scheme in conjunction with the enhanced HELLO message to establish the PTK and use the established PTK to distribute GTK to the node's one-hop neighbours throughout the entire network. We also identify various attacking scenarios specifically in AODV and present security analysis to prove that our proposed SEAODV is able to effectively defend against most of those identified attacks. Moreover, our scheme is lightweight and computationally efficient since only symmetric cryptographic operations (e.g., MAC) are involved.

The rest of the paper is organized as follows. Section 2 discusses related work. Details of our SEAODV protocol is presented in section 3. Section 4 identifies various potential attack scenarios in SEAODV and presents the security analysis. The performance evaluation is explained in section 5. Finally, section 6 concludes the paper.

2 Related Work

Research work of secure routing protocols has been explored on ad hoc networks. However, most of them are not efficient or still vulnerable to various types of attacks. Furthermore, they are designed for ad hoc networks and do not provide specific security features (e.g., hop-by-hop authentication) for mesh networks.

SAODV [4] is a secure variant of AODV [5]. SAODV defends against impersonation attacks and modification of hop count and sequence number attacks. However, it does not support hop-by-hop authentication. The intermediate nodes on the path cannot verify the authenticity of the messages from their predecessors.

ARAN (Authenticate Routing for Ad hoc Networks) [6] adopts digital signature to ensure hop-by-hop authentication and routing message integrity; However, it experiences significant computation overheads and route acquisition latency. Each node needs to verify signatures every time it receives the signed messages, removes the certificate and signature of its predecessor, use its own private key to sign the message originally broadcast by the source and appends its own certificate before rebroadcasting it to its one-hop neighbours. Further, both SAODV and ARAN cannot defend against the DoS attack. When a node receives a routing message, it has to verify signatures. Adversaries can utilize this flaw to inject a large number of faked signed messages and to intentionally make the designated node repeatedly verify the signatures.

Ariande is a secure on-demand source routing protocol [7] in which TESLA[8], digital signatures and MAC are employed to ensure source authentication. The drawback of this scheme is that the route request message is not authenticated until it reaches the destination which makes it possible for adversary to initiate route request flooding attack. In endairA [9], a variant of Ariande, instead of signing the route request message, intermediate nodes sign the route reply. This scheme experiences less cryptographic computation, but is still vulnerable to malicious route request flooding attack.

Hybrid Wireless Mesh Protocol (HWMP) [10][11][12] allows two MPs (Mesh Point) to communicate using peer-to-peer paths. This approach is primarily used by nodes that experience a changing environment and when there is no root MP configured. While the proactive tree building mode can be an efficient choice for nodes in a fixed network topology, HWMP does not address the security issues and suffers from different type of attacking scenarios that will be described in section 4.

LHAP [13] is a lightweight transparent authentication protocol for ad hoc networks. LHAP uses TESLA to maintain the trust relationship among nodes, which is not a realistic approach due to the delayed key disclosure period in TESLA. Furthermore, in LHAP, simply attaching the traffic key right after the raw message is not secure since the traffic key has no relationship with the message being transmitted.

3   Security Enhanced AODV Routing

SEAODV requires each node in the network to maintain two key hierarchies: a broadcast key hierarchy and a unicast key hierarchy. The broadcast key hierarchy of a node stores the broadcast keys shared with its one hop neighbours while the unicast key hierarchy stores all their secret pairwise keys. Besides, SEAODV employs the Blom's key pre-distribution scheme and an enhanced HELLO message to compute a pairwise transient key (PTK), which is subsequently used to distribute the group transient key (GTK). PTK and GTK are employed to secure the unicast and broadcast routing messages respectively. In addition, PTK and GTK provide SEAODV with a hop-by-hop authentication routing solution in which routing messages are protected at every hop during the route setup process.

3.1   Enhanced HELLO Message

The variables and notations given in the table 1 illustrate the notations of the proposed secure routing protocol and relevant cryptography operations.

3.1.1   HELLO RREQ
HELLO message in AODV is broadcasted to its one-hop neighbours to maintain the updated local connections. In SEAODV, each node embeds Blom's column of the public G matrix [9] into its HELLO RREQ message [14]. Since each column of the public known matrix G can be regenerated by applying a seed from each node, every node only needs to store the seed in order to exchange their public information of the matrix G.

Assume node A is the originator of the HELLO RREQ message, the enhanced HELLO RREQ is shown in the following format.

$$EN_{HRREQ}: [M_{type}, M_{ID}, IP*, Seed_G, IP_A, OSN_A]$$

Node A needs to broadcast the enhanced HELLO RREQ to its one-hop neighbours periodically whenever the node A needs

1. To announce to its one-hop neighbours that A is in active mode
2. To let its one-hop neighbours know A's seed of the column of public Matrix G
3. To trigger its one-hop neighbours to send their encrypted GTK back to A

Table 1. Variables and Notations

EN_HRREQ	Enhanced hello RREQ message	OSN_A	Originator sequence number of node A
EN_HRREP	Enhanced hello RREP message	DSN_B	Destination sequence number of node B
GTK_A	Group Transient Key of Node A	IP_A	IP address of node A
PTK_A	Pairwise Transient Key of Node A	IP*	Broadcast IP address
{d}GTK_A	Encryption of data d with key GTK_A	NC_A	Nonce issued by node A
{d}PTK_A	Encryption of data d with key PTK_A	Seed_G_A	Seed of column of Public Matrix G, Node A
MAC(K,M)	Computation of MAC over message M with key K	P_Row_A_A[ ]	Row of Private Matrix A, Node A
T	Timestamp	M_type	Message type
M	All the elements before the MAC field in SEAODV routing message	M_ID	Message ID

3.1.2 HELLO RREP

In the enhanced HELLO RREP, the hop count field is replaced with zero and the lifetime field is equal to the value of ALLOWED_HELLO_LOSS × HELLO_INTERVAL.

Assume node B, a one-hop neighbour of node A, has received the HELLO RREQ. To respond A, B unicasts an enhanced HELLO RREP message which is shown as follows

$$EN_{HRREP}: [M_{type}, NC_B, IP_B, DSN_B, IP_A, Lifetime, \{GTK_B, IP_B, NC_B\}PTK_B]$$

B needs to send the enhanced HELLO RREP message to node A under the following situations

1. Acknowledges node A that B has already received the public key of A.
2. Derives the secret pairwise key it shares with A and use this shared key to encrypt its own group key and unicasts it back to A.

Upon receiving the HELLO RREP from B, A can confirm that there is a bi-directional wireless link between A and B. A can also know the GTK of B by using A's PTK to decrypt the received HELLO RREP.

3.2 Exchange Public Seed_G and GTK Using Enhanced HELLO Message

During the key pre-distribution phase, every legitimate node in the wireless mesh network stores its public known Seed_G (seed of the column of public G matrix) and the corresponding private row of the generated A matrix. The Seed_G and GTK key of each node is exchanged among nodes in the WMN using the enhanced HELLO RREQ and HELLO RREP message and the exchange process are depicted in the following three steps.

Step 1: Exchange of Seed_G of the public G matrix

Suppose B is a one-hop neighbour of A in Figure 1 where EN_HRREQ represents the enhanced hello RREQ message and EN_HRREP denotes the enhanced hello RREP message, when A and B exchange their Seed_G, they picks them up from their key pool, and broadcasts the enhanced HELLO RREQ to their one-hop neighbours.

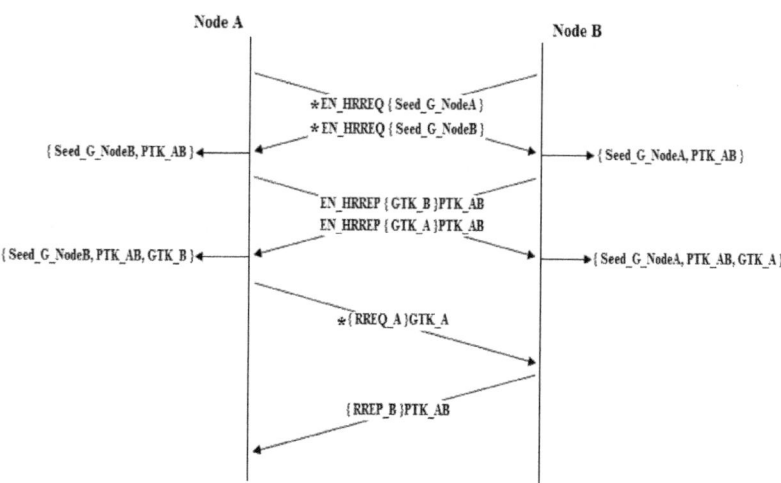

Fig. 1. Public Seed and GTK Key Exchange Process

Step 2: Derivation of PTK (Pairwise Transient Key)

On receiving the public Seed_G of its one-hop neighbours, the node uses both Seed_G it received from its one-hop neighbour and its corresponding private row of matrix A to compute the unique PTK that it shares with every one-hop neighbour with Blom's scheme.

Initially node A has A(i) and seed for G(i), and node B has A(j) and seed for G(j). After exchanging the seeds, node A regenerates G(j) and node B regenerates G(i). The pairwise secret key of nodes A and B, Kij and Kji, can be computed by both nodes independently with the following equation.

$$K_{ij} = K_{ji} = A_i \times G_j = A_j \times G_i$$

Upon finishing step 2, each node has stored the Seed_G of its one-hop neighbours and derived unique PTK pairwise key shares with its one-hop neighbours. Every node now can encrypt its GTK key with its PTK key and unicast it back to the originator of the HELLO RREQ message with the HELLO RREP message.

Step 3: Exchange of GTK (Group Transient Key) through HELLO RREP

The unicast HELLO RREP message from node A and B can be expressed as follows

$$A \rightarrow : EN_{HRREP}: [M_{type}, NC_A, IP_A, DSN_A, IP_B, Lifetime, \{GTK_A, IP_A, NC_A\}PTK_A]$$

$$B \rightarrow : EN_{HRREP}: [M_{type}, NC_B, IP_B, DSN_B, IP_A, Lifetime, \{GTK_B, IP_B, NC_B\}PTK_B]$$

After receiving HELLO RREQ from A, B unicasts a HELLO RREP message to A. The encrypted GTK_B is also attached within the unicast HELLO RREP message. Once A receives HELLO RREP, it decrypts the GTK_B with its private PTK_A and stores it in its database. Every node now stores the Seed_G, a group of PTK and GTK pairwise keys from its one-hop neighbours.

3.3   Securing Route Discovery

To implement a hop-by-hop authentication, each node must verify the incoming message from its one-hop neighbours before re-broadcast or unicast it. The trust relationship between each pair of nodes relies on their shared GTK and PTK keys, which have been obtained during the key exchange process.

In SEAODV, the route discovery process is similar to that of standard AODV, but a MAC extension is appended to the end of the AODV routing message. The format of the modified RREQ is given in Figure 2.

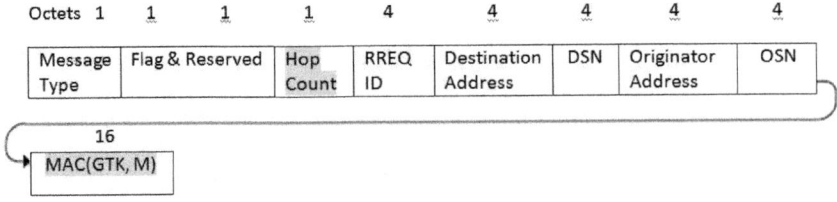

Fig. 2. The Modified RREQ

where the first shadowed box is a mutable field representing the hop count. The second shadowed box is the MAC field that has been appended at the end of the RREQ routing message. The MAC is computed over message M using the key GTK of the node who broadcasts a RREQ to its one-hop neighbours. Message M refers to all the elements before the MAC field in the RREQ message. As an example, if a node A wants to broadcast a modified RREQ, it sends the following message

$$A \rightarrow * : RREQ : [M, MAC(GTK, \ M)]$$

Whenever a node needs to discover a route to a designated destination, it broadcasts the modified RREQ message to its neighbours. Upon receiving the broadcast RREQ, each neighbour checks whether it possesses the GTK key of the sender by checking its GTK group. If there is a match, the receiving node computes the corresponding MAC with the received message and the GTK. If the MAC value is verified, the received RREQ is considered authentic and unaltered. The receiving node will then update the hop-count in RREQ, its routing table, and subsequently sets up the reverse path back to the source by recording the neighbour from which it received the RREQ. If the node is the destination, it will respond a RREP with a new MAC(PTK, M) affixed to the end of the RREP and unicast the RREP back to its next hop of its reverse path towards the source. The appended MAC(PTK, M) is computed on the RREP message with the PTK key the node shares with its next hop, to which the RREP is going to be forwarded. In the case the node is an intermediate node, the node applies its own GTK key on the updated RREQ to compute the new MAC(GTK, M) and attaches it to the end of the RREQ before it re-broadcasts the new RREQ to its one-hop neighbours. The receiving node will simply discard RREQ if the node does not have the GTK key of the sender or if their MAC value does not match.

3.4 Securing Route Setup

Eventually, the RREQ message reaches the destination or an intermediate node that already has a fresh route to the destination. A destination or an intermediate node can generate a modified RREP and unicast it back to the next hop from which it received the RREQ towards the originator of the RREQ. Since RREP routing message is authenticated at each hop, adversaries have no opportunity to re-direct the traffic.

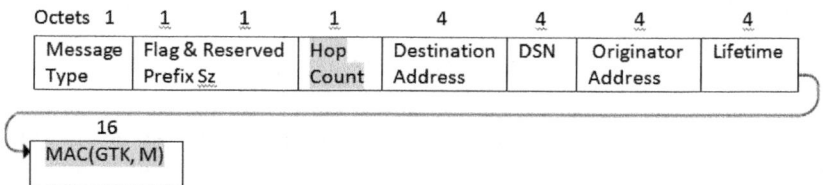

Fig. 3. The Modified RREP

The format of a modified RREP is shown in Figure 3. One more field, MAC(PTK, M), is attached to the end of the AODV RREP. This MAC is computed with the PTK key that the node secretly shares with the one to which the RREP is going to be forwarded.

Before unicasting the modified RREP back to its originator, the node needs to check its routing table to identify the next hop from which it received the broadcast RREQ; the node then applies the PTK key it shares with the identified next hop to compute the MAC(PTK, M) and affixes this MAC to the end of RREP. For instance, suppose a node B who needs to unicast a modified RREP back to Node A, it sends the following message.

$$B \rightarrow * : RREP : [M, MAC(PTK_{BA}, M)]$$

where $PTK_{AB} = PTK_{BA}$.

Upon receiving RREP from node B, A checks whether PTK_{BA} is in its PTK group. If there is a match, A verifies MAC'(PTK_AB, M). Node A will then update the hop-count field in the RREP and its own routing table, sets up the forwarding path towards the destination. Besides, A also searches the appropriate PTK key it shares with its next hop to which the new RREP is going to be forwarded towards the source. A then uses the PTK key to construct a new MAC and attaches it at the end of the new RREP. The received RREP is deemed to be unauthentic and will be discarded if the two MACs are not equal.

3.5 Securing Route Maintenance

In AODV, a node generates a RERR message under the following three situations:

1. If a node receives data packet destined to another node for which it does not have an active route in its routing table.
2. Whenever there is a broken link for the next hop of an active route has been detected by a node, the node will initiate a RERR message to all its precursors that may use the broken next hop towards to their destinations.
3. On receiving a RERR from a neighbour for one or more active routes.
    The following figure shows the format of a modified RERR message

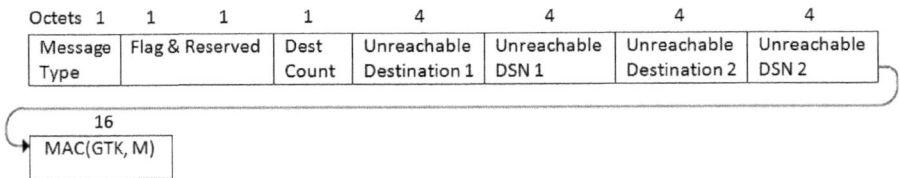

Octets 1	1	1	1	4	4	4	4
Message Type	Flag & Reserved	Dest Count	Unreachable Destination 1	Unreachable DSN 1	Unreachable Destination 2	Unreachable DSN 2	

16
MAC(GTK, M)

Fig. 4. The Modified RERR

where a MAC(GTK, M) is appended to the end of the RERR. The modified RERR is broadcast to all its one-hop neighbours in order to deliver the notification of the unreachable destinations. Suppose A is a one-hop neighbour of B and B broadcasts the following modified RERR message

$$B \rightarrow * : RERR : [M, MAC(GTK_B, M)]$$

where GTK_B is the GTK key that node B secretly shares with all its one-hop neighbours and the MAC field is generated with node B's GTK_B. Upon receiving the broadcast message from B, node A checks whether it has GTK_B and further computes the MAC'(GTK_B, M'). The MAC'(GTK_B, M') is to be compared against the received MAC(GTK_B, M). If they match, A searches its routing table and tries to identify the affected routes which use node B as their next-hop. If no routes are affected, A simply drops the RERR and listens to the channel again. A may also discards the RERR if the GTK_B is not found or the MAC'(GTK_B, M') is not consistent with the one from RERR. If A identifies that at least one route is to be affected and these routes satisfy (1) the route must be active; (2) the route uses B as its next-hop; (3) the destinations of the affected route in A's routing table are members of the unreachable destinations list specified in the received RERR message from B. A takes the following actions:

1. Marks the affected routes as invalid in its routing table.
2. Updates the RERR message. For example, the number of affected destinations in Node A's routing table might not be the same as that of the received one from node B and it has to be smaller or identical to that of in RERR obtained from node B.
3. Computes the new MAC(GTK_A, M) by using its own GTK_A and the updated RERR obtained from step 2 and attaches it to the end of the updated RERR.
4. Broadcasts the new RERR to all its one-hop neighbours and starts listening to the channel.

4 Security Analysis

We analyze the security of SEAODV and compare it with ARAN, SAODV and LHAP.

RREQ Flooding
ARAN suffers from the expensive digital signature verification operations, while SAODV also incurs massive overhead in signature verification process. Contrarily, LHAP offers better immunity due to its light-weight nature by using one-way hash chain and only authenticates RREQ from its one-hop neighbours. The number of hash operations required to verify the authenticity of a message is from single hash operation up to maximum number of tolerance in terms of packet loss. SEAODV only authenticates RREQ from nodes that are in the list of its active one-hop neighbours. Hash operations are required in SEAODV and re-creation of MAC is simple, fast and one time only.

RREP Routing Loop
In ARAN, the signed routing message makes impersonation and modification of sequence numbers impossible. SAODV does not support hop-by-hop authentication. Also, being a source-destination authentication protocol, any intermediate nodes in SAODV could be impersonated during the fly of RREP. LHAP uses one-way hash chain to protect the message by simply appending traffic key right after the raw message. Malicious node can simply block the wireless transmission between two neighbouring nodes, modifies the messages, put the corresponding intercepted traffic keys right after the messages and send them back to the wireless channel. SEAODV supports hop-by-hop authentication. GTK and PTK keys are used to secure the broadcast and unicast routing messages respectively. The entire routing message is MACed, and thus the possibilities of impersonation and modification are eliminated.

Route Re-direction
Both ARAN and SEAODV defend against this type of attack. ARAN employs digital signature to sign every routing message in a hop-by-hop fashion, while in SEAODV, GTK and PTK keys are used to compute the MAC, which secures all the fields in the entire routing message. SAODV cannot effectively prevent the hop count from being increased by malicious nodes. This increases the chances of the route being de-selected from the potential candidate routes, which is another form of route re-direction attack. In LHAP, again malicious nodes can use the exact technique described in RREP Routing Loop to create this type of attack.

Formation of Routing Loops
Two conditions need to be satisfied in order to launch this attack. The malicious node has to impersonate a legitimate node in the network and is able to modify the metric such as hop count to be a better value in terms of less hop count in this case. SAODV is able to prevent the hop-count from decreasing, and thus avoid this attack. ARAN and SEAODV can also defeat this attack due to its hop-by-hop authentication. However, in LHAP, as long as the malicious node gets a chance to intercept the effective traffic keys and re-use them in a timely manner, there is a possibility to launch this type of attack.

RERR Fabrication
In ARAN, messages can only be fabricated by nodes with valid certificates and ARAN offers non-repudiation. Nodes keep sending fabricated routing messages might get excluded from the future route computation. While in SAODV, malicious node may simply impersonate nodes other than the one initiates the original RERR and forward the signed RERR to other nodes in the network. By doing do, malicious nodes can not only deplete the energy of the nodes, but also successfully defeat the routing protocol. LHAP also suffers from this type of attack; malicious node could use the captured traffic key to be attached after the modified RERR as long as the captured traffic keys are still "fresh" to be authenticated by the receivers. SEAODV experiences least negative impact against this attack since a receiving node only authenticates the RERR that comes from its active one-hop neighbours. This forces malicious node can only forward the replayed RERRs come from the receiving nodes' one-hop neighbours in order to launch this type of attack.

Tunnelling
ARAN uses the total time consumed as the metric to seek a route which does not guarantee the shortest path in terms of hop count but does offer the quickest path. However, it still cannot defeat the tunnelling attack because malicious nodes can simply adopt high-power gain transceiver to tunnel the routing messages such as RREP in order to make the source believe that the "tunnelled path" is the quickest one. As a consequence, malicious nodes would have been included on the final route towards destination and gained all the subsequent data packets passed through them. Similar methodology would be taken by malicious nodes to launch this attack on SAODV and SEAODV with the difference that now the actual routing metric is mi-srepresented in terms of hop counts. LHAP only authenticates messages from its one-hop neighbours, it makes tunnelling attack become more tougher to be launched since malicious nodes now have to intercept the "fresh enough" traffic keys at both ends of the tunnel. Summaries for each routing protocol in terms of defending against those identified attacks are presented in table 2.

Table 2. Vulnerabilities of Various Routing Protocols

Attack	AODV	ARAN	SAODV	LHAP	SEAODV
RREQ Flooding	Yes	Yes	Yes	Yes	Yes
RREP Routing Loop	Yes	No	Yes	Yes	No
Route Re-direction	Yes	No	Yes	Yes	No
Formation of Routing Loops	Yes	No	No	Yes	No
RERR Fabrication	Yes	Yes	Yes	Yes	Yes
Tunnelling	Yes	Yes	Yes	Yes	Yes

5 Performance Evaluation

Performance evaluation is presented to prove that SEAODV is superior against ARAN and SAODV in terms of computation cost and route acquisition latency.

5.1 Computation Cost

Since each node in WMN is considered as both a sender and a receiver, the total computation cost incurred at each node should be the cost of the node as a sender and a receiver. This methodology is applied to the evaluation of the computation cost for ARAN, SAODV and SEAODV.

Variables and notations used for the computing and communication costs is shown in table 3.

Table 3. Variables and Notations

$Signature_{Gen}$	Signature generation cost
$Signature_{Ver}$	Signature verification cost
H	Hash operation cost
MAC	Cost for computing a MAC
$Max_{Hop_{Count}}$	Maximum hop count
Hop_{Count}	Number of hop count
N	Total number of nodes on the established route
$Broadcast$	Broadcast routing message
$Unicast$	Unicast routing message

ARAN

In ARAN, if the receiver is one hop away from the originator, two signature verifications are required. The first verifies the certificate of the originator of RREQ or RREP and obtains the public key of the originator. The second verifies the signature of the originator by using the public key of the originator. Moreover, the node needs to perform four signature verifications should the routing message come from the node other than the originator of RREQ or RREP. If a node is multi-hops away from the originator, it experiences four signature verifications when receives a RREQ or RREP from its one-hop neighbour.

The total computation cost for a final established route with N nodes between source S and destination D is described as follows

$$2 \times (N-4) \times (Signature_{Gen} + 4 \times Signature_{Ver}) + 2 \times [(Signature_{Gen} + 2 \times Signature_{Ver}) + (Signature_{Gen} + 4 \times Signature_{Ver})] + 2 \times (Signature_{Gen} + 4 \times Signature_{Ver})$$

The computation cost of N nodes indicates that as the number of nodes on the established final route increases, the number of intermediate nodes who are at least two hops away from the originator of RREQ or RREP also rises, and hence the total computational cost of all the nodes on the final route are going to boost up.

SAODV

SAODV offers two types of signature extensions: single signature and double signature extensions. To evaluate its computation cost, we consider the single signature extension in which intermediate nodes cannot reply to a RREQ message due to its unable to properly sign its RREP message. The only node that can reply to a RREQ is the destination itself. Before rebroadcasting a RREQ or forwarding a RREP, a node needs to apply the hash function to the hash value of the signature extension so as to account for the new hop. If the node is the originator of RREQ or RREP, the generation of digital signature is requires but the hash operation is not needed. Upon receiving the RREQ or RREP, the receiver apply the hash function $h(Max_{HopCount} - Hop_{Count})$ times to the value in the hash field in order to secure the hop count. Apart from that, the receiving node also needs to verify the signature generated by the originator of the RREQ or RREP.

The computation cost of a node being as a sender or a receiver in SAODV is given below.

Sender (originator of RREQ):	$Signature_{Gen}$
Sender (intermediate node):	H
Receiver:	$H \times (Max_{HopCount} - Hop_{Count}) + 2 \times Signature_{Ver}$

Therefore, the computation cost of N nodes is as follows

$$2 \times \left[\sum_{i=0}^{N-3} H + H \times (Max_{HopCount} - i) + 2 \times Signature_{Ver}\right] + 2 \times \{Signature_{Gen} + H \times Max_{HopCount} + H \times [Max_{HopCount} - (N-2)] + 2 \times Signature_{Ver}\}$$

The calculation of computation cost of N nodes in SAODV shows that as the number of nodes on the finalized route increase, more hash operations and signature verifications are required during the route set up process.

SEAODV

The computation cost of SEAODV is simple and straightforward in contrast to that of ARAN and SAODV. In SEAODV, the computation cost of each node on the route is exactly the same whenever the node acts to be a sender or a receiver. The total computation cost for a finalized route of N nodes can be deduced as $2 \times (N - 2) \times 2 \times MAC + 2 \times 2 \times MAC$ where MAC stands for the cost of computing a MAC using a GTK or PTK key. Figure 5 shows the computation cost of ARAN, SAODV and SEAODV which is calculated in millisecond and the result demonstrates that the computation cost of SEAODV is much less in contrast to SAODV and ARAN.

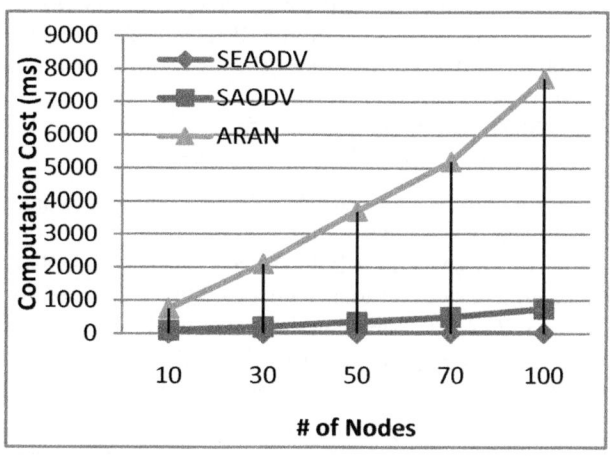

Fig. 5. Computation Cost of ARAN, SAODV and SEAODV

5.2 Communication Cost

ARAN, SAODV and SEAODV are similar in the way of discovering routes. RREQ is broadcast by the originator of the on-demand node towards the destination. Upon receiving the RREQ, the destination unicasts the RREP back to the source from which the RREQ is generated by using the reverse path which has been set up during the flooding of the RREQ. All of these routing protocols apply the same methodology in their routing mechanisms. Therefore, the communication cost involved in ARAN, SAODV and SEAODV are the same for N nodes which is $(N - 1) \times Broadcast + Hop_{Count} \times Unicast$. However, ARAN, SAODV and SEAODV experience various number of control bytes within every routing message (e.g., RREQ and RREP). The more control bytes incurred in a single routing message, the larger the entire routing message. Therefore, routing message with bigger size tends to have a lower probability of successful reception at the destination and suffer longer delay.

Before computing the latency produced by the communication overhead for each of the routing protocols mentioned above, the following assumptions are made:

1. Network throughput is 400kbps for a single flow (in this case, a single pair of source and destination) ;
2. 1024 bit RSA algorithm is used for ARAN and SAODV;
3. HMAC is the MAC algorithm for SEAODV;
4. In ARAN, the size of RREQ or RREP generated by the source or destination is smaller than those forwarded by intermediate nodes, which include two signatures and two certificates. While the RREQ or RREP originates by either the source or destination is only comprised of one signature and one certificate. Presume that the route discovery packet (RDP) in ARAN is the same size as that of used in AODV, which is 24 bytes. Therefore, for RREQ and RREP with single signature and certificate, the total size is 312 bytes. For RREQ or RREP with double signatures and double certificates, the total size is extended to 568 bytes;
5. In SAODV, 312 bytes in total for both RREQ and RREP, which include original AODV message (24 bytes), signature (128 bytes), top hash (16 bytes), hash (16 bytes) and certificate (128 bytes);
6. In SEAODV, there are totally 40 bytes for either RREQ or RREP. The AODV message costs 24 bytes and the HMAC is 16 bytes.

There are only two routing messages in ARAN with single signature, others are double signatures. The total number of bytes required to be transmitted in order to ensure a secure route set up is $568 \times (N - 2) \times 2 + 312 \times 2$ bytes. In SAODV, all routing messages are the same size, hence the total number of bytes required to be transmitted is $312 \times (N - 2) \times 2 + 312 \times 2$ bytes. Similarly, the total number of bytes of SEAODV incurred during the route set up process is $40 \times (N - 2) \times 2 + 40 \times 2$ bytes.

The communication cost for transmitting the required bits can be calculated below

$$Communication\ cost = \frac{Total\ No.\ of\ bits\ need\ to\ be\ transmitted\ (bits)}{Network\ throughput}$$

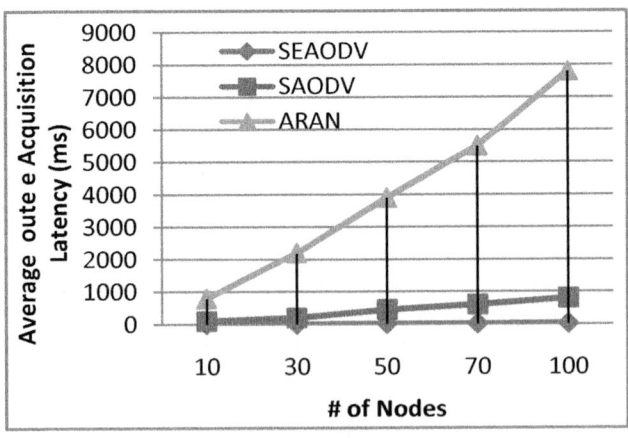

Fig. 6. Average Route Acquisition Latency for ARAN, SAODV and SEAODV(ms)

The average route acquisition latency (the total of computation and communication cost) is defined as the average delay between the sending of a RREQ packet by a source for discovering a route to a destination and the receipt of the first corresponding RREP. Figure 6 shows that the average route acquisition latency (in millisecond) of our SEAODV is much less in contrast to SAODV and ARAN due to the use of MAC and the smaller size of routing messages.

6 Conclusion

In this paper, we present a security enhanced AODV routing protocol (SEAODV). In SEAODV, Blom's key pre-distribution scheme is used to establish keys to ensure that every two nodes in the network uniquely share the pairwise keys. Our scheme adds secure AODV extensions to the original AODV routing messages. Each node in the network possesses two types of keys: PTK and GTK, where PTK is computed using Blom's scheme, node makes use of PTK to accomplish the distribution of GTK. PTK protects the unicast routing message and every pair of nodes secretly shares their own PTK while GTK secures the broadcast routing message and shares privately between the node and its one-hop neighbours. Depending on the type of the routing message, MAC is computed using either PTK or GTK. Evaluation results shows that SEAODV does offer superior performance in terms of computation cost and route acquisition latency as compares to other two secure routing protocols, ARAN and SAODV.

References

1. Sichitiu, M.L.: Wireless mesh networks: opportunities and challenges. In: Proceedings of the Wireless World Congress (2005)
2. Akyildiz, I.F., Wang, X., Wang, W.: Wireless mesh networks: A survey. Computer Networks (2005)
3. Bruno, R., Conti, M., Gregori, E.: Mesh networks: Commodity multihop ad hoc networks. IEEE Communications Magazine 43, 63–71 (2005)
4. Zapata, M., Asokan, N.: Securing ad-hoc routing protocols. In: Proceedings of ACM Workshop on Wireless Security (WiSe) (2002)
5. Perkins, C.E., Belding Royer, E., Das, S.R.: Ad hoc on-demand distance vector routing. In: IETF RFC 3561 (2003)
6. Sangiri, K., Dahil, B.: A Secure Routing Protocol for Ad Hoc Networks. In: Proceedings of 10th IEEE International Conference on Network Protocols (2002)
7. Hu, Y.-C., Perrig, A., Johnson, D.B.: Ariadne, A Secure On-Demand Routing Protocol for Ad Hoc Networks. In: Proceedings of MobiCom, Atlanta, GA (2002)
8. Perrig, A., Canetti, R., Tygar, J.D., Song, D.: Efficient authentication and signing of multicast streams over lossy channels. In: Proceedings of IEEE Symposium on Security and Privacy, pp. 56–73 (2000)
9. Du, W., Deng, J., Han, Y.S., Varshney, P.K.: A Pairwise Key Pre-distribution Scheme for Wireless Sensor Networks. ACM, New York (2003)

10. IEEE 802.11s Task Group. Draft Amendment to Standard for Information Technology Telecommunications and Information Exchange Between Systems – LAN/MAN Specific Requirements – Part 4: Wireless Medium Access Control (MAC) and physical layer (PHY) specifications: Amendment: ESS Mesh Networking. IEEE 802.11s/D1.06 (2007)
11. Bahr, M.: Proposed Routing for IEEE 802.11s WLAN Mesh Networks. In: 2nd Annual International Wireless Internet Conference (WICON), Boston, MA, USA (2006)
12. Bahr, M.: Update on the Hybrid Wireless Mesh protocol of 802.11s. In: Proceedings of IEEE International Conference on Mobile Adhoc and Sensor Systems, MASS, pp. 1–6 (2007)
13. Zhu, S., Xu, S., Setia, S., Jajodia, S.: LHAP: A Lightweight Hop-by-Hop Authentication Protocol for Ad-Hoc Networks. In: ICDCS International Workshop on Mobile and Wireless Network, Providence, Rodhe Island (2003)
14. Jing, X., Lee, M.J.: Energy-Aware Algorithms for AODV in Ad Hoc Networks. In: Proceedings of Mobile Computing and Ubiquitous Networking, Yokosuka, Japan (2004)

Auto-generation of Least Privileges Access Control Policies for Applications Supported by User Input Recognition

Sven Lachmund and Gregor Hengst

DOCOMO Euro-Labs, Munich, Germany

Abstract. Applications are typically executed in the security context of the user. Nonetheless, they do not need all the access rights granted. Executing applications with minimal rights (least privileges) is desirable. In case of an attack, only a fraction of resources can be accessed. The state-of-the-art on application-based access control policy generation has limitations: existing work does not generate least privileges policies, policies are not always complete and the process requires complex manual interaction. This paper presents an almost fully automated approach which counters these limitations. It achieves this by (1) extending a static analysis approach by user input recognition, by (2) introducing a new runtime approach on user input recognition which is based on information tracking and Aspect-Oriented Programming and by (3) combining the other two contributions with some of the existing work. The combined approaches are integrated into the software development life cycle and thus, policy generation becomes practicable. A prototype of the runtime approach is implemented which proves feasibility and scalability.

1 Introduction

In today's mainstream operating systems applications are typically executed with the security context of the user. Since applications are used for a specific purpose, they do not need all the access rights of the user. Applications should rather be executed with only those access rights they actually need (least privileges [1]).

If the application has a vulnerability which can be exploited by an attacker, allowing the attacker to control the application, the attacker is able to access the resources which the application is permitted to access. If the application has restricted access, potential damage of the attack can be confined. Due to complexity and extensibility of today's software, applications typically have vulnerabilities [2,3].

To execute applications with least privileges, applications have to be assigned access rights individually, as the purpose of applications and the resources they need to access vary significantly. Generic policies and protection domains are not specific enough. If applications have their individual access rights, limited to the minimum, they can *execute normally*, i.e. they have all the access rights

M.L. Gavrilova et al. (Eds.): Trans. on Comput. Sci. XI, LNCS 6480, pp. 17–38, 2010.

they need to carry out their operations, but not more. Applications access two categories of resources. The first category – the *system resources* – comprises those resources which the application accesses to carry out its operations independently of any user interaction. The second category – the *user resources* – comprises resources chosen by the user while interacting with the application. For example, a library that is loaded by an application to invoke a function represents a system resource and a text document accessed by a user in a text editor represents a user resource. It is not known prior to execution which user resources a user will access at runtime. In contrast, access to system resources can be derived from the application's code.

There is existing work, both on research level [4,5,6,7] and product level [8] (see Section 2), that automates generation of policies individually per application. While this existing work generates least privileges policies for system resources, it fails doing so for user resources. It fails due to permitting access to all user resources which might be accessed by the user during execution by adding *generic permissions* to the policy. An example for these generic permissions is to permit access to the entire home directory of a user. This overapproximation of access rights violates the principle of least privileges.

The objective is to reduce the set of access rights by discarding access rights to user resources in order to generate a least privileges policy. Treating system resources differently from user resources is the key. In this paper, access rights for system resources are collected and policies are generated using the existing work. However, for treating access to user resources, user input is identified and its propagation through the control flow of the application is analysed. If data that is input by the user is used as resource identifier at a permission check, the access is considered as access to a user resource. User-initiated resource access is determined that way. Permissions for accessing user resources are not added to the policy. The generated policy – the *application policy* – only consists of all the necessary access rights to system resources. The entire process is automated in order to minimise the involvement of the developer. Policy generation is performed at development time as a kind of side-task during implementation and testing. Technically, user input recognition is based on static analysis and runtime observations of the application's code.

The contributions of this paper are:

– Improving a static analysis-based approach by integrating user input recognition,
– introducing a new scheme for user input recognition based on user input tracking with aspect-oriented programming and
– Combining existing work on static analysis and runtime observation with the two other contributions.

Combining all these contributions eliminates major drawbacks of the existing work on policy auto-generation: overapproximation is eliminated, completeness of the policies is given and manual user interactions are eliminated. A prototype has been implemented which proves feasibility and scalability of the taint tracking approach.

Executing the application with this application policy being enforced at runtime would prevent the user from accessing any user resource in the application. Therefore the application policy is adapted on the target system where the application is executed. This can be done dynamically at runtime upon user interactions. Whenever a user chooses a resource in the application, the corresponding access right is added to the policy. Consequently, the application can only access user resources chosen by the user. This satisfies the principle of least privileges. The user perception is the same as in systems based on the object-capability security model (see Section 2.4). An alternative is to adapt the policy statically at deployment time or at load time, by specifying which user resources are accessible. We already elaborated various approaches for this adaptation. Some of them are similar to the user input recognition at development time (see Section 3). However, other approaches are beyond the focus of this paper.

Since the policy is generated by the developer and augmented on the target system, responsibilities are split: the developer defines the access rights the application needs and the target system only has to define which resources the user should be able to access. This results in policy generation being practicable for all the involved stakeholders. In contrast, existing work involves the user in complex manual tasks, as all the policy is generated on the target system. It is difficult for the user to determine which access rights to system resources an application needs. Consequently, existing work is not widely used in practice.

The paper is organised as follows. Section 2 discusses related work. Section 3 contributes distinction of access to system resources and user resources. It also presents the automated application policy generation process. The prototype implementation of the runtime observation approach is addressed by Section 4. Section 5 illustrates the contribution applied on an application. An evaluation of the contribution and the prototype is provided by Section 6. Section 7 addresses issues to be considered, such as embedding the presented work in the software development life cycle (SDLC) and future work. Section 8 concludes the paper.

This work has been carried out based on the object-oriented programming (OOP) paradigm [9,10]. It is assumed that the programming language and its execution environment are entirely object-oriented. Terms, such as *class*, *object*, *method*, *field*, *member*, *type* and *modifier* are used in their OOP context throughout the paper.

2   Related Work

This section is organised in line with the contributions. Section 2.1 describes the static analysis-based approach which is improved in this paper. Therefore it is discussed in depth. Section 2.2 covers relevant runtime observation approaches for generating application policies. Further approaches which are interesting but not used in this paper are briefly discussed in Section 2.3 Other related work is addressed thereafter.

2.1   Static Analysis by Call Graph

Policy generation by Koved, Centonze, Pistoia, et al. [4,5,6] is based on static analysis. It creates a call graph of applications written in Java which represents methods as nodes and method calls as edges. The call graph is used to determine which method calls result in permission checks. For each of these permission checks, the allocation site of the involved Permission class (representing access rights in Java [11]) is determined. Each Permission class contains an identifier which represents the accessed resource. The values of all of these identifiers are collected. They are put in the policy, as these resources will be accessed by the applications during execution. Libraries and applications are analysed differently.

For libraries a *summary* is created for all possible paths in the call graph of the library which start at any permission check node and end at any protected or public method. *Data flow analysis* [4,5] is applied to determine the paths. Each of the end point methods causes a permission check in the library when being invoked by the application. For the applications, these methods are entry points into the library. The summary contains all the required permissions for these calls.

Application analysis is limited to the paths in the call graph of the application that go from a start node to a node that is an entry point of a library for which a summary has been created. A set containing all the entry point nodes of all the libraries used by the application is created. It is partitioned in three subsets depending on properties of the resource identifiers. Paths are treated differently in the analysis, depending on the partition to which the entry point node where the path ends is assigned.

The first subset contains all those methods that use a string constant defined in the library to define the resource identifier. These methods are processed by a data flow analysis, like in the library analysis.

The second subset contains all those entry point methods that receive one or more String argument(s) when being invoked by the application. These arguments are allocated by the application and used as resource identifiers for the permission check in the library. These methods are processed with a more complex algorithm. The complete algorithm is presented in [6]. It starts with the *string analysis* described below to determine all possible resource identifiers in the application code that are used as arguments of entry point methods. *Slices* [12] are created for each of these String arguments to determine their propagation through the application. The slices identify the propagation paths in the call graph that belong to the String argument, without introducing paths that do not exist in the application's control flow.

The *string analysis* is a processing step of the application analysis where String objects of the application are analysed. The string analysis is capable of tracking all instantiations and modifications (e.g. concatenation) of strings representing resource identifiers. *Transforms* [6] of String modifying operations are created to determine the output String when an input String is provided. The string analysis creates a context-free language representing possible values for input

Strings and output Strings, derived from all modifications that are applied to the given string in the application. String objects are labelled to document their allocation site and all subsequent modifications. The labels map nodes in the call graph of the application to literals in the context-free language. After the string analysis, each character carries its own history of modifications from allocation to the site where they are used as arguments for allocation of a Permission object. Fig. 1 illustrates this on an example. Each character of the string is assigned a list of labels, where each label describes the operation or allocation site. String analysis increases precision over data flow analysis, as resource identifiers can be determined in cases where the work in [4] and [5] only generate generic permissions.

The third subset contains all those entry point methods that have non-string arguments containing String objects. These String objects are used as resource identifiers. For each of these non-string objects rules are predefined that describe how to obtain the resource identifier from the contained String object(s). If no rule is predefined for a specific non-string object, a generic permission that matches the Permission type used for the corresponding permission check is used, which permits access to any resource of that type. Once the resource identifiers have been extracted, the analysis continues with the one for the second subset.

Once the analyses are done for an entry point, the permissions needed when calling the corresponding method are determined. They are added to the corresponding node in the call graph. After all entry points are analysed, all the collected permissions are propagated backwards through the call graph to the start nodes. In nodes which are join-points of paths, permissions are combined with set unions. After the backward propagation, the start nodes contain all the permissions the application needs. The policy is created from these permissions.

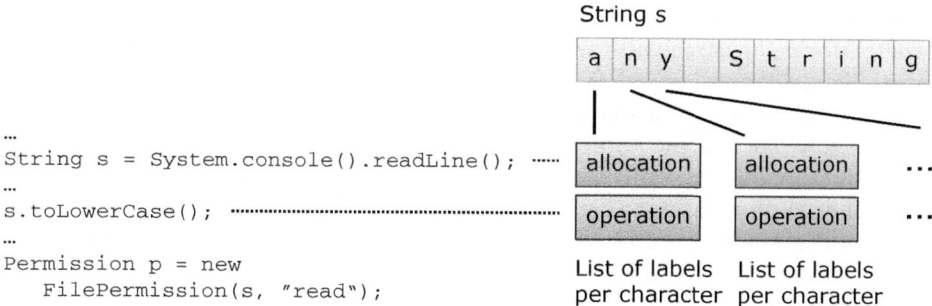

Fig. 1. Example of labelling characters of a string in the string analysis

The drawback of call graph-based analysis (as introduced in [4,5]) is that the call graph overapproximates [13]: it contains paths which do not correspond to program flow of the code. This overapproximation – we call it *call graph-based overapproximation* – results in access rights in the policy which the application actually does not need. Therefore the analyses which are sketched here are introduced in [6] in order to eliminate call graph-based overapproximation.

A different sort of overapproximation – we call it *indecisiveness overapproximation* – remains in [6], however. With all the analyses, many access rights which the application needs can be obtained statically. The rest of the access rights can only be determined on execution, as the involved resource identifiers are only defined at runtime. Thus they are unknown at the time of static analysis. In [5], static analysis is combined with dynamic analysis, but this is to overcome call graph-based overapproximation. For permissions that can only be determined at runtime, generic permissions are added to the policy in [6] (as well as in [5]). These generic permissions are added for all user resources and for those system resources which are not specified in the source code of the application (e.g. the file separator character in Java). Some resource identifiers are defined by operations at runtime. These operations receive various arguments which also may partially be available at runtime only. In such a case, overapproximation is countered in [6]: the string analysis provides transforms for operations which modify input that is used as resource identifier. These transforms are used to determine permissions statically. Generic permissions are only added if for certain operations no transforms are defined.

The improvement of the algorithm presented in Section 3.3 further reduces indecisiveness overapproximation of this approach by avoiding generic permissions for user resources in the generated policy.

2.2   Runtime Observation

Cowan's AppArmor [14,8] and Provos' Systrace [7] are runtime observation approaches. System calls are recognised and recorded, based on the Linux Security Modules (LSM). An application is executed several times in learning mode. All the performed system calls are written to a policy. Finally, the policy is examined manually and applied. Any future execution of the application is controlled within the bounds of this policy.

The drawback of the approach is the manual policy examination. It is left to the user. In addition, a general drawback of runtime approaches is that completeness of execution coverage cannot be determined. Only if all the functionality of the application is executed, a complete policy is generated. As a consequence, generated policies are likely to be incomplete. This is closely related to the need of generic permissions for user resources. They are needed as it cannot be determined whether during execution all possible user resources have been accessed. Due to these generic permissions, runtime observations overapproximate.

However, the approach is an important work on auto-generation of policies. It has been chosen for creating the observation records in Section 3.4. In order to benefit from this approach, it is combined with the other contributions of this paper. This eliminates the involvement of the user for manual policy examination and it avoids generic permissions for user resources by filtering permission checks which are performed by the user.

2.3   Further Analysis Approaches

Wagner and Dean [15] present a combination of static analysis and runtime monitoring. Using static analysis, a model of the application is created which is represented by an automaton. This automaton models the order in which system calls are made by the application. Each system call of the application initiates a state transition of the automaton. Any system call which the application makes when executing normally transfers the automaton from a valid state to another. Any system call that is normally not made by the application leads to an error state. Using this automaton at runtime allows recognising illegal state transitions of the application, which is used to terminate the application. The advantage is that this approach takes the history of system calls into consideration. However, while this approach keeps track of the application's control flow, it does not provide fine grained access control for resources. Once a system call is permitted, the resources on which the system call operates are not further restricted. Thus, this approach should always be combined with other access control models.

Polymer [16] also follows a two step approach. The bytecode of Java applications and libraries is instrumented with jumps into the Policy object, which performs policy enforcement. The policy is a compiled Java object. At runtime, the instrumented code and the Policy object act as runtime monitors to perform access control on the level of method calls. Polymer does not support generating the policies. This is a purely manual task. It also provides limited dynamic policy adaptation at runtime.

2.4   Other Related Work

Systems based on the object-capability security model [17] provide the application with a reference after the user has chosen a resource. The application itself does not have access rights for the resource, but via the reference, the application can access the resource. Thus, the user transparently provides the necessary access rights which the application needs. These object-capability-based systems and the contribution of this paper have the same user perception in common. It is the user's responsibility to carefully choose resources the application should operate on.

Taint tracking [18,19,20,21] is used to filter potentially dangerous user input before it is used for sensitive operations. There are different approaches, but they all have in common that user input is tracked when propagating through an application. Some work recognises intrusions when user input is used as arguments for certain sensitive operations, as these operations are normally only executed with arguments that are not specified by the user. If user input reaches such sensitive operations, the application will be terminated. In other work, control characters are filtered from user input. For an SQL statement, for example, characters, such as semicolon and quote characters, would be filtered. This prevents the user from rewriting the SQL statement. In this paper, taint tracking is used for tracking user defined resource identifiers.

3   Policy Generation with User Interaction Recognition

When analysing an application in order to generate an access control policy for it, all the control flow that leads to resource access is of interest. Each access to a resource initiates a permission check. The permission check determines whether access is permitted or prohibited. Control flow analysis starts at the function that starts the application and it ends at methods which perform a permission check. If all these control flows of an application are captured along with the corresponding access rights, a complete description of the application's access behaviour is available which can be used to generate an access control policy. Existing work does that by static code analysis [4,5,6], as described in Section 2.1 and by runtime observations [5,8,7] (see Section 2.2). Static analysis models stack inspection-based access control [22] statically and runtime observations collect all the access attempts of an executed application when they occur. This paper uses a combination of static analysis and runtime observations in order to eliminate either one's limitations. Since static analysis covers the entire code of the application, the generated policy is complete, i.e. it contains at least all the access rights the application needs. Runtime observations are incomplete, but they can determine permissions which cannot be determined statically prior to execution.

As explained in Section 2 and as further elaborated upon in Section 6.2, existing work suffers from different types of overapproximation. In order to eliminate overapproximation, the different types of overapproximation are to be treated differently. This requires distinguishing them. This distinction can be achieved by recognising user input. Indecisiveness overapproximation can be eliminated for user resources if all user-initiated access is discarded from the generated policy. Therefore this section integrates user input recognition into the policy auto-generation process. How to eliminate indecisiveness overapproximation for system resources is addressed by Section 6.2.

Integrating user input recognition into the policy generation process allows determining if an access is initiated by the user. Two promising approaches of user input tracking were found by the authors: (1) using information tracking along with Aspect-Oriented Programming (AOP) [23] to track user input during execution (known as *taint tracking*; see Section 2) and (2) extending the call graph-based static analysis. This paper uses both these approaches in combination. The contribution of this section extends the static analysis approach explained in depth in Section 2.1 and it uses ideas from the runtime observation approach by Cowan [14,8] (see Section 2.2).

Extending policy generation by user input recognition leads to the process depicted in Fig. 2. All the steps of the process are integrated in the development phase of the application. Policy generation itself takes place in step 5, where the policy is generated from all the input which is collected in the steps 1, 3 and 4. These four steps (marked by slightly darker background colour in Fig. 2) are discussed in this section.

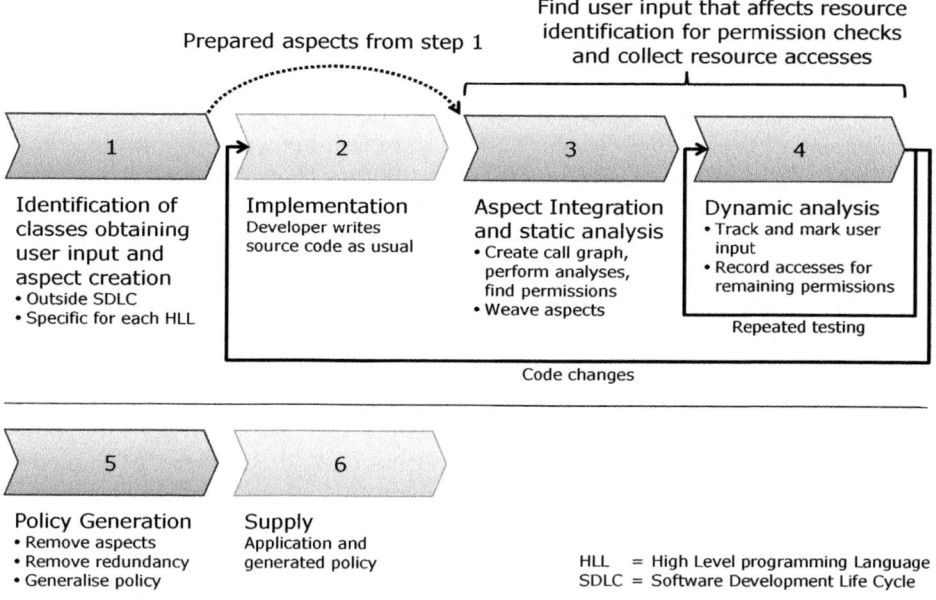

Fig. 2. The process of auto-generating the application policy including recognition of user interactions

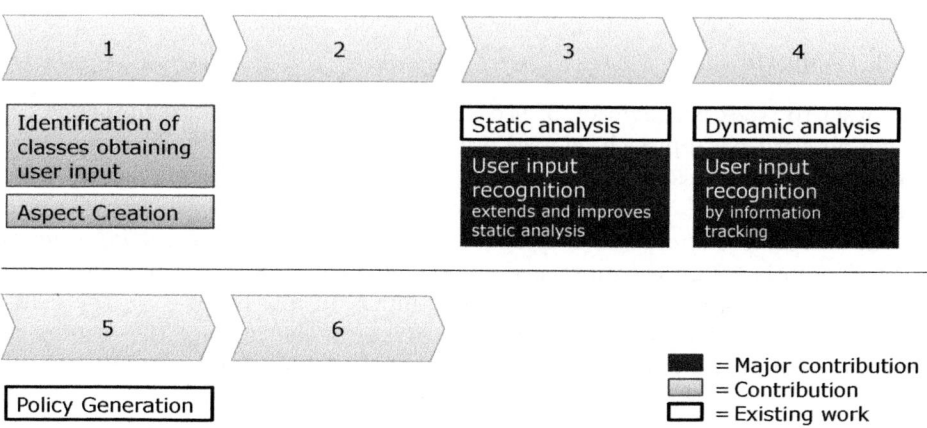

Fig. 3. Combination of existing work and contributions

Fig. 3 illustrates in which way existing work and contributions are combined in the steps of Fig. 2.

3.1 Classes Obtaining User Input

Tracking propagation of user input starts at those classes which obtain user input. These classes are collected in Step 1 of Fig. 2.

All the classes that come with a programming language and its execution environment – the so called *system library* – are well-known and finite in number and size. Thus, all the classes of the system library that obtain user input comprise a subset of the system library which is also finite. A list of all classes that obtain user input is compiled statically individually per programming language. The list is used as starting point for tracking user input through the application. If all entry points, such as console input, Graphical User Interface (GUI) input, network input, file input, database input and others, are taken into account, a complete list can be compiled. Some external libraries (e.g. windowing toolkits) are to be added separately if they are not part of the system library. As the object-oriented programming paradigm follows the idea of composing components, applications typically reuse components from the libraries to implement their functionality. Applications do rather not create own classes covering commonly used low level functionality, such as obtaining user input. Consequently, the list is not to be changed by the developer when implementing an application. It is only to be extended if an external library is used that does not provide its own list.

For some classes of the list, all instances obtain user input (e.g. a text box of a windowing toolkit). With other classes, only certain instances obtain user input (e.g. the input stream in Java that connects keyboard input to standard-in, but not necessarily any other input stream).

3.2 Aspects

In order to track propagation of user input through the application in the runtime observation approach, classes are augmented by aspects. This subsection discusses which classes are augmented and what the aspects do (Steps 1 and 3 in Fig. 2).

The aspects augment classes by a new field that stores *taint information*. When data is assigned to instance objects of these classes, the new field is set *tainted* if the data is obtained from user input. In any other case, the field is set *not tainted*. The aspects observe all operations that change the state of objects containing taint information. They update taint information accordingly. If data is propagated to other objects, taint information is also propagated by the aspects. Consequently, all the classes that are involved in user input propagation are augmented. The objects that perform permission checks, finally, receive the usual data they need for permission check, i.e. the resource identifier and the requested access right. They also receive taint information that is stored in the object containing the resource identifier. This allows aspects in the objects that perform permission checks to distinguish user-initiated access requests from application-initiated requests.

When classes deal with user input, apart from obtaining it, they can store, modify or transform this user input. Methods of the class perform this functionality. They are augmented by aspects to set taint information accordingly. A class that stores user input also needs to store taint information. A class that modifies user input also needs to modify taint information. A class that transforms user input into another type also needs to provide the target class with taint information. The target class needs to receive and store this taint information. Methods that perform other functionality that does not affect user input need no augmentation.

User input is data that is stored in a class either by calling a method of the class or by assigning the data directly to a field. In whatever way the state of the field is changed, taint information is to be set accordingly. If a method changes the state of the field, the method is augmented by the necessary aspect. If the field is accessed directly, the member class which represents the field is augmented.

Aspects are prepared statically outside the process of policy auto-generation (Step 1 in Fig. 2). The prepared aspects are specific for each programming language. An example aspect for Java is listed in Section 4.2. The aspects are weaved into the application's code during development (Step 3 in Fig. 2). Many classes (mainly high level classes) are augmented with generic aspects, i.e. aspects with pointcut definitions which apply to a wide range of classes. Such a generic aspect applies, for instance, to all methods of all classes that return a value of type String. Some classes (mainly low level classes) are augmented with individual aspects to cover all their specific data propagation possibilities.

3.3   Call Graph and String Analysis

In order to distinguish user independent actions of the application from user interactions, values and allocation sites of resource identifiers are determined in the static analysis (Step 3 in Fig. 2). As soon as a resource identifier is allocated by a class from the list (see Section 3.1), the resource identifier is known to be defined by the user. The string analysis (see Section 2.1) is extended by integrating this distinction. The labels are analysed to find all the characters of which the resource identifier is composed and their allocation sites. Each of the allocation sites is looked up in the list of user input obtaining classes. If the allocation site is listed, the allocation label is tagged as originating from user input. All the other labels are analysed for their string operations. If the operations keep the original content obtained from the allocation site, their corresponding labels are also tagged. If the content is changed, e.g. by using a substring, the user input tag is discarded. This is to prevent the application from composing a resource identifier from parts of user input to gain access to arbitrary resources at runtime. The transforms (see Section 2.1) are extended by describing whether user input tags shall be dropped when the corresponding methods are applied.

If all the labels of a character get the user input tag, the character itself is tagged as user input. The string representing the resource identifier can either consist of (1) only characters that are tagged as user input, (2) characters that

are tagged as user input and characters that are not tagged as user input and
(3) no character that is tagged as user input. Treatment of those strings that
only consist of tagged or untagged characters is easy: the strings are tagged
according to their character tags. In the mixed case, the string is tagged as user
input. In all potentially dangerous cases, the transforms of string operations
removed the user input tag before. Thus, it is safe to treat the mixed case as
user input. If the string is tagged, the permission object which uses the string
as resource identifier is also tagged. The tagged permission object indicates that
the permission it represents is defined by user input.

Fig. 4 depicts the extended analysis on an example: the allocated string originates from user input and the endsWith operation does not remove this property;
thus, the characters used in the permission check originate from user input.

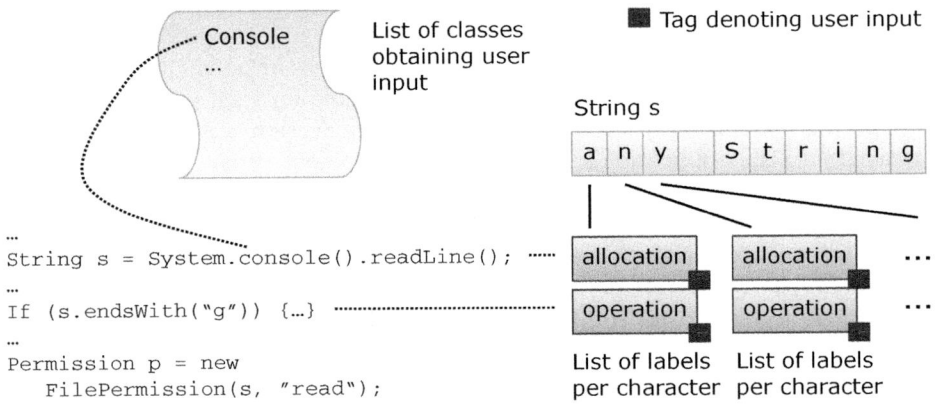

Fig. 4. Example of library-client application analysis extended by tagging labels of
those characters of a string that originate from user input

The application analysis of Section 2.1 is extended by adding this user input
analysis. Since subset 1 of the partitioned set of entry points is independent of
user input, no extensions are applied. Both subset 2 and subset 3 entry points
require extensions. After all the labelled strings are available, they are used for
user input analysis. The extended algorithm collects permissions as before, but it
tags those permissions that contain resource identifiers obtained from user input.
This makes them distinguishable from the others which have resource identifiers
not obtained from user input.

There is no need to extend the library analysis, as the application analysis
considers all input that comes from outside the library when the library is used
by an application.

The approach presented here further reduces overapproximation of the call
graph-based approach discussed in Section 2.1. All the resource identifiers that
are defined by user input, which cannot be determined prior to runtime, are

not added to the application policy at all. Thus, there is no need for generic permissions as a result of the static analysis. Cases in which generic permissions are still required are limited to system values of the runtime (e.g. the file separator character in Java) and to resource identifiers that are read from other sources (e.g. database or file). Therefore overapproximation is reduced, but not eliminated. These cases of remaining overapproximation can be treated by defining sets of possible values to further reduce overapproximation, as discussed in Section 6.2.

3.4   Dynamic Analysis

During dynamic analysis (Step 4 in Fig. 2), the augmented classes are capable of tracking user input through the application. The application is executed repeatedly for software testing. During these executions, all access requests are recorded and stored during permission checks to generate the application policy from these records [5,8] (see Section 2). The records are stored together with the corresponding taint information for each resource identifier. If a resource identifier originates from user input, the corresponding record is discarded. For processing tainted data, content and semantics of the data is irrelevant. Only the propagation of data together with its taint information is relevant.

3.5   Policy Generation

The policy is generated in Step 5 of Fig. 2. After static analysis collected permissions and after execution has been finished, the access control policy is generated from all the acquired permissions.

For policy generation, redundant permissions are removed, as they are useless. Permissions with specific access rights that are implied by more generic permissions are removed as well. Finally, the policy is generated from the remaining records.

Since aspects are only needed for policy generation, they are removed in the policy generation step. The deployed application does not differ from an application for which no policy was generated.

4   Prototype Implementation

The prototype of the runtime observation approach with taint tracking [24] has been implemented in Java [25] and AspectJ [26]. Java's modular design allows replacing components easily. The access control model of Java [11] is flexible and advanced. Fine-grained access control is possible. We extend Java's access control facilities by replacing the default Policy Enforcement Point (PEP), i.e. the SecurityManager class. Access control is performed on the level of the Java Virtual Machine (JVM), based on the various subclasses of the Permission class.

The list of user input obtaining classes contains core classes like, for example, java.io.Console and java.io.InputStream. Direct user input is obtained from standard-in using these classes. To obtain user input from arguments given when starting the application, any class containing a main method is added to the list. Among the windowing toolkits only Swing is exemplified here. Text input fields (e.g. javax.swing.TextComponent) and dialogue boxes (e.g. javax.swing.JFileChooser) are added to the list. The classes Byte, Character, Integer and String are also added, as the InputStream uses the first three to store the input it gets from its data source. These classes are also involved in other String operations, processing user input until it is stored in a String representation.[1] In order to analyse tracked taint information, the SecurityManager.checkPermission method is added as well.

Since Java does not only know objects, but also support primitive types when adding aspects to classes, there are two categories of classes to distinguish. Category 1 represents all the low level classes that correspond to primitive types: they store data in their fields in primitive types. Category 1 classes access their fields directly. These classes are the end of the hierarchy of member classes. Category 2 comprises all the classes that store their data in fields that are classes by themselves. Category 2 classes need to call methods of their member classes whenever they store or read data therein. There can be an arbitrary hierarchy of member classes of category 2. All classes which are not category 1 are category 2.

Category 1 classes need to be augmented in any case. Each time, data is stored in these classes, taint information is to be set according to the origin of that data. If, for instance, data is obtained from the console (i.e. from System.in), it is known that this instance of InputStream always produces user input. The array of int in which the system library class InputStream stores the user input, needs to set its data tainted.

Category 2 classes do not necessarily need to be augmented. Their member classes refer to other objects which already may contain taint information. However, if a category 2 class is capable of transforming its content to another type, the corresponding methods need to set taint information of the target type according to the source type. Therefore such a category 2 class needs to be augmented.

Native methods are augmented using around advice. Taint is tracked on return values. When the method is called, its arguments are analysed and when it returns, its return value can be set tainted. This requires understanding the semantics of the method to some extent.

4.1 Implementation Details

We implemented a plug-in for the integrated development environment *eclipse* [27]. This plug-in accompanies the software development and testing process by augmenting classes with aspects and by auto-generating the policy. The plug-in

---

[1] Java actually uses primitive types in InputStream and other low level classes.

contains the aspects. The developer does not need to write their own aspects, unless the developer writes code that is capable of obtaining user input directly without using existing Java classes.

The application is always executed using the replacement SecurityManager – called ObservingSecurityManager – and a replacement policy provider – called PolicyObserver. The PolicyObserver always returns false for each permission check without consulting any policy. This causes the AccessControlContext, which is involved in policy enforcement, to throw a security exception [11]. This exception is caught by the ObservingSecurityManager and forwarded to a monitor class, which analyses and stores its contents. That way, the subject, the subclass of the Permission class, the resource identifier, the action, the entire call stack and the code base are obtained by the monitor class. Due to the aspect by which the ObservingSecurityManager is augmented, the monitor class also obtains taint information and stores it in the records.

The ObservingSecurityManager suppresses the caught exception and returns silently. Consequently, the application gets all access attempts permitted. This allows testing any application feature without being hindered by security constraints. As this takes place in the development phase, there is no threat for the system where the application is deployed.

The policy is generated from all the records. Filtering of user-initiated access and removal of redundancy is done as described in Section 3.

In order to keep the ObservingSecurityManager small and independent of analysing the exception, the ObservingSecurityManager sends the exception it caught on a permission check to a server process using RMI. This server process generates the policy.

4.2   Aspects

As described in Section 3.2, there are generic and specific aspects. For the prototype, it is advisable to classify them in three groups: group 1 consists of aspects that are needed to store, read and transfer taint information in classes. They provide methods to set and get taint information and they add a taint bit that stores taint information. All classes that obtain or process user input need to be augmented by these aspects. Classes are augmented by *inter-type declaration*, i.e. classes inherit from both the aspect and the Object class (or a sub-class). Augmented classes can then store and change their own taint information. Pointcuts of Group 1 aspects define which classes are to be augmented by that functionality. Aspects in group 2 are generic aspects that specify in which cases group 1 aspects' functionality is to be called in order to update the taint bit. This is the case when data in Group 1 classes is set or modified. These aspects are generic, as they apply to multiple classes satisfying some common properties. Group 3 aspects are all the specific aspects that deal with peculiarities of certain library classes. They have the same purpose as Group 2 aspects, but they are specifically written to track the taint bit in a particular class. Listing 1 shows one of the Group 2 aspects.

Listing 1. Advice augmenting main method

```
1 before(String[] args): execution(public static void *.main(String[]
2     || String...) && args(arg) {
3   for (String string : arg) ((Taint)string).setTainted(true);
4 }
```

5   Example

In the following, the prototype is used to generate the policy for the *UMU XACML-Editor* (Version 1.3) exemplarily. The results are compared to the state-of-the-art and evaluated. The UMU XACML-Editor [28] is a GUI-based XACML file editor written using Swing. At first, all the resource access has been collected using the ObservingSecurityManager. Table 1 lists all the access attempts which occur when the application is executed. Access attempts with numbers 3, 4 and 9 in the table are initiated by the user. When applying existing work, the same access attempts are collected, as the ObservingSecurityManager performs the same analysis. However, creating a policy from the collected access attempts is of little avail. It permits the application to access all the system resources it needs to execute normally, but it only permits the application to access those user resources (i.e. XACML files in the example) the user has chosen when the access attempts were recorded. Therefore existing work involves the user to manually inspect the policy. Thereby, the user ought to add a generic permission for file access which permits the application to access all the XACML files which may be opened in the XACML-Editor in the future. In order to prevent adding this generic permission, our prototype identifies user interactions. In the case of the UMU XACML-Editor, all user interactions are initiated via file dialogues. They are all identified and marked in the records. They are discarded for policy generation. Consequently, the generated policy does not contain access rights with numbers 3, 4 and 9 from Table 1, but all the other access rights. In contrast to existing work, the resulting policy does not overapproximate. Therefore it is the least privileges policy for the application. A modified file dialogue can then augment the policy at runtime upon user interaction (as described in Section 7.2).

6   Evaluation

6.1   Prototype

The AOP aspects of the prototype instrument the system library. They identify user-initiated resource access correctly. As depicted by Table 2, the overall number of necessary aspects is kept in a manageable range. This is due to the generic aspects which affect a large number of classes. For production, external libraries need to be augmented as well. From the feedback of the prototype implementation, this is a scaling task with respect to the size of the external libraries.

Table 1. All access attempts of the UMU XACML-Editor. Duplicates are omitted. Access attempts with numbers 3, 4 and 9 (bold font) are initiated by the user.

No.	Permission	Resource	Action
1	java.awt.AWTPermission	accessEventQueue	
2	java.awt.AWTPermission	showWindowWithoutWarningBanner	
3	java.io.FilePermission	/user/policy	read
4	java.io.FilePermission	/user/policy.xml	write
5	java.io.FilePermission	/UMU-XACML-Editor/bin/icons/cara.gif	read
6	java.io.FilePermission	/UMU-XACML-Editor/bin/icons/nube.gif	read
7	java.io.FilePermission	/UMU-XACML-Editor/bin/icons/target.gif	read
8	java.io.FilePermission	/UMU-XACML-Editor/bin/icons/verde.gif	read
9	java.io.FilePermission	/user	write
10	java.lang.RuntimePermission	exitVM	
11	java.lang.RuntimePermission	modifyThreadGroup	
12	java.util.PropertyPermission	elementAttributeLimit	read
13	java.util.PropertyPermission	entityExpansionLimit	read
14	java.util.PropertyPermission	maxOccurLimit	read
15	java.util.PropertyPermission	os.name	read
16	java.util.PropertyPermission	os.version	read
17	java.util.PropertyPermission	user.dir	read

Table 2. Number of aspects in the prototype implementation. The figures are limited to the packages java.lang, java.io, java.net and javax.swing. Aspects needed for compensating primitive types are not considered. For inter-type declarations, the number of affected classes is limited to directly affected classes. Through inheritance more classes become affected.

Group		Advice	Named Pointcuts	Affected Classes
Inter-type declarations	1	1	0	11
Generic aspects	2	8	8	41
Specific aspects	3	28	36	26

6.2 Elimination of Overapproximation

Existing work on policy generation suffers from limitations, as discussed earlier. Runtime observations are incomplete (see Section 2.2) and static analysis suffers from indecisiveness overapproximation (see Section 2.1).

Combining observations and static analysis (as done by Centonze et al. [5], see Section 2.1) counters the drawback of incompleteness of runtime observations. The combination can further reduce overapproximation of static analysis, as generic permissions can be more precise and minimised to a set of valid values in some cases, but they still remain.

The contributions of this paper further reduce overapproximation. By discarding user-initiated resource access, all access to user resources is omitted from the policy. There is no generic permission in the policy and the content of the policy does also not depend on the user resources which the tester has chosen during runtime observations. Access to user resources is the major cause for over-approximation in existing work. Thus, the primary source of overapproximation is eliminated.

There are still special cases where overapproximation remains: in cases where no transform is defined for an operation which is analysed by the string analysis, as well as in cases where no set of possible values for a resource identifier which is set by the execution environment is defined. Both these cases can be countered by ensuring that all the transforms and sets are defined. The policy generation process can be implemented in a way that it identifies missing transforms and sets. The sets can then be defined directly in the policy generation process, for instance by adding annotations to the code. Consequently, overapproximation is eliminated which results in policies that are complete and that also represent the least privileges of the corresponding application. This is a major benefit over the state-of-the-art. Table 3 summarises the differences and the gains by comparing the state-of-the-art, the individual contributions and the combined contributions of this paper.

Table 3. Comparison of state-of-the-art and contributions

	Static analysis [6] (see Section 2.1)	Dynamic analysis [7] (see Section 2.2)	Static analysis extended by contributions from Section 3.3	Dynamic analysis extensions by contributions from Sections 3.2 and 3.4	Combination of all contributions of Section 3
Over-approximation	Overapproximates on permissions only known on execution (indecisiveness overapprox.)	Overapproximates on user resources	Reduced indecisiveness overapproximation to system resources only	Does not overapproximate as no user resources are captured	Does not overapproximate
Completeness	Complete	Incomplete	Complete	Incomplete	Complete
Scalability	Scales due to separated library analysis	Scales due to the ability of combining permissions in include files	Scales due to separated library analysis	Scales as the number of aspects to be created is in a manageable range	Scales as the individual solutions combined here scale and as the combination does not add non-linear complexity
Automation	Automated to large extent	Manual inspection of generated policies required	Automated with a few exceptions	Aspects are created manually per programming language; Policy generation is fully automatic	Aspects are created manually per programming language; Policy generation is fully automatic
Requirements	Source code or object code	Complete test coverage	Source code	Complete test coverage	Source code

7   Discussion

7.1   Threat Model

The policy generation process contributed in this paper is meant for protecting systems against applications that misbehave due to programming errors and due to being exploitable by attackers. Since the policy is generated by the developer, a developer of a malicious application can generate a policy that permits all the malicious access. The contribution does not protect against malicious applications.

7.2 The Big Picture

If no further measures are taken in the deployment phase and/or in the execution phase, the generated policy is of little avail. If the application is deployed together with its application policy, manifold measures are advisable, as discussed next.

For deployment, the application policy can be checked against the policy of the system on which the application is to be deployed, to see if they do not contradict. This can be done manually by examining the policy or automatically. The European research project *Security of Software and Services for Mobile Systems (S3MS)* [29,30,31] provides means to prove that the application policy matches the application and that the application policy does not contradict the system policy.

As mentioned in Section 1, it is not sufficient to only apply the application policy at runtime. The policy needs to be adapted.

7.3 Future Work

The work presented here is limited to *volatile user input*. This is input that is only relevant for the currently executed instance of an application. It is used, for example, to open a file the user intends to edit in the application. We will also address *persistent user input* in our future work, which persists over multiple executions of the application. This is the case, for instance, if the user specifies the path and name of a configuration file which is read each time the application is started. Once specified by the user, the application should have access rights for future executions.

Means are needed to handle new classes that are capable of obtaining user input by themselves without relying on classes from the system library. In such a rare case, aspects are to be auto-generated from aspect templates. This way, the developer is not required to write aspects in order to apply user input tracking to these new classes.

The approach presented here relies on the availability of source code of the application. Applicability on intermediate language code (Java bytecode or .NET CIL) is to be elaborated.

In some cases, additional information is required for generating the policy. In these cases the developer needs to specify meta information. Investigations on integrating this meta information specification into the development process with little developer involvement are required.

8 Conclusion

This paper presents means to auto-generate least privileges access control policies for applications. While existing work is used for the process of retrieving the contents for the policy by static and dynamic analyses, this paper introduces a way to distinguish resource access performed by the application from resource access initiated by the user. This distinction allows generating the application policy that satisfies the principle of least privileges. The application policy does not contain any access right to user resources, whereas existing work permits generic access to user resources. Access rights users need to access resources in the application are later added on the target system.

Two approaches are presented here. The static analysis approach uses a call graph of the application and performs various analyses to determine and tag resource identifiers that are defined by the user. The policy is generated without adding permissions that are based on tagged resource identifiers. The runtime observation approach tracks user input through the application using taint tracking and aspect-oriented programming. If user input propagates to a permission check where the resource identifier is specified by the user, the corresponding access is treated as user-initiated. A prototype is implemented in Java. Its implementation shows that the total number of aspects is kept in a manageable range. It suffices to augment those classes by aspects which play a key role in processing user input. As a result, the approach is feasible, it scales with respect to the size of instrumented libraries and it reduces overapproximation of existing approaches significantly. However it requires a fully object-oriented programming language, as AOP cannot be applied on primitive data types.

If both approaches are combined, a complete and sound policy is generated and overapproximation is eliminated. As the policy is auto-generated, the effort for the developer is low. The resulting application policy can be used on the target system to execute the application in its bounds. The target system only needs to specify access rights for user resources. Thus, the effort is also low there. As a consequence, policy generation becomes practical. An outlook of three obvious possibilities to apply the contributions in practice concludes the paper:

The mobile phone is an appealing target, as the mobile industry controls most of the phases of the SDLC. Development environments can be extended by the analyses, the generated policy can be included in the supply chain of applications and the execution environments on the mobile phones can be adapted to perform policy adaptation. Controlling access on a mobile phone to user resources, such as the phone book or the agenda, means controlling access to personal data which not all the applications need. Effective tailored access control on a per-application basis is possible and practical that way.

Execution environments, such as the Java Virtual Machine and the .NET CLR, provide a fine-grained and flexible security architecture that allows enforcing tailored access control policies. The problem in practice, however, is that it is complex to write proper policies. Integrating the policy auto-generation process into the SDLC reduces this effort to a minimum.

The two execution environments .NET CLR and Android allow for defining the access rights an application needs in a configuration file which is supplied together with the application. This is used at install time in order to assign the right access rights. However, there is no support in collecting all the needed access rights. The policy auto-generation process can be used to close this gap by filling the section of required permissions in the configuration file.

References

1. Saltzer, J.H., Schroeder, M.D.: The Protection of Information in Computer Systems. Proceedings of the IEEE 63(9), 1278–1308 (1975)
2. McGraw, G.: Software Security - Building Security. Addison-Wesley, USA (2006)
3. National Institute of Standards and Technology: National vulnerability database statistics, http://nvd.nist.gov/statistics.cfm (last checked: August 2010)
4. Koved, L., Pistoia, M., Kershenbaum, A.: Access rights analysis for java. In: OOPSLA 2002: Proceedings of the 17th ACM SIGPLAN Conference on Object-Oriented Programming, Systems, Languages, and Applications, pp. 359–372. ACM, New York (2002)
5. Centonze, P., Flynn, R., Pistoia, M.: Combining Static and Dynamic Analysis for Automatic Identification of Precise Access-Control Policies. In: Proceedings of the 23rd Annual Computer Security Applications Conference, ACSAC 2007, pp. 292–303 (December 2007)
6. Geay, E., Pistoia, M., Tateishi, T., Ryder, B.G., Dolby, J.: Modular String-Sensitive Permission Analysis with Demand-Driven Precision. In: Proceedings of the 31st International Conference on Software Engineering, pp. 177–187. IEEE, Los Alamitos (May 2009)
7. Provos, N.: Improving host security with system call policies. In: SSYM 2003: Proceedings of the 12th conference on USENIX Security Symposium, Berkeley, CA, USA, pp. 18–18. USENIX Association (2003)
8. Novell, Inc.: AppArmor, http://en.opensuse.org/AppArmor/ (last checked: August 2010)
9. Goldberg, A., Kay, A.: Smalltalk-72 Instruction Manual. Technical Report SSL 76-6, Learning Research Group, Xerox Palo Alto Research Center, California, USA (1976)
10. Eckel, B.: Thinking in Java, 3rd edn. Prentice Hall, Nwe Jersey (2003)
11. Gong, L., Ellison, G., Dagenforde, M.: Inside Java 2 Platform Security, 2nd edn. Addison-Wesley, Reading (2003)
12. Horwitz, S., Reps, T., Binkley, D.: Interprocedural Slicing Using Dependence Graphs. In: PLDI 1988: Proceedings of the ACM SIGPLAN 1988 Conference on Programming Language Design and Implementation, pp. 35–46. ACM, New York (1988)
13. Shivers, O.: Control flow analysis in scheme. In: Proceedings of the ACM SIGPLAN Conference on Programming Language Design and Implementation, pp. 164–174 (1988)
14. Cowan, C., Wright, C., Smalley, S., Morris, J., Kroah-Hartman, G.: Linux security modules: General security support for the linux kernel. In: Proceedings of the 11th USENIX Security Symposium, San Francisco, CA, USA (August 2002)
15. Wagner, D., Dean, D.: Intrusion detection via static analysis. In: Proceedings of the 22nd IEEE Symposium on Security and Privacy, pp. 156–169 (May 2001)
16. Bauer, L., Ligatti, J., Walker, D.: Composing security policies with Polymer. In: Proceedings of the 2005 ACM SIGPLAN Conference on Programming Language Design and Implementation (PLDI 2005), Chicago, IL, USA, pp. 305–314 (2005)

17. Miller, M.S.: Robust Composition - Towards a Unified Approach to Access Control and Concurrency Control. PhD thesis, Johns Hopkins University, Baltimore, MD, USA (May 2006)
18. Xu, W., Bhatkar, E., Sekar, R.: Taint-enhanced policy enforcement: A practical approach to defeat a wide range of attacks. In: 15th USENIX Security Symposium, pp. 121–136 (2006)
19. Nguyen-Tuong, A., Guarnieri, S., Greene, D., Shirley, J., Evans, D.: Automatically hardening web applications using precise tainting. In: 20th IFIP International Information Security Conference (SEC), pp. 372–382 (2005)
20. Sabelfeld, A., Myers, A.C.: Language-based information-flow security. IEEE Journal on Selected Areas in Communications 21 (2003)
21. Denning, D.E., Denning, P.J.: Certification of programs for secure information flow. Communications of the ACM 20(7), 504–513 (1977)
22. Wallach, D.S., Felten, E.W.: Understanding java stack inspection. In: Proceedings of the 1998 IEEE Symposium on Security and Privacy, pp. 52–63 (1998)
23. Kiczales, G., Lamping, J., Mendhekar, A., Maeda, C., Lopes, C., Loingtier, J.M., Irwin, J.: Aspect-Oriented Programming. In: Liu, Y., Auletta, V. (eds.) ECOOP 1997. LNCS, vol. 1241, pp. 220–242. Springer, Heidelberg (1997)
24. Hengst, G.: Auto-generation of access-control policies - elaboration of an information tracking approach and its prototype implementation. Bachelor's thesis, Munich University of Applied Sciences (September 2009)
25. Sun Microsystems Inc.: Java Technology, http://java.sun.com/ (last checked: August 2010)
26. Eclipse Foundation: Aspectj, http://www.eclipse.org/aspectj/ (last checked: August 2010)
27. Eclipse Foundation: eclipse, http://www.eclipse.org (last checked: August 2010)
28. Dólera Tormo, G., Martinez Perez, G.: UMU XACML-Editor, http://sourceforge.net/projects/umu-xacmleditor/ (last checked: August 2010)
29. S3MS project consortium: Security of Software and Services for Mobile Systems (S3MS), European research project, http://www.s3ms.org/ (last checked: August 2010)
30. Dragoni, N., Martinelli, F., Massacci, F., Mori, P., Schaefer, C., Walter, T., Vetillard, E.: Security-by-Contract (SxC) for Software and Services of Mobile Systems. In: Nitto, E.D., Sassen, A.M., Traverso, P., Zwegers, A. (eds.) At Your Service-Oriented Computing From an EU Perspective, pp. 429–455. MIT Press, Cambridge (2009)
31. Aktug, I., Naliuka, K.: ConSpec - a formal language for policy specification. In: First International Workshop on Run Time Enforcement for Mobile and Distributed Systems (REM 2007), Dresden, Germany (September 27, 2007)

Impossibility Results for RFID Privacy Notions

Frederik Armknecht[1], Ahmad-Reza Sadeghi[2], Alessandra Scafuro[3],
Ivan Visconti[3], and Christian Wachsmann[2]

[1] University of Mannheim, Germany
armknecht@informatik.uni-mannheim.de
[2] Horst Görtz Institute for IT-Security (HGI), Ruhr-University Bochum, Germany
{ahmad.sadeghi,christian.wachsmann}@trust.rub.de
[3] Dipartimento di Informatica ed Applicazioni, University of Salerno, Italy
{scafuro,visconti}@dia.unisa.it

Abstract. RFID systems have become increasingly popular and are already used in many real-life applications. Although very useful, RFIDs introduce privacy risks since they carry identifying information that can be traced. Hence, several RFID privacy models have been proposed. However, they are often incomparable and in part do not reflect the capabilities of real-world adversaries. Recently, Paise and Vaudenay presented a general RFID security and privacy model that abstracts and unifies most previous approaches. This model defines mutual authentication (between RFID tags and readers) and several privacy notions that capture adversaries with different tag corruption behavior and capabilities.

In this paper, we revisit the model proposed by Paise and Vaudenay and investigate some subtle issues such as tag corruption aspects. We show that in their formal definitions tag corruption discloses the temporary memory of tags and leads to the impossibility of achieving both mutual authentication and any reasonable notion of RFID privacy in their model. Moreover, we show that the strongest privacy notion (narrow-strong privacy) cannot be achieved simultaneously with reader authentication even under the strong assumption that tag corruption does not disclose temporary tag states. Further, we show other impossibility results that hold if the adversary can manipulate an RFID tag such that it resets its state or when tags are stateless.

Although our results are shown on the privacy definition by Paise and Vaudenay, they give insight to the difficulties of setting up a mature security and privacy model for RFID systems that aims at fulfilling the sophisticated requirements of real-life applications.

Keywords: RFID, Privacy, Authentication, Security, Resettability.

1 Introduction

Radio Frequency Identification (RFID) enables RFID *readers* to perform fully automatic wireless identification of objects that are labeled with RFID *tags*, and is widely deployed to many applications (e.g., access control [2,29], electronic tickets [31,29], and e-passports [19]). As pointed out in previous publications

M.L. Gavrilova et al. (Eds.): Trans. on Comput. Sci. XI, LNCS 6480, pp. 39–63, 2010.
© Springer-Verlag Berlin Heidelberg 2010

(see, e.g., [38,20,34]), this prevalence of RFID technology introduces various risks, in particular concerning the privacy of its users and holders. The most deterrent privacy risk concerns the tracking of users, which allows the creation and misuse of detailed user profiles. Thus, it is desired that an RFID system provides *anonymity* (confidentiality of the tag identity) as well as *untraceability* (unlinkability of the communication of a tag), even in case the state (e.g., the secret) of a tag has been disclosed.

The design of a secure privacy-preserving RFID scheme requires a careful analysis in an appropriate formal model. There is a large body of literature on security and privacy models for RFID (see, e.g., [3,21,8,37,30,12]). Existing solutions often do not consider important aspects like adversaries with access to auxiliary information, e.g., on whether the identification of a tag was successful, or the privacy of corrupted tags whose state has been disclosed. In particular, tag corruption is usually considered to happen only *before* and/or *after* but *not during* a protocol-run. However, in practice there are a variety of side-channel attacks (see., e.g., [24,18,22]) that extract the state of a tag based on the observation of, e.g., the power consumption of the tag *while* it is executing a protocol with the reader. Since RFID tags are usually cost-effective devices without expensive tamper-proof mechanisms [2,29], tag corruption is an important aspect to be covered by the underlying (formal) security model. Though in literature, tag corruption during protocol execution is rarely considered. To the best of our knowledge, the security and privacy model in [8] is the only one that considers corruption of tags during protocol executions and proposes a protocol in this model. However, this model does not consider issues like the privacy of tags after they have been corrupted and privacy against adversaries with access to auxiliary information. Moreover, [8] only provides an informal security analysis of the proposed protocol. Recently, tag corruption during protocol-runs has been informally discussed in [12]. However, the formal RFID security and privacy model proposed in [12] assumes that such attacks cannot occur. Moreover, [12] indicates informally without giving formal proofs that tag corruption during protocol execution may have an impact on the formal definitions of [37,30], which are basis for many subsequent works (see, e.g., [26,25,7,33,32,11,10,36,35]). The first papers addressing tag corruption during protocol runs in the model of [37] are [11,10], where it is shown that privacy can be achieved under the assumption that tag corruption during protocol execution can be detected by the tag.

In this paper, we focus on the security and privacy model by Paise and Vaudenay [30] (that is based on [37]), which we call the *PV-Model* (Paise-Vaudenay Model) in the following. The PV-Model is one of the most comprehensive RFID security and privacy models up to date since it captures many aspects of real world RFID systems and aims at abstracting most previous works in a single concise framework. It defines mutual authentication between RFID tags and readers and several privacy notions that correspond to adversaries with different tag corruption abilities. However, as we show in this paper, the PV-Model suffers from subtle deficiencies and weaknesses that are mainly caused by tag corruption aspects: in the PV-Model, each tag maintains a state that can be divided into a persistent and a temporary part.[1] The *persistent state* subsumes all information that must be available to the tag in more than one interaction with the reader (e.g., the authentication secret of the tag) and can be updated during the interaction with the reader. The *temporary state* consists of all ephemeral information that is discarded by the tag after each interaction with the reader (e.g., the randomness used by the tag). As discussed in [30], in the PV-Model it is impossible to achieve any notion of privacy that allows tag corruption if the adversary can obtain *both* the persistent *and* the temporary tag state by tag corruption. This issue is addressed by the PV-Model by the assumption that each tag erases its temporary state each time it gets out of the reading range of the adversary. However, this assumption leaves open the possibility to corrupt a tag *while* it is in the reading range of the adversary, i.e., *before* its temporary state is erased. In particular, the PV-Model allows the adversary to corrupt a tag *while* it is executing the authentication protocol with the reader.

Moreover, an adversary in practice could physically tamper with a tag such that the tag resets its state and randomness to a previous value. This form of physical attack is not considered in the PV-Model and thus, the study of privacy notions done in [30] does not address these attacks.

Contribution. In this paper, we point out subtle weaknesses and deficiencies in the PV-Model. First, we show that the assumption of erasing temporary tag states whenever a tag gets out of the reading range of the adversary made by the PV-Model is not strong enough. We prove that, even under this assumption, it is *impossible* to achieve reader authentication and simultaneously *any* notion of privacy that allows tag corruption. This implies that the PV-Model cannot provide privacy along with mutual authentication without relying on tamper-proof hardware, which is unrealistic for low-cost RFID tags. Consequently, two of the three schemes presented in [30] do not satisfy their claimed properties.

Our second contribution is to show that even under the strong assumption that the temporary tag state is not subject to tag corruption attacks, some privacy notions still remain impossible in the PV-Model. This implies that the third protocol of [30] has another conceptually different weakness.

Finally, we show that by extending the model of [30] to capture reset attacks on tag states and randomness, no privacy can be achieved, and, more interestingly, when tags are stateless (i.e., when tags cannot update their persistent state), then destructive privacy is impossible. Although our results are shown on the privacy model by Paise and Vaudenay, we believe that our work is helpful for developing a mature security and privacy model for RFID systems that fulfills the sophisticated requirements of real-life applications.

Outline. We first informally discuss the general RFID scenario on a high level in Section 2. Then we focus on the formalization of the relevant aspects by revisiting the RFID security and privacy model by Paise and Vaudenay (PV-Model) [30]

---

[1] During a protocol execution tags could store some temporary information that allows them to verify the response of the reader.

in Section 3. In Section 4 we present our first result while our second result is shown in Section 5. In Section 6 we show our third impossibility result based on resettable and stateless tags. Finally, we conclude in Section 7.

2   RFID System and Requirement Analysis

System model. An RFID system consists of at least an operator $\mathcal{I}$, a reader $\mathcal{R}$ and a tag $\mathcal{T}$. $\mathcal{I}$ is the entity that enrolls and maintains the RFID system. Hence, $\mathcal{I}$ initializes $\mathcal{T}$ and $\mathcal{R}$ before they are deployed in the system. $\mathcal{T}$ and $\mathcal{R}$ are called *legitimate* if they have been initialized by $\mathcal{I}$. In many applications $\mathcal{T}$ is a hardware token with constrained computing and memory capabilities that is equipped with a radio interface [2,29]. All information, e.g., secrets and data that is stored on $\mathcal{T}$ is denoted as the *state* of $\mathcal{T}$. Usually $\mathcal{T}$ is attached to some object or carried by a user of the RFID system [14,28]. $\mathcal{R}$ is a stationary or mobile computing device that interacts with $\mathcal{T}$ when $\mathcal{T}$ gets into the reading range of $\mathcal{R}$. The main purpose of this interaction usually is the authentication of $\mathcal{T}$ to $\mathcal{R}$. Depending on the use case, $\mathcal{R}$ may also authenticate to $\mathcal{T}$ and/or obtain additional information like the identity of $\mathcal{T}$. $\mathcal{R}$ can have a sporadic or permanent online connection to some backend system $\mathcal{D}$, which typically is a database maintaining detailed information on all tags in the system. $\mathcal{D}$ is initialized and maintained by $\mathcal{I}$ and can be read and updated by $\mathcal{R}$.

Trust and adversary model. The operator $\mathcal{I}$ maintains the RFID system and is considered to behave correctly. However, $\mathcal{I}$ may be curious and collect user information. Since $\mathcal{T}$ and $\mathcal{R}$ communicate over a radio link, any entity can eavesdrop and manipulate this communication, even from outside the nominal reading range of $\mathcal{R}$ and $\mathcal{T}$ [23]. Thus, the adversary $\mathcal{A}$ can be every (potentially unknown) entity. Besides the communication between $\mathcal{T}$ and $\mathcal{R}$, $\mathcal{A}$ can also obtain useful auxiliary information (e.g., by visual observation) on whether $\mathcal{R}$ accepted $\mathcal{T}$ as a legitimate tag [21,37]. Most commercial RFID tags are cost-efficient devices without expensive protection mechanisms against physical tampering [2,29]. Hence, $\mathcal{A}$ can physically attack (*corrupt*) $\mathcal{T}$ and obtain its state, e.g., its secrets. In practice, RFID readers are embedded devices that can be integrated into mobile devices (e.g., mobile phones or PDAs) or computers. The resulting complexity exposes readers to sophisticated hard- and software attacks, e.g., viruses and Trojans. This problem aggravates for mobile readers that can easily be lost or stolen. Hence, $\mathcal{A}$ can get full control over $\mathcal{R}$ [4,16,27].

Security and privacy objectives. The most deterrent privacy risk concerns the *tracking* of tag users, which allows the creation and misuse of detailed user profiles in an RFID system [20]. For instance, detailed movement profiles can leak sensitive information on the personal habits and interests of the tag user. The major security threats are to create illegitimate (*forge*) tags that are accepted by honest readers, to simulate (*impersonate*) or to copy (*clone*) legitimate tags, and to permanently prevent users from using the RFID system (*denial-of-service*) [8].

Thus, an RFID system should provide *anonymity* as well as *untraceability* of a tag T even when the state of T has been disclosed. Anonymity means the confidentiality of the identity of T whereas untraceability refers to the unlinkability of the communication of T. The main security objective is to ensure that only legitimate tags are accepted by honest readers (*tag authentication*). Most use cases (like access control systems) additionally require $\mathcal{R}$ to determine the authentic tag identity (*tag identification*). Moreover, there are several applications (e.g., electronic tickets) where reader authentication is a fundamental security property. However, there are also use cases (e.g., electronic product labels) that do not require reader authentication.

3   The PV-Model

In this section, we recall the RFID security and privacy model by Paise and Vaudenay (PV-Model) [30] that refines the model in [37]. We give a more formal specification of this model, which is one of the most comprehensive RFID privacy and security models up to date. We start by specifying our notation.

General notation. For a finite set S, $|S|$ denotes the size of S whereas for an integer (or a bit-string) n the term $|n|$ means the bit-length of n. The term $s \in_R S$ means the assignment of a uniformly chosen element of S to variable s. Let A be a probabilistic algorithm. Then $y \leftarrow \mathsf{A}(x)$ means that on input x, algorithm A assigns its output to variable y. The term $[\mathsf{A}(x)]$ denotes the set of all possible outputs of A on input x. $\mathsf{A}_K(x)$ means that the output of A depends on x and some additional parameter K (e.g., a secret key). The term $\mathsf{Prot}[\mathsf{A}:x_\mathsf{A};\ \mathsf{B}:x_\mathsf{B};\ *:x_{pub}] \rightarrow [\mathsf{A}:y_\mathsf{A};\ \mathsf{B}:y_\mathsf{B}]$ denotes an interactive protocol Prot between two probabilistic algorithms A and B. Hereby, A (resp. B) gets a private input x_A (resp. x_B) and a public input x_{pub}. While A (resp. B) is operating, it can interact with B (resp. A). After the protocol terminates, A (resp. B) returns y_A (resp. y_B). Let E be some event (e.g., the result of a security experiment), then $\Pr[E]$ denotes the probability that E occurs. Probability $\epsilon(l)$ is called *negligible* if for all polynomials f it holds that $\epsilon(l) \leq 1/f(l)$ for all sufficiently large l. Probability $1 - \epsilon(l)$ is called *overwhelming* if $\epsilon(l)$ is negligible.

3.1   System Model

The PV-Model considers RFID systems that consist of a single operator $\mathcal{I}$, a single reader $\mathcal{R}$ and a polynomial number of tags T. Note that the PV-Model does not explicitly define an entity that corresponds to the operator $\mathcal{I}$ but implies the existence of such an entity. $\mathcal{R}$ is assumed to be capable of performing public-key cryptography and of handling multiple instances of the mutual authentication protocol with different tags in parallel. Each tag T is a passive device, i.e., it does not have its own power supply but is powered by the electromagnetic field of $\mathcal{R}$. Hence, T cannot initiate communication, has a narrow communication range (i.e., a few centimeters to meters) and erases its temporary state (i.e., all session-specific information and randomness) after it gets out of the reading range of $\mathcal{R}$. Each $\mathcal{T}$ is assumed to be capable of computing basic cryptographic functions like hashing, random number generation and symmetric-key encryption. The authors of [37,30] also use public-key encryption, although it exceeds the capabilities of most currently available RFID tags [2,29].

Security and privacy objectives. The main security objective of the PV-Model is mutual authentication. More precisely, $\mathcal{R}$ should only accept legitimate tags and must be able to identify them, while each legitimate tag $\mathcal{T}$ should only accept $\mathcal{R}$. Availability and protection against cloning are not captured by the PV-Model. The privacy objectives are anonymity and unlinkability.

Definitions. The operator $\mathcal{I}$ sets up $\mathcal{R}$ and all tags $\mathcal{T}$. Hence, there are two setup algorithms where $\mathcal{R}$ and $\mathcal{T}$ are initialized and their system parameters (e.g., keys) are generated and defined. A protocol between $\mathcal{T}$ and $\mathcal{R}$ covers mutual authentication.

Definition 1 (RFID System [30]). *An RFID system is a tuple of probabilistic polynomial time (p.p.t.) algorithms* $(\mathcal{R}, \mathcal{T}, \mathsf{SetupReader}, \mathsf{SetupTag}, \mathsf{Ident})$ *that are defined as follows:*

$\mathsf{SetupReader}(1^l) \to (sk_\mathcal{R}, pk_\mathcal{R}, \mathrm{DB})$ *On input of a security parameter l, this algorithm creates the public parameters $pk_\mathcal{R}$ that are known to all entities. Moreover, it creates the secret parameters $sk_\mathcal{R}$ and a database DB that can only be accessed by $\mathcal{R}$.*

$\mathsf{SetupTag}_{pk_\mathcal{R}}(\mathrm{ID}) \to (K, S)$ *uses $pk_\mathcal{R}$ to generate a tag secret K and tag state S, initializes $\mathcal{T}_{\mathrm{ID}}$ with S, and stores (ID, K) in DB.*

$\mathsf{Ident}[\mathcal{T}_{\mathrm{ID}} : S; \ \mathcal{R} : sk_\mathcal{R}, \mathrm{DB}; \ * : pk_\mathcal{R}] \to [\mathcal{T}_{\mathrm{ID}} : out_{\mathcal{T}_{\mathrm{ID}}}; \ \mathcal{R} : out_\mathcal{R}]$ *is an interactive protocol between $\mathcal{T}_{\mathrm{ID}}$ and $\mathcal{R}$. $\mathcal{T}_{\mathrm{ID}}$ takes as input its current state S while $\mathcal{R}$ has input $sk_\mathcal{R}$ and DB. The common input to all parties is $pk_\mathcal{R}$. After the protocol terminates, $\mathcal{R}$ returns either the identity ID of $\mathcal{T}_{\mathrm{ID}}$ or $\perp$ to indicate that $\mathcal{T}_{\mathrm{ID}}$ is not a legitimate tag. $\mathcal{T}_{\mathrm{ID}}$ returns either ok to indicate that $\mathcal{R}$ is legitimate or $\perp$ otherwise.*

Definition 2 (Correctness [30]). *An RFID system (Definition 1) is correct if $\forall \ l$, $\forall \ (sk_\mathcal{R}, pk_\mathcal{R}, \mathrm{DB}) \in [\mathsf{SetupReader}(1^l)]$, and $\forall \ (K, S) \in [\mathsf{SetupTag}_{pk_\mathcal{R}}(\mathrm{ID})]$ $\mathsf{Ident}[\mathcal{T}_{\mathrm{ID}} : S; \ \mathcal{R} : sk_\mathcal{R}, \mathrm{DB}; \ * : pk_\mathcal{R}] \to [\mathcal{T}_{\mathrm{ID}} : \mathsf{ok}; \ \mathcal{R} : \mathrm{ID}]$ holds with overwhelming probability.*

3.2 Trust and Adversary Model

In the PV-Model, the issuer $\mathcal{I}$, the backend database $\mathcal{D}$ and the readers are assumed to be trusted whereas a tag $\mathcal{T}$ can be compromised. All readers and $\mathcal{D}$ are subsumed to *one single* reader entity $\mathcal{R}$ that cannot be corrupted. This implies that all readers are assumed to be tamper-resistant devices that have a permanent online connection to $\mathcal{D}$.[2] The PV-Model defines privacy and security

---

[2] Depending on the use case, this assumption can be problematic in practice, e.g., for mobile readers that usually have only a sporadic or no online connection and that are subject to a variety of soft- and hardware attacks.

as security experiments, where a p.p.t. adversary $\mathcal{A}$ can interact with a set of oracles that model the capabilities of $\mathcal{A}$. These oracles are:

CreateTag^b(ID) Allows $\mathcal{A}$ to set up a tag $\mathcal{T}_{\text{ID}}$ with identifier ID by internally calling SetupTag$_{pk_{\mathcal{R}}}$(ID) to create (K, S) for $\mathcal{T}_{\text{ID}}$. If input $b = 1$, then (ID, K) is added to DB. If $b = 0$, then (ID, K) is *not* added to DB.

Draw(δ) $\rightarrow$ $(vtag_1, b_1, \dots, vtag_n, b_n)$ Initially, $\mathcal{A}$ cannot interact with any tag but must query Draw to get access to a set of tags chosen according to a probability distribution δ. $\mathcal{A}$ knows the tags it can interact with by some temporary tag identifiers $vtag_1, \dots, vtag_n$. Draw manages a secret look-up table Γ that keeps track of the real tag identifier ID_i associated with each temporary tag identifier $vtag_i$, i.e., $\Gamma[vtag_i] = \text{ID}_i$. Moreover, Draw also provides $\mathcal{A}$ with information on whether the tags are legitimate ($b_i = 1$) or not ($b_i = 0$).

Free($vtag$) Makes tag $vtag$ inaccessible to $\mathcal{A}$ such that $\mathcal{A}$ cannot interact with $vtag$ until it is made accessible again under a new temporary identifier $vtag'$ by another Draw query.

Launch( ) $\rightarrow$ π Makes $\mathcal{R}$ to start a new instance π of the Ident protocol.

SendReader(m, π) $\rightarrow$ m' Sends a message m to instance π of the Ident protocol that is running on $\mathcal{R}$. $\mathcal{R}$ interprets m as a protocol message of instance π of the Ident protocol and responds with a message m'.

SendTag($m, vtag$) $\rightarrow$ m' Sends a message m to the tag $vtag$, which interprets m as a protocol message of the Ident protocol and responds with a message m'.

Result(π) Returns 1 if instance π of the Ident protocol has been completed and the tag $\mathcal{T}_{\text{ID}}$ that participated in instance π has been accepted by $\mathcal{R}$. Otherwise Result returns 0.

Corrupt($vtag$) $\rightarrow$ S Returns the current state S (i.e., all information stored in the memory) of the tag $vtag$ to $\mathcal{A}$.

The PV-Model distinguishes eight adversary classes, which differ in (i) their ability to corrupt tags and (ii) the availability of auxiliary information, i.e., the ability to access the Corrupt and Result oracle, respectively.

Definition 3 (Adversary Classes [30]). *An adversary is a p.p.t. algorithm that has arbitrary access to all oracles described in Section 3.2. Weak adversaries cannot access the Corrupt oracle. Forward adversaries cannot query any other oracle than Corrupt after they made the first Corrupt query. Destructive adversaries cannot query any oracle for vtag again after they made a Corrupt(vtag) query. Strong adversaries have no restrictions on the use of the Corrupt oracle. Narrow adversaries cannot access the Result oracle.*

Tag corruption aspects. Depending on the concrete scenario, the temporary tag state is disclosed under tag corruption. In general, any concrete scenario will range between the following two extremes: (i) corruption discloses the full temporary tag state, or (ii) corruption does not disclose any information on the temporary tag state. In Section 4 and 5, we will prove that in both cases some privacy notions are impossible to achieve in the PV-Model. Thus, *independently* of any possible interpretation of tag corruption, impossibility results exist that contradict the claims of [30].

3.3 Security Definition

The security definition of the PV-Model focuses on attacks where the adversary aims to impersonate or forge a legitimate tag $\mathcal{T}$ or the reader $\mathcal{R}$. It does *not* capture availability and security against cloning.

Tag authentication. The definition of tag authentication is based on a security experiment $\mathbf{Exp}_{\mathcal{A}_{\text{sec}}}^{\mathcal{T}\text{-aut}}$ where a strong adversary $\mathcal{A}_{\text{sec}}$ (Definition 3) must make $\mathcal{R}$ to identify some tag $\mathcal{T}_{\text{ID}}$ in some instance π of the Ident protocol. To exclude trivial attacks (e.g., relay attacks), $\mathcal{A}_{\text{sec}}$ is not allowed to simply forward all the messages from $\mathcal{T}_{\text{ID}}$ to $\mathcal{R}$ in instance π nor to corrupt $\mathcal{T}_{\text{ID}}$. This means that at least some of the protocol messages that made $\mathcal{R}$ to return ID must have been computed by $\mathcal{A}_{\text{sec}}$ without knowing the secrets of $\mathcal{T}_{\text{ID}}$. With $\mathbf{Exp}_{\mathcal{A}_{\text{sec}}}^{\mathcal{T}\text{-aut}} = 1$ we denote the case where $\mathcal{A}_{\text{sec}}$ wins the security experiment.

Definition 4 (Tag Authentication [30]). *An RFID system (Definition 1) achieves tag authentication if for every strong adversary $\mathcal{A}_{\text{sec}}$ (Definition 3) $\Pr[\mathbf{Exp}_{\mathcal{A}_{\text{sec}}}^{\mathcal{T}\text{-aut}} = 1]$ is negligible.*

Reader Authentication. The definition of reader authentication is based on a security experiment $\mathbf{Exp}_{\mathcal{A}_{\text{sec}}}^{\mathcal{R}\text{-aut}}$ where a strong adversary $\mathcal{A}_{\text{sec}}$ (Definition 3) must successfully impersonate $\mathcal{R}$ to a legitimate tag $\mathcal{T}_{\text{ID}}$. Also here, to exclude trivial attacks, $\mathcal{A}_{\text{sec}}$ must achieve this without simply forwarding the protocol messages from $\mathcal{R}$ to $\mathcal{T}_{\text{ID}}$. This means that $\mathcal{A}_{\text{sec}}$ must have computed at least some of the protocol messages that made $\mathcal{T}_{\text{ID}}$ to return ok. With $\mathbf{Exp}_{\mathcal{A}_{\text{sec}}}^{\mathcal{R}\text{-aut}} = 1$ we denote the case where $\mathcal{A}_{\text{sec}}$ wins the security experiment.

Definition 5 (Reader Authentication [30]). *An RFID system (Definition 1) achieves reader authentication if for every strong adversary $\mathcal{A}_{\text{sec}}$ (Definition 3) $\Pr[\mathbf{Exp}_{\mathcal{A}_{\text{sec}}}^{\mathcal{R}\text{-aut}} = 1]$ is negligible.*

Note that both tag and reader authentication are critical properties that must be preserved even against strong adversaries.

3.4 Privacy Definition

The privacy definition of the PV-Model is very flexible and, dependent on the adversary class (see Definition 3), it covers different notions of privacy. It captures anonymity and unlinkability and focuses on the privacy leakage of the communication of tags with the reader. It is based on the existence of a simulator $\mathcal{B}$, called *blinder*, that can simulate $\mathcal{R}$ and any tag $\mathcal{T}$ without knowing their secrets such that an adversary $\mathcal{A}_{\text{prv}}$ cannot distinguish whether it is interacting with the real or the simulated RFID system. The rationale behind this simulation-based definition is that the communication of $\mathcal{T}$ and $\mathcal{R}$ does not leak any information about $\mathcal{T}$. Hence, everything $\mathcal{A}_{\text{prv}}$ observes from the interaction with $\mathcal{T}$ and $\mathcal{R}$ appears to be independent of $\mathcal{T}$ and consequently, $\mathcal{A}_{\text{prv}}$ cannot distinguish different tags based on their communication.

This privacy definition can be formalized by the following privacy experiment $\mathbf{Exp}_{\mathcal{A}_{prv}}^{prv\text{-}b} = b'$: let $\mathcal{A}_{prv}$ be an adversary according to Definition 3, l be a given security parameter and $b \in_R \{0, 1\}$. In the first phase of the experiment, $\mathcal{R}$ is initialized with $(sk_{\mathcal{R}}, pk_{\mathcal{R}}, DB) \leftarrow \mathsf{SetupReader}(1^l)$. The public key $pk_{\mathcal{R}}$ is given to $\mathcal{A}_{prv}$ and $\mathcal{B}$. Now, $\mathcal{A}_{prv}$ is allowed to arbitrarily interact with all oracles defined in Section 3.2. Hereby, $\mathcal{A}_{prv}$ is subject to the restrictions of its corresponding adversary class (see Definition 3). If $b = 1$, all queries to the Launch, SendReader, SendTag and Result oracles are redirected to and answered by $\mathcal{B}$. Hereby, $\mathcal{B}$ can observe all queries $\mathcal{A}_{prv}$ makes to all other oracles that are not simulated by $\mathcal{B}$ and the corresponding responses ("$\mathcal{B}$ sees what $\mathcal{A}_{prv}$ sees"). After a polynomial number of oracle queries, the second phase of the experiment starts. In this second stage, $\mathcal{A}_{prv}$ cannot interact with the oracles but is given the secret table Γ of the Draw oracle. Finally, $\mathcal{A}_{prv}$ returns a bit b'.

Definition 6 (Privacy [37]). *Let C be one of the adversary classes according to Definition 3. An RFID system (Definition 1) is C-private if for every adversary $\mathcal{A}_{prv}$ of C there exists a p.p.t. algorithm $\mathcal{B}$ (blinder) such that the advantage $\mathbf{Adv}_{\mathcal{A}_{prv}}^{prv} = \left| \Pr\left[\mathbf{Exp}_{\mathcal{A}_{prv}}^{prv\text{-}0} = 1\right] - \Pr\left[\mathbf{Exp}_{\mathcal{A}_{prv}}^{prv\text{-}1} = 1\right] \right|$ of $\mathcal{A}_{prv}$ is negligible. $\mathcal{B}$ simulates the Launch, SendReader, SendTag and Result oracles to $\mathcal{A}_{prv}$ without having access to $sk_{\mathcal{R}}$ and DB. Hereby, all oracle queries $\mathcal{A}_{prv}$ makes and their corresponding responses are also sent to $\mathcal{B}$.*

All privacy notions defined in the PV-Model are summarized in Figure 1, which also shows their relations. It has been shown that strong privacy is impossible [37] while the technical feasibility of destructive privacy currently is an open problem.

$$\begin{array}{ccccccc}
\text{Strong} & \Rightarrow & \text{Destructive} & \Rightarrow & \text{Forward} & \Rightarrow & \text{Weak} \\
\Downarrow & & \Downarrow & & \Downarrow & & \Downarrow \\
\text{Narrow-Strong} & \Rightarrow & \text{Narrow-Destructive} & \Rightarrow & \text{Narrow-Forward} & \Rightarrow & \text{Narrow-Weak}
\end{array}$$

Fig. 1. Privacy notions defined in the PV-Model and their relations

4   Corruption with Temporary State Disclosure

We now point out a subtle weakness of the PV-Model. We show that in the PV-Model it is *impossible* to achieve *any* notion of privacy simultaneously with reader authentication (under temporary state disclosure) except for the weak and narrow-weak privacy notions. As a consequence, two of the protocols given in [30] do not achieve their claimed privacy properties.

We stress that this impossibility result is due to the fact that, according to the formal definitions of the PV-Model, the adversary can obtain the *full* state including the temporary memory of a tag by corrupting the tag *while* it is executing a protocol with the reader. Such attacks are a serious threat in practice, in particular to low-cost RFID tags, and hence must be formally considered.

Although [30] informally discusses an issue related to tag corruption during protocol execution, we show that such attacks are *not* adequately captured by the formal definitions of the PV-Model. Hence, the only achievable privacy notions are those where the adversary is not allowed to corrupt tags at all. Since in practice tag corruption is realistic, this implies that using the PV-Model is not helpful when reader authentication and a reasonable notion of privacy are needed.

Impossibility of narrow-forward privacy. To prove our first impossibility result, we need the following lemma, which we will prove in detail further below:

Lemma 1. *If there is a blinder $\mathcal{B}$ for every narrow-forward adversary $\mathcal{A}_{\mathrm{prv}}$ such that $\mathbf{Adv}_{\mathcal{A}_{\mathrm{prv}}}^{\mathrm{prv}}$ is negligible (Definition 6), then $\mathcal{B}$ can be used to construct an adversary $\mathcal{A}_{\mathrm{sec}}^{\mathcal{B}}$ such that $\Pr[\mathbf{Exp}_{\mathcal{A}_{\mathrm{sec}}^{\mathcal{B}}}^{\mathcal{R}\text{-aut}} = 1]$ is non-negligible (Definition 5).*

Based on this lemma, we set up the following theorem, which we need later to prove our main impossibility result:

Theorem 1. *There is no RFID system (Definition 1) that achieves both reader authentication (Definition 5) and narrow-forward privacy (Definition 6) under temporary tag state disclosure.*

Proof (Theorem 1). Let $\mathcal{A}_{\mathrm{prv}}$ be a narrow-forward adversary (Definition 3). Definition 6 requires the existence of a blinder $\mathcal{B}$ such that $\mathcal{A}_{\mathrm{prv}}$ cannot distinguish $\mathcal{B}$ from the real oracles. From Lemma 1 it follows that such a $\mathcal{B}$ can be used to impersonate $\mathcal{R}$ to any legitimate tag $\mathcal{T}_{\mathrm{ID}}$ with non-negligible probability. Hence, the existence of $\mathcal{B}$ contradicts reader authentication (Definition 5). □

Proof (Lemma 1). First, we show how to construct $\mathcal{A}_{\mathrm{sec}}^{\mathcal{B}}$ from $\mathcal{B}$. Second, we prove that $\mathcal{A}_{\mathrm{sec}}^{\mathcal{B}}$ violates reader authentication (Definition 5) if $\mathcal{B}$ is such that $\mathbf{Adv}_{\mathcal{A}_{\mathrm{prv}}}^{\mathrm{prv}}$ is negligible for every narrow-forward $\mathcal{A}_{\mathrm{prv}}$ (Definition 3).

Let $q_{\mathcal{R}} \in \mathbb{N}$ with $q_{\mathcal{R}} > 0$ be the (expected) number of SendReader queries as specified by the Ident protocol and let $S_i^{\mathcal{R}}$ be the state of $\mathcal{R}$ after processing the i-th SendReader query. The initial reader state $S_0^{\mathcal{R}}$ includes the public key $pk_{\mathcal{R}}$ and the secret key $sk_{\mathcal{R}}$ of $\mathcal{R}$ as well as a pointer to the credentials database DB. Note that during the processing of a SendReader query, $\mathcal{R}$ can update DB. $\mathcal{R}$ can be considered as a tuple of algorithms $(\mathcal{R}_\pi^{(1)}, \ldots, \mathcal{R}_\pi^{(q_{\mathcal{R}})})$, where $\mathcal{R}_\pi^{(i)}$ represents the computation done by $\mathcal{R}$ when processing the i-th SendReader query in instance π of the Ident protocol. More formally: $(S_1^{\mathcal{R}}, m_1) \leftarrow \mathcal{R}_\pi^{(0)}(S_0^{\mathcal{R}})$ and $(S_{i+1}^{\mathcal{R}}, m_{2i+1}) \leftarrow \mathcal{R}_\pi^{(i)}(S_i^{\mathcal{R}}, m_{2i})$ for $1 \leq i < q_{\mathcal{R}}$. Since tags are passive devices that cannot initiate communication $\mathcal{R}$ must send the first protocol message. Thus, $\mathcal{R}$ generates all protocol messages with odd indices whereas the tag $\mathcal{T}$ generates all messages with even indices. In case the Ident protocol specifies that $\mathcal{T}$ sends the last protocol message, then $m_{2q_{\mathcal{R}}-1}$ is the empty string.

Let $q_{\mathcal{T}} \in \mathbb{N}$ with $q_{\mathcal{T}} > 0$ be the (expected) number of SendTag queries as specified by the Ident protocol and let $S_i^{\mathcal{T}}$ be the state of $\mathcal{T}$ after processing the i-th SendTag query. $\mathcal{T}$ can be represented as a tuple of algorithms $(\mathcal{T}^{(1)}, \ldots, \mathcal{T}^{(q_{\mathcal{T}})})$ where $\mathcal{T}^{(i)}$ means the computation done by $\mathcal{T}$ when processing the i-th SendTag

---

Alg. 1. Adversary $\mathcal{A}_{\text{sec}}^{\mathcal{B}}$ violating reader authentication

---

1: CreateTag(ID)
2: $vtag \leftarrow \text{Draw}(\Pr[\text{ID}] = 1)$
3: $\pi \leftarrow \text{Launch}(\ )$ $\triangleright$ simulated by $\mathcal{B}$
4: $m_1 \leftarrow \text{SendReader}(-, \pi)$ $\triangleright$ simulated by $\mathcal{B}$
5: $i \leftarrow 1$
6: **while** $i < q_{\mathcal{R}}$ **do**
7:      **if** $i \leq q_{\mathcal{T}}$ **then** $m_{2i} \leftarrow \text{SendTag}(m_{2i-1}, vtag)$ $\triangleright$ simulated by $\mathcal{B}$
8:      **end if**
9:      $m_{2i+1} \leftarrow \text{SendReader}(m_{2i}, \pi)$ $\triangleright$ simulated by $\mathcal{B}$
10:      $i \leftarrow i + 1$
11: **end while**
12: $out_{\mathcal{T}_{\text{ID}}} \leftarrow \text{SendTag}(m_{2q_{\mathcal{R}}-1}, vtag)$ $\triangleright$ computed by $\mathcal{T}_{\text{ID}}$

---

query in an instance of the Ident protocol that involves $\mathcal{T}$. More formally: $(S_{i+1}^{\mathcal{T}}, m_{2i}) \leftarrow \mathcal{T}^{(i)}(S_i^{\mathcal{T}}, m_{2i-1})$ for $1 \leq i \leq q_{\mathcal{T}}$. Note that $m_{2q_{\mathcal{T}}}$ is the empty string if Ident specifies that $\mathcal{R}$ must send the last protocol message.

The idea of $\mathcal{A}_{\text{sec}}^{\mathcal{B}}$ is to internally use $\mathcal{B}$ as a black-box to simulate the final protocol message of $\mathcal{R}$ that makes each legitimate tag $\mathcal{T}_{\text{ID}}$ to accept $\mathcal{A}_{\text{sec}}^{\mathcal{B}}$ as $\mathcal{R}$. The construction of $\mathcal{A}_{\text{sec}}^{\mathcal{B}}$ is shown in Algorithm 1. First, $\mathcal{A}_{\text{sec}}^{\mathcal{B}}$ creates a legitimate tag $\mathcal{T}_{\text{ID}}$ (step 1) and makes it accessible (step 2). Both steps are also shown to $\mathcal{B}$, which expects to observe all oracle queries. Then, $\mathcal{A}_{\text{sec}}^{\mathcal{B}}$ makes $\mathcal{B}$ to start a new instance π of the Ident protocol with $\mathcal{T}_{\text{ID}}$ (step 3) and obtains the first protocol message m_1 generated by $\mathcal{B}$ (step 4). Now, $\mathcal{A}_{\text{sec}}^{\mathcal{B}}$ internally runs $\mathcal{B}$ that simulates both $\mathcal{T}_{\text{ID}}$ and $\mathcal{R}$ until $\mathcal{B}$ returns the final reader message $m_{2q_{\mathcal{R}}-1}$ (steps 5–11). Finally, $\mathcal{A}_{\text{sec}}^{\mathcal{B}}$ sends $m_{2q_{\mathcal{R}}-1}$ to the real tag $\mathcal{T}_{\text{ID}}$ (step 12). $\mathcal{A}_{\text{sec}}^{\mathcal{B}}$ succeeds if $\mathcal{T}_{\text{ID}}$ accepts $\mathcal{B}$ as $\mathcal{R}$. More formally, this means that:

$$\Pr\left[\mathbf{Exp}_{\mathcal{A}_{\text{sec}}^{\mathcal{B}}}^{\mathcal{R}\text{-aut}} = 1\right] = \Pr\left[\text{Ident}\left[\mathcal{T}_{\text{ID}}\!:\!S_0^{\mathcal{T}_{\text{ID}}};\ \mathcal{A}_{\text{sec}}^{\mathcal{B}}\!:\!-;\ *\!:\!pk_{\mathcal{R}}\right] \to \left[\mathcal{T}_{\text{ID}}\!:\!\text{ok};\ \mathcal{A}_{\text{sec}}^{\mathcal{B}}\!:\!\cdot\right]\right] \quad (1)$$

We stress that this indeed is a valid attack w.r.t. Definition 5 since $\mathcal{A}_{\text{sec}}$ does not just forward the protocol messages between $\mathcal{R}$ and $\mathcal{T}_{\text{ID}}$.

Next, we show that narrow-forward privacy (Definition 6) ensures that $\mathcal{A}_{\text{sec}}^{\mathcal{B}}$ succeeds with non-negligible probability, i.e., that Eq. 1 is non-negligible. Note that in case Eq. 1 is negligible, this implies that with non-negligible probability $p_\perp$ message $m_{2q_{\mathcal{R}}-1}$ generated by $\mathcal{B}$ makes $\mathcal{T}_{\text{ID}}$ to return $out_{\mathcal{T}_{\text{ID}}} = \perp$. In the following, we show that if $p_\perp$ is non-negligible, then there is a narrow-forward adversary $\mathcal{A}_{\text{prv}}$ that has non-negligible advantage $\mathbf{Adv}_{\mathcal{A}_{\text{prv}}}^{\text{prv}}$ to distinguish $\mathcal{B}$ form the real oracles, which contradicts narrow-forward privacy (Definition 6). The construction of $\mathcal{A}_{\text{prv}}$ is shown in Algorithm 2. First, $\mathcal{A}_{\text{prv}}$ creates a legitimate tag $\mathcal{T}_{\text{ID}}$ (step 1) and makes it accessible (step 2). Then, $\mathcal{A}_{\text{prv}}$ makes $\mathcal{R}$ to start a new instance π of the Ident protocol with $\mathcal{T}_{\text{ID}}$ (step 3) and obtains the first protocol message m_1 from $\mathcal{R}$ (step 4). Now, $\mathcal{A}_{\text{prv}}$ eavesdrops on the execution of the Ident protocol up to to the point *after* $\mathcal{R}$ has sent its last protocol message $m_{2q_{\mathcal{R}}-1}$ (steps 5–11) and corrupts $\mathcal{T}_{\text{ID}}$ just *before* $\mathcal{T}_{\text{ID}}$ received $m_{2q_{\mathcal{R}}-1}$ (step 12). Next, $\mathcal{A}_{\text{prv}}$ performs the computation $\mathcal{T}_{\text{ID}}$ would have done on receipt of $m_{2q_{\mathcal{R}}-1}$

Alg. 2. Narrow-forward adversary $\mathcal{A}_{\mathrm{prv}}$

1: CreateTag(ID)
2: $vtag \leftarrow \mathsf{Draw}(\Pr[\mathtt{ID}] = 1)$
3: $\pi \leftarrow \mathsf{Launch}(\,)$
4: $m_1 \leftarrow \mathsf{SendReader}(-, \pi)$
5: $i \leftarrow 1$
6: **while** $i < q_{\mathcal{R}}$ **do**
7:     **if** $i \le q_{\mathcal{T}}$ **then** $m_{2i} \leftarrow \mathsf{SendTag}(m_{2i-1}, vtag)$
8:     **end if**
9:     $m_{2i+1} \leftarrow \mathsf{SendReader}(m_{2i}, \pi)$
10:     $i \leftarrow i + 1$
11: **end while**
12: $S_{q_{\mathcal{R}}}^{\mathcal{T}_{\mathtt{ID}}} \leftarrow \mathsf{Corrupt}(vtag)$
13: $out_{\mathcal{T}_{\mathtt{ID}}} \leftarrow \mathcal{T}_{\mathtt{ID}}{}^{(q_{\mathcal{R}})}(S_{q_{\mathcal{R}}}^{\mathcal{T}_{\mathtt{ID}}}, m_{2q_{\mathcal{R}}-1})$
14: **if** $out_{\mathcal{T}_{\mathtt{ID}}} = \mathrm{ok}$ **then return** 0
15: **else return** 1
16: **end if**

(step 13). If this computation results in $out_{\mathcal{T}_{\mathtt{ID}}} = \mathrm{ok}$, $\mathcal{A}_{\mathrm{prv}}$ returns 0 to indicate that it interacted with the real oracles (step 14). Otherwise, $\mathcal{A}_{\mathrm{prv}}$ indicates the presence of $\mathcal{B}$ by returning 1 (step 15). Note that $\mathcal{A}_{\mathrm{prv}}$ indeed is a narrow-forward adversary (Definition 3) since $\mathcal{A}_{\mathrm{prv}}$ never queries Result and none of the oracles defined in Section 3.2 after corrupting $\mathcal{T}_{\mathtt{ID}}$.

Next, we show that $\mathcal{A}_{\mathrm{prv}}$ has non-negligible advantage $\mathbf{Adv}_{\mathcal{A}_{\mathrm{prv}}}^{\mathrm{prv}}$ if $p_\perp$ is non-negligible. Therefore, we first consider the case where $\mathcal{A}_{\mathrm{prv}}$ interacts with the real oracles. Since $\mathcal{T}_{\mathtt{ID}}$ is legitimate, it follows from correctness (Definition 2) that $out_{\mathcal{T}_{\mathtt{ID}}} = \mathrm{ok}$ with overwhelming probability p_{ok}. Hence, $\Pr\left[\mathbf{Exp}_{\mathcal{A}_{\mathrm{prv}}}^{\mathrm{prv}\text{-}0} = 1\right] = 1 - p_{\mathrm{ok}}$ is negligible. Now, consider the case where $\mathcal{A}_{\mathrm{prv}}$ interacts with $\mathcal{B}$. Note that by the contradicting hypothesis, $\mathcal{B}$ generates a protocol message $m_{2q_{\mathcal{R}}-1}$ that makes $\mathcal{T}_{\mathtt{ID}}$ to return $out_{\mathcal{T}_{\mathtt{ID}}} = \perp$ with non-negligible probability $p_\perp$. Thus, we have $\Pr\left[\mathbf{Exp}_{\mathcal{A}_{\mathrm{prv}}}^{\mathrm{prv}\text{-}1} = 1\right] = p_\perp$. Hence, it follows that $\mathbf{Adv}_{\mathcal{A}_{\mathrm{prv}}}^{\mathrm{prv}} = \left|1 - p_{\mathrm{ok}} - p_\perp\right|$. Note that due to correctness both p_{ok} is overwhelming and by assumption $p_\perp$ is non-negligible. Hence, $\mathbf{Adv}_{\mathcal{A}_{\mathrm{prv}}}^{\mathrm{prv}}$ is non-negligible, which contradicts narrow-forward privacy (Definition 6). In turn, this means that narrow-forward privacy ensures that Eq. 1 is non-negligible, which finishes the proof.   □

Since the impossibility of narrow-forward privacy (Theorem 1), implies the impossibility of all other stronger privacy notions (see Figure 1), we have the following corollary, which corresponds to the first main claim of this paper:

Corollary 1. *In the PV-Model there is no RFID system (Definition 1) that achieves both reader authentication (Definition 5) and any privacy notion that is different from weak and narrow-weak privacy (Definition 6) under temporary state disclosure.*

5   Corruption without Temporary State Disclosure

Our first impossibility result shows that the PV-Model requires further assumptions to evaluate the privacy properties of RFID systems where tag corruption is of concern. A natural question therefore is, whether one can achieve mutual authentication along with some form of privacy, if the temporary tag state is *not* disclosed. Hence, in this section we consider the case where corruption *only* reveals the persistent tag state but *no* information on the temporary tag state.

The attack and the impossibility result shown in Section 4 critically use the fact that in the PV-Model an adversary $\mathcal{A}_{\mathrm{prv}}$ can learn the temporary state of a tag during the Ident protocol. This allows $\mathcal{A}_{\mathrm{prv}}$ to verify the response of $\mathcal{R}$ (that may have been simulated by $\mathcal{B}$) and hence, due to reader authentication (Definition 5), $\mathcal{A}_{\mathrm{prv}}$ can distinguish with non-negligible advantage between the real oracles and $\mathcal{B}$. However, if $\mathcal{A}_{\mathrm{prv}}$ cannot obtain temporary tag states, it cannot perform this verification. Hence, the impossibility result we proved in Section 4 does not necessarily hold if the temporary state is safe to corruption.

Impossibility of narrow-strong privacy. We now show our second impossibility result: in the PV-Model, it is *impossible* to achieve narrow-strong privacy along with reader authentication. This means that even in case the adversary cannot obtain the temporary tag state, the most challenging privacy notion defined in [30] (narrow-strong privacy) still remains unachievable. This implies a conceptually different weakness of the claimed narrow-strong private protocol in [30].

Theorem 2. *In the PV-Model there is no RFID system (Definition 1) that fulfills both reader authentication (Definition 5) and narrow-strong privacy (Definition 6).*

Proof (Theorem 2). Narrow-strong privacy (Definition 6) requires the existence of a blinder $\mathcal{B}$ that simulates the Launch, SendReader and SendTag oracles such that every narrow-strong adversary $\mathcal{A}_{\mathrm{prv}}$ has negligible advantage $\mathbf{Adv}_{\mathcal{A}_{\mathrm{prv}}}^{\mathrm{prv}}$ to distinguish $\mathcal{B}$ from the real oracles. We now show that $\mathcal{B}$ can be used to construct an algorithm $\mathcal{A}_{\mathrm{sec}}^{\mathcal{B}}$ that violates reader authentication (Definition 5).

The construction of $\mathcal{A}_{\mathrm{sec}}^{\mathcal{B}}$ is as shown in Algorithm 3. First, $\mathcal{A}_{\mathrm{sec}}^{\mathcal{B}}$ creates a legitimate tag $\mathcal{T}_{\mathrm{ID}}$ (step 1), makes it accessible (step 2), and corrupts it (step 3). These three steps are also shown to $\mathcal{B}$, which expects to observe all oracle queries. Then, $\mathcal{A}_{\mathrm{sec}}^{\mathcal{B}}$ makes $\mathcal{B}$ to start a new instance π of the Ident protocol with $\mathcal{T}_{\mathrm{ID}}$ (step 4) and obtains the first protocol message m_1 generated by $\mathcal{B}$ (step 5). Now, $\mathcal{A}_{\mathrm{sec}}^{\mathcal{B}}$ internally runs $\mathcal{B}$ that simulates *vtag* and $\mathcal{R}$ until $\mathcal{B}$ returns the final reader message $m_{2q_{\mathcal{R}}-1}$ (steps 6–12). Finally, $\mathcal{A}_{\mathrm{sec}}^{\mathcal{B}}$ sends $m_{2q_{\mathcal{R}}-1}$ to the real tag $\mathcal{T}_{\mathrm{ID}}$ (step 13). $\mathcal{A}_{\mathrm{sec}}^{\mathcal{B}}$ succeeds if $\mathcal{T}_{\mathrm{ID}}$ accepts $m_{2q_{\mathcal{R}}-1}$ and returns $out_{\mathcal{T}_{\mathrm{ID}}} = \mathrm{ok}$, which means that $\mathcal{T}_{\mathrm{ID}}$ accepts $\mathcal{B}$ as $\mathcal{R}$. More formally, this means that:

$$\Pr\left[\mathbf{Exp}_{\mathcal{A}_{\mathrm{sec}}^{\mathcal{B}}}^{\mathcal{R}\text{-}\mathrm{aut}} = 1\right] = \Pr\left[\mathsf{Ident}\left[\mathcal{T}_{\mathrm{ID}} : S_0^{\mathcal{T}_{\mathrm{ID}}};\ \mathcal{A}_{\mathrm{sec}}^{\mathcal{B}} : -;\ * : pk_{\mathcal{R}}\right] \rightarrow \left[\mathcal{T}_{\mathrm{ID}} : \mathrm{ok};\ \mathcal{A}_{\mathrm{sec}}^{\mathcal{B}} : \cdot\right]\right] \quad (2)$$

We stress that this indeed is a valid attack w.r.t. Definition 5 since $\mathcal{A}_{\mathrm{sec}}$ does not just forward the protocol messages between $\mathcal{R}$ and $\mathcal{T}_{\mathrm{ID}}$.

Alg. 3. Adversary $\mathcal{A}_{\text{sec}}^{\mathcal{B}}$ violating reader authentication

1: CreateTag(ID)
2: $vtag \leftarrow \text{Draw}(\Pr[\text{ID}] = 1)$
3: $S_0^{\mathcal{T}_{\text{ID}}} \leftarrow \text{Corrupt}(vtag)$
4: $\pi \leftarrow \text{Launch}(\ )$ $\triangleright$ simulated by $\mathcal{B}$
5: $m_1 \leftarrow \text{SendReader}(-, \pi)$ $\triangleright$ simulated by $\mathcal{B}$
6: $i \leftarrow 1$
7: **while** $i < q_{\mathcal{R}}$ **do**
8:    **if** $i \leq q_{\mathcal{T}}$ **then** $m_{2i} \leftarrow \text{SendTag}(m_{2i-1}, vtag)$
9:    **end if**
10:    $m_{2i+1} \leftarrow \text{SendReader}(m_{2i}, \pi)$ $\triangleright$ simulated by $\mathcal{B}$
11:    $i \leftarrow i + 1$
12: **end while**
13: $out_{\mathcal{T}_{\text{ID}}} \leftarrow \text{SendTag}(m_{2q_{\mathcal{R}}-1}, vtag)$ $\triangleright$ computed by $\mathcal{T}_{\text{ID}}$

From reader authentication (Definition 5) it follows that Eq. 2 must be negligible. However, this implies that with overwhelming probability $\mathcal{B}$ generates at least one protocol message that makes $\mathcal{T}_{\text{ID}}$ to finally return $out_{\mathcal{T}_{\text{ID}}} = \bot$. Let p_t be the probability that this is the case for message m_{2t-1} for some $t \in \{1, \ldots, q_{\mathcal{T}}\}$. We now show a narrow-strong adversary $\mathcal{A}_{\text{prv}}$ that succeeds with non-negligible advantage $\mathbf{Adv}_{\mathcal{A}_{\text{prv}}}^{\text{prv}}$ if p_t is non-negligible, which contradicts narrow-strong privacy (Definition 6). The construction of $\mathcal{A}_{\text{prv}}$ is shown in Algorithm 4. First, $\mathcal{A}_{\text{prv}}$ creates a legitimate tag $\mathcal{T}_{\text{ID}}$ (step 1), makes it accessible (step 2), and corrupts it (step 3). Note that by a Corrupt query, $\mathcal{A}_{\text{prv}}$ only learns the persistent tag state $S_0^{\mathcal{T}_{\text{ID}}}$ of $\mathcal{T}_{\text{ID}}$. Then, $\mathcal{A}_{\text{prv}}$ makes $\mathcal{R}$ to start an instance π of the Ident protocol with $\mathcal{T}_{\text{ID}}$ (step 4) and obtains the first protocol message m_1 from $\mathcal{R}$ (step 5). Now, $\mathcal{A}_{\text{prv}}$ guesses t (step 6) and simulates $\mathcal{T}_{\text{ID}}$ (using $S_0^{\mathcal{T}_{\text{ID}}}$) in the Ident protocol up to the point where SendReader returns message m_{2t-1} (steps 7–13). Next, $\mathcal{A}_{\text{prv}}$ performs the computation $\mathcal{T}_{\text{ID}}$ would have done on receipt of message m_{2t-1} (step 14). Finally, $\mathcal{A}_{\text{prv}}$ returns either 0 to indicate that it interacted with the real oracles (step 15) or 1 to indicate the presence of $\mathcal{B}$ (step 16).

Next, we show that $\mathcal{A}_{\text{prv}}$ has non-negligible $\mathbf{Adv}_{\mathcal{A}_{\text{prv}}}^{\text{prv}}$ if $p_{\bot}$ is non-negligible. Therefore, we first consider the case where $\mathcal{A}_{\text{prv}}$ interacts with the real oracles. Since $\mathcal{T}_{\text{ID}}$ is legitimate, it follows form correctness (Definition 2) that $out_{\mathcal{T}_{\text{ID}}} = \text{ok}$ holds with overwhelming probability p_{ok}. This means that $\Pr\left[\mathbf{Exp}_{\mathcal{A}_{\text{prv}}}^{\text{prv-0}} = 1\right] = 1 - p_{\text{ok}}$ is negligible. Now, consider the case where $\mathcal{A}_{\text{prv}}$ interacts with $\mathcal{B}$. Note that by the contradicting hypothesis, with non-negligible probability p_t $\mathcal{B}$ generates a message m_{2t-1} that makes $\mathcal{T}_{\text{ID}}$ to return $out_{\mathcal{T}_{\text{ID}}} = \bot$. Moreover, $\mathcal{A}_{\text{prv}}$ guesses t with probability of at least $1/q_{\mathcal{T}}$. Thus, we have $\Pr\left[\mathbf{Exp}_{\mathcal{A}_{\text{prv}}}^{\text{prv-1}} = 1\right] \geq \frac{p_t}{q_{\mathcal{T}}}$. Hence, it follows that $\mathbf{Adv}_{\mathcal{A}_{\text{prv}}}^{\text{prv}} \geq |1 - p_{\text{ok}} - \frac{p_t}{q_{\mathcal{T}}}|$. Note that due to correctness p_{ok} is overwhelming while p_t is non-negligible by assumption and $q_{\mathcal{T}}$ is polynomially bounded. Hence, $\mathbf{Adv}_{\mathcal{A}_{\text{prv}}}^{\text{prv}}$ is non-negligible, which contradicts narrow-strong privacy (Definition 6). $\qquad\square$

Alg. 4. Narrow-strong adversary $\mathcal{A}_{\text{prv}}$

1: CreateTag(ID)
2: $vtag \leftarrow \text{Draw}(\Pr[\text{ID}] = 1)$
3: $S_0^{\mathcal{T}_{\text{ID}}} \leftarrow \text{Corrupt}(vtag)$
4: $\pi \leftarrow \text{Launch}()$
5: $m_1 \leftarrow \text{SendReader}(-, \pi)$
6: $t \in \{1, \ldots, q_{\mathcal{T}}\}$
7: $i \leftarrow 1$
8: **while** $i < t$ **do**
9:     $(S_{i+1}^{\mathcal{T}_{\text{ID}}}, m_{2i}) \leftarrow \mathcal{T}_{\text{ID}}^{(i)}(S_i^{\mathcal{T}_{\text{ID}}}, m_{2i-1})$
10:     **if** $i < q_{\mathcal{R}}$ **then** $m_{2i+1} \leftarrow \text{SendReader}(m_{2i}, \pi)$
11:     **end if**
12:     $i \leftarrow i + 1$
13: **end while**
14: $out_{\mathcal{T}_{\text{ID}}} \leftarrow \mathcal{T}_{\text{ID}}^{(t)}(S_t^{\mathcal{T}_{\text{ID}}}, m_{2t-1})$
15: **if** $out_{\mathcal{T}_{\text{ID}}} = \text{ok}$ **then return** 0
16: **else return** 1
17: **end if**

6 Impossibility Results for Resettable and Stateless Tags

It is well known (see [9] for details and in particular [5] for identification schemes) that standard security notions do not work anymore when the adversary can manipulate the device that is running an honest party protocol, in particular when the adversary can reset the internal state of the device. To face this security issue, Canetti et al. [9] considered the concept of *resettability* for obtaining a security notion that is resilient to "reset attacks", e.g., attacks where the adversary can force a device to reuse the same randomness. The crucial importance of this notion is proved by several results (see, e.g., [5,9,13,6,17]) with the focus on obtaining feasibility results and efficient constructions for proof systems and identification schemes in such hostile settings. *Reset attacks* have been motivated in particular by the use of smart cards since some specific smart cards, when disconnected from power, go back to their initial state and perform their computations using the same randomness they already used before. However, the concept of a reset attacks can have a wider applicability. In particular reset attacks are always possible when the adversary controls the environment and can therefore force a stateless device to use the same randomness in different executions of a protocol.

As discussed in Section 2, most RFID tags in practice are low-cost devices that are usually not protected against physical tampering. Moreover, the randomness generator of a real-life RFID tag has already been successfully attacked [15]. Therefore, it is interesting to investigate the impact of reset attacks on the security and privacy of RFID systems.

In this section, we focus on the effect of reset attacks on privacy as defined in both the PV-Model [30] and the model it is based on [37]. Therefore, we first extend the formal adversary model in [37,30] to capture reset attacks. Then, we show that any privacy notion as defined in Definition 6 is spoiled when an adversary is able to launch reset attacks. We finally show that, when restricting the power of the adversary to the capability of resetting *only* the persistent state of a tag, i.e., the randomness of the tag is out of the control of the adversary, it is impossible to achieve destructive privacy.

6.1   Impossibility of Narrow-Weak Privacy under Reset Attacks

In order to extend the model in [37,30] to capture reset attacks, we add an additional oracle Reset(*vtag*) to the adversary model shown in Section 3.2. This oracle allows the adversary to reset the randomness and the state of a tag *vtag* to their initial values. We stress out that resetting a tag is a mere adversarial action and is never performed by honest parties. Thus we do not require that such an action must be carried out efficiently, instead according to the result showed in [5,9] we assume that it can be carried out in polynomial time. Note that, as for the Corrupt oracle, the Reset oracle is not simulated by the blinder $\mathcal{B}$ (see Definition 6) but is observed by it.

Now we are ready to formalize the impossibility of achieving any privacy notion in the extended model of [37,30] when the adversary can perform reset attacks against tags.

Theorem 3. *In the model of [37,30], no privacy notion (Definition 6) is achievable if the adversary is allowed to query the* Reset *oracle.*

Proof (Theorem 3). We show a narrow-weak adversary $\mathcal{A}_{\mathrm{prv}}$ that can distinguish with non-negligible advantage $\mathbf{Adv}^{\mathrm{prv}}_{\mathcal{A}_{\mathrm{prv}}}$ whether it is interacting with the real oracles or a blinder $\mathcal{B}$. The construction of $\mathcal{A}_{\mathrm{prv}}$ is shown in Algorithm 5. First, $\mathcal{A}_{\mathrm{prv}}$ creates two legitimate tags $\mathcal{T}_{\mathrm{ID}0}, \mathcal{T}_{\mathrm{ID}1}$ (steps 1–2) and makes one of them accessible (step 3). Then $\mathcal{A}_{\mathrm{prv}}$ eavesdrops a complete execution protocol of the Ident protocol between *vtag* and $\mathcal{R}$ (steps 4-11). We define τ as the complete transcript of the protocol execution. Note that τ contains the messages sent by both $\mathcal{R}$ and *vtag*. Now, $\mathcal{A}_{\mathrm{prv}}$ resets the state of *vtag* by querying the Reset oracle (step 12) and makes *vtag* inaccessible again by querying the Free oracle (step 13). Next, $\mathcal{A}_{\mathrm{prv}}$ makes a randomly chosen tag *vtag'* accessible (step 14) and then executes a complete run of the Ident protocol with *vtag'* simulating $\mathcal{R}$ (steps 15–18). To simulate $\mathcal{R}$, $\mathcal{A}_{\mathrm{prv}}$ uses the messages that have been sent by $\mathcal{R}$ in the previous execution according to the transcript τ. Finally, $\mathcal{A}_{\mathrm{prv}}$ obtains a new protocol transcript τ'. If the same tag has played both times, then $\mathcal{A}_{\mathrm{prv}}$ expects that the transcripts τ and τ' are the same due to the Reset oracle. The idea is that $\mathcal{B}$ has no information about which tag has been drawn in step 14 (the resetted one or the other one). Thus, $\mathcal{B}$ can at most guess which tag has been chosen when answering the SendTag query in the second protocol execution.

In the following we show that $\mathcal{A}_{\mathrm{prv}}$ has non-negligible advantage $\mathbf{Adv}^{\mathrm{prv}}_{\mathcal{A}_{\mathrm{prv}}}$ of distinguishing between $\mathcal{B}$ and real oracles, which violates narrow-weak privacy. First, we consider the case where $\mathcal{A}_{\mathrm{prv}}$ interacts with the real oracles. It is easy to see that in this case the attack is always successful. Indeed, if $\mathcal{A}_{\mathrm{prv}}$ interacts with the same tag in both executions of the Ident protocol, then, due to the

Alg. 5. Experiment with a narrow-weak adversary $\mathcal{A}_{\mathrm{prv}}$

```
1:  CreateTag(ID₀)
2:  CreateTag(ID₁)
3:  vtag ← Draw(Pr[ID₀] = ½, Pr[ID₁] = ½)
4:  m₁ ← SendReader(−, π)
5:  i ← 1
6:  while i < q_R do
7:      if i ≤ q_T then m_{2i} ← SendTag(m_{2i−1}, vtag)
8:      end if
9:      m_{2i+1} ← SendReader(m_{2i}, π)
10:     i ← i + 1
11: end while
12: Reset(vtag)
13: Free(vtag)
14: vtag′ ← Draw(Pr[ID₀] = ½, Pr[ID₁] = ½)
15: i ← 1
16: while i ≤ q_T do m_{2i} ← SendTag(m_{2i−1}, vtag′)
17:     i ← i + 1
18: end while
19: if τ = τ′ then out_A ← 1
20: else out_A ← 0
21: end if
22: return (Γ[vtag] = Γ[vtag′] ∧ out_A) ∨ (Γ[vtag] ≠ Γ[vtag′] ∧ out̄_A)
```

Reset query, challenging $vtag'$ with the *same* messages must generate the same protocol transcript. Thus, after $\mathcal{A}_{\mathrm{prv}}$ is given the hidden table Γ, one of the two conditions must hold: either $\mathcal{A}_{\mathrm{prv}}$ has (i) interacted with the same tag twice and the transcripts match (which is always true in case $\Gamma[vtag] = \Gamma[vtag']$), or (ii) the tag involved in the second execution of Ident is not the resetted tag and the protocol transcripts are different (which holds with overwhelming probability in case $\Gamma[vtag] \neq \Gamma[vtag']$ due to tag authentication, since otherwise $\mathcal{A}_{\mathrm{prv}}$ can create a faked tag state that can be used to generate the messages of a legitimate tag with non-negligible probability). Hence, $\mathcal{A}_{\mathrm{prv}}$ succeeds in $\mathbf{Exp}^{\mathrm{prv-0}}_{\mathcal{A}_{\mathrm{prv}}}$ with probability $1 - \epsilon(l)$ where ϵ is a negligible function in the security parameter l. Formally, $\Pr\left[\mathbf{Exp}^{\mathrm{prv-0}}_{\mathcal{A}_{\mathrm{prv}}} = 1\right] = \Pr\left[(\Gamma[vtag] = \Gamma[vtag']) \wedge out_A\right] + \Pr\left[(\Gamma[vtag] \neq \Gamma[vtag']) \wedge \overline{out_A}\right] = \frac{1}{2} \cdot 1 + \frac{1}{2} \cdot (1 - \epsilon(l)) = 1 - \epsilon(l)/2$. Next we consider the case where the SendTag oracle is simulated by $\mathcal{B}$. In this case any $\mathcal{B}$ can at most guess which tag has been selected by Draw. Hence, the probability that $\mathcal{A}_{\mathrm{prv}}$ wins the experiment $\mathbf{Exp}^{\mathrm{prv-1}}_{\mathcal{A}_{\mathrm{prv}}}$ is at most $\Pr\left[\mathbf{Exp}^{\mathrm{prv-1}}_{\mathcal{A}_{\mathrm{prv}}} = 1\right] = \Pr\left[(\Gamma[vtag] = \Gamma[vtag']) \wedge out_A\right] + \Pr\left[(\Gamma[vtag] \neq \Gamma[vtag']) \wedge \overline{out_A}\right] \leq \frac{1}{2} \cdot \frac{1}{2} + \frac{1}{2} \cdot \frac{1}{2} = \frac{1}{2}$. According to Definition 6, from the above probability it follows that $\mathcal{A}_{\mathrm{prv}}$ has non-negligible advantage $\mathbf{Adv}^{\mathrm{prv}}_{\mathcal{A}_{\mathrm{prv}}} \geq 1 - \epsilon(l)/2 - \frac{1}{2}$ to distinguish between $\mathcal{B}$ and the real oracles. $\qquad\square$

Alg. 6. Narrow-forward adversary $\mathcal{A}_{\mathrm{prv}}$

1: CreateTag(ID)
2: $vtag \leftarrow$ Draw($\Pr[\mathrm{ID}] = 1$)
3: Free($vtag$)
4: $vtag \leftarrow$ Draw($\Pr[\mathrm{ID}] = 1$)
5: $t \in \{1, \ldots, q_{\mathcal{T}}\}$
6: $m_1 \leftarrow$ SendReader($-, \pi$)
7: $i \leftarrow 1$
8: **while** $i \leq t$ **do**
9:    $m_{2i} \leftarrow$ SendTag($m_{2i-1}, vtag$)
10:    $m_{2i+1} \leftarrow$ SendReader(m_{2i}, π)
11:    $i \leftarrow i + 1$
12: **end while**
13: $S \leftarrow$ Corrupt($vtag$)
14: **return** 1 if and only if the temporary state in S is empty

---

6.2   Impossibility of Destructive Privacy with Stateless Tags

In this section we show that destructive privacy is impossible to achieve in the model of [37,30] when tags are *stateless*, i.e., when their persistent state cannot be updated. This implies that destructive privacy is impossible when an adversary can reset the persistent state of a tag to its original value: by resetting a tag, the adversary can interact with a tag that uses the same state several times, which corresponds to an experiment with a stateless tag. We stress that in a stateless RFID scheme the Free oracle erases any temporary information stored on the tag. Otherwise there would be an updatable information that survives even when a tag is not powered, and thus the tag would be stateful.

We recall that in our previous notation we associate $S_i^{\mathcal{T}}$ to the full state (including both the persistent *and* temporary state) of a tag when playing the i-th message from the moment it has been drawn, i.e., powered on. We start by giving a useful preliminary lemma.

Lemma 2. *In any stateless narrow-forward RFID scheme the temporary tag state is always empty.*

Proof (Lemma 2). To prove the lemma we show in Algorithm 6 that if there exists a non-empty temporary tag state, then there exists a narrow-forward adversary $\mathcal{A}_{\mathrm{prv}}$ that distinguishes between the real oracles and $\mathcal{B}$. We stress that for a stateless tag, due to the Free query, the output S returned by a Corrupt($vtag$) query played immediately after a Draw query corresponds to the persistent state generated by the CreateTag oracle. Clearly, when interacting with the real oracles the output of $\mathcal{A}_{\mathrm{prv}}$ is different than 1 with non-negligible probability. Indeed, since stateless tags are allowed to have some non-empty temporary state, there exists at least one round, which can be guessed with non-negligible probability by the selection of t, that, when followed by the Corrupt query, reveals to $\mathcal{A}_{\mathrm{prv}}$ that the temporary state of the tag is not empty.

During interaction with the blinder $\mathcal{B}$ the tag does not play any round, as all SendTag queries are simulated by $\mathcal{B}$. Therefore, the output of the above experiment is always equal to 1, which shows that $\mathcal{A}_{\mathrm{prv}}$ is successful and the claim holds. ☐

Due to Lemma 2 we can assume that $(S^{\mathcal{T}_{\mathrm{ID}}}, \cdot) \leftarrow \mathcal{T}_{\mathrm{ID}}^{(i)}(S^{\mathcal{T}_{\mathrm{ID}}}, \cdot)$, i.e., the new state after each round is always identical to the previous one. Recall that an RFID scheme is stateless if the persistent tag state is not allowed to change over time. In this section we show that when the tag state does not change, then achieving destructive privacy is impossible.

Theorem 4. *There is no stateless RFID system (Definition 1) that achieves destructive privacy (Definition 6).*

Proof (Theorem 4). Recall that destructive privacy implies forward privacy (see Figure 1). We prove that a stateless RFID system cannot achieve destructive and narrow-forward privacy at the same time. The proof is by contradiction. Note that a destructive private stateless RFID system implies the existence of a blinder $\mathcal{B}$ such that $\mathcal{A}_{\mathrm{prv}}$ fails in distinguishing the real oracles from their simulation by $\mathcal{B}$ with overwhelming probability. Thus, we first show a destructive adversary $\mathcal{A}_{\mathrm{prv}}$ for which there must exist a successful blinder, that we denote by $\mathcal{B}_D$. Then, we construct a narrow-forward adversary $\mathcal{A}_{\mathrm{prv}}^{\mathcal{B}_D}$ that internally uses $\mathcal{B}_D$ to violate forward privacy. Hence, we obtain a contradiction.

Since we are considering *stateless* tags, we assume that at each step of the tag algorithm the persistent state remains unchanged. Formally, this means that $\mathcal{T}$ can be represented as a tuple of algorithms $(\mathcal{T}^{(1)}, \ldots, \mathcal{T}^{(q_{\mathcal{T}})})$ where $\mathcal{T}^{(i)}$ means the computation done by $\mathcal{T}$ when processing the i-th SendTag query in an instance of the Ident protocol that involves $\mathcal{T}$. We have $m_{2i} \leftarrow \mathcal{T}^{(i)}(S^{\mathcal{T}}, m_{2i-1})$ for $1 \leq i \leq q_{\mathcal{T}}$ where $q_{\mathcal{T}}$ is an upper bound on the number of messages sent by $\mathcal{T}$ during the protocol.

Let $\mathcal{A}_{\mathrm{prv}}$ be the destructive adversary defined in Algorithm 7. Informally, the attack is the following: $\mathcal{A}_{\mathrm{prv}}$ faithfully forwards the messages generated by $\mathcal{R}$ and $\mathcal{T}$, up to a certain (randomly chosen) round t of the Ident protocol execution. Then $\mathcal{A}_{\mathrm{prv}}$ corrupts $\mathcal{T}$ and gets its state. Since $\mathcal{A}_{\mathrm{prv}}$ is destructive, it is not allowed to query any other oracle for $\mathcal{T}$ after corrupting $\mathcal{T}$ but $\mathcal{A}_{\mathrm{prv}}$ can still compute the remaining protocol messages of $\mathcal{T}$ by running the tag algorithm with the state obtained by corruption. Then $\mathcal{A}_{\mathrm{prv}}$ picks a state S with the same distribution used by CreateTag (i.e., SetupTag) with the purpose of distinguishing if it is interacting with the real oracles or $\mathcal{B}_D$. Then $\mathcal{A}_{\mathrm{prv}}$ randomly selects one of the two states and continues the protocol execution running the tag algorithm with the chosen state until the end of the protocol. The main idea is that when $\mathcal{A}_{\mathrm{prv}}$ runs the tag algorithm with the state obtained through the Corrupt query, then, due to correctness (Definition 2), $\mathcal{R}$ will accept, i.e., the Result query outputs 1 with overwhelming probability, while $\mathcal{R}$ will reject otherwise.

Formally, $\mathcal{A}_{\mathrm{prv}}$ behaves as follows: First, $\mathcal{A}_{\mathrm{prv}}$ creates two legitimate tags $\mathcal{T}_{\mathrm{ID}}$ (step 1 and step 2) and makes one of them accessible (step 3). Then, $\mathcal{A}_{\mathrm{prv}}$ asks

Alg. 7. Destructive adversary $\mathcal{A}_{\mathrm{prv}}$

1: CreateTag(ID)
2: CreateTag(ID′)
3: $vtag \leftarrow \mathsf{Draw}(\Pr[\mathtt{ID}] = \frac{1}{2}, \Pr[\mathtt{ID}'] = \frac{1}{2})$
4: $\pi \leftarrow \mathsf{Launch}()$
5: $m_1 \leftarrow \mathsf{SendReader}(-, \pi)$
6: $j_{\mathcal{R}} \in_R \{1, \ldots, q_{\mathcal{R}}\}$
7: $i \leftarrow 1$
8: **while** $i < j_{\mathcal{R}}$ **do** $m_{2i} \leftarrow \mathsf{SendTag}(m_{2i-1}, vtag)$
9:     $m_{2i+1} \leftarrow \mathsf{SendReader}(m_{2i}, \pi)$
10:     $i \leftarrow i + 1$
11: **end while**
12: $S^{\mathcal{T}_{\mathtt{ID}}} \leftarrow \mathsf{Corrupt}(vtag)$
13: $b \in_R \{0, 1\}$
14: **if** $b = 1$ **then**
15:     $m_{2j_{\mathcal{R}}} \leftarrow \mathcal{T}_{\mathtt{ID}}^{(j_{\mathcal{R}})}(S^{\mathcal{T}_{\mathtt{ID}}}, m_{2j_{\mathcal{R}}-1})$
16: **else**
17:     pick a state S with the same distribution used by CreateTag (i.e., SetupTag)
18:     $S^{\mathcal{T}_{\mathtt{ID}}} \leftarrow S$
19:     $m_{2j_{\mathcal{R}}} \leftarrow \mathcal{T}_{\mathtt{ID}}^{(j_{\mathcal{R}})}(S^{\mathcal{T}_{\mathtt{ID}}}, m_{2j_{\mathcal{R}}-1})$
20: **end if**
21: $m_{2j_{\mathcal{R}}+1} \leftarrow \mathsf{SendReader}(m_{2j_{\mathcal{R}}}, \pi)$
22: $i \leftarrow j_{\mathcal{R}} + 1$
23: **while** $i < q_{\mathcal{R}}$ **do**
24:     **if** $i \leq q_{\mathcal{T}}$ **then** $m_{2i} \leftarrow \mathcal{T}_{\mathtt{ID}}^{(i)}(S^{\mathcal{T}_{\mathtt{ID}}}, m_{2i-1})$
25:     **end if**
26:     $m_{2i+1} \leftarrow \mathsf{SendReader}(m_{2i}, \pi)$
27:     $i \leftarrow i + 1$
28: **end while**
29: **return** $\left(\mathsf{Result}(\pi) \wedge b\right) \vee \left(\overline{\mathsf{Result}(\pi)} \wedge \bar{b}\right)$

$\mathcal{R}$ to start a new instance π of the Ident protocol with $\mathcal{T}_{\mathtt{ID}}$ (step 4) and obtains the first protocol message m_1 from $\mathcal{R}$ (step 5). Then $\mathcal{A}_{\mathrm{prv}}$ randomly chooses a protocol round $j_{\mathcal{R}}$ (step 6) and starts eavesdropping on the execution of the Ident protocol up to the point after $\mathcal{R}$ has sent protocol message $m_{2j_{\mathcal{R}}-1}$ (steps 7–11). Then $\mathcal{A}_{\mathrm{prv}}$ gets the tag state $S^{\mathcal{T}_{\mathtt{ID}}}$ by querying the Corrupt oracle, just before $\mathcal{T}_{\mathtt{ID}}$ receives $m_{2j_{\mathcal{R}}-1}$ (step 12). Now $\mathcal{A}_{\mathrm{prv}}$ chooses a random bit b (step 13) to decide how to complete the protocol execution. In case $b = 1$, $\mathcal{A}_{\mathrm{prv}}$ continues by simulating $vtag$ using the state $S^{\mathcal{T}_{\mathtt{ID}}}$ obtained by the Corrupt query (steps 14–15). In case $b = 0$, $\mathcal{A}_{\mathrm{prv}}$ sets $S^{\mathcal{T}_{\mathtt{ID}}}$ to a new state generated on the fly (steps 16–19). Hereafter, $\mathcal{A}_{\mathrm{prv}}$ simulates the tag by running the algorithm $\mathcal{T}^{(i)}$ with the state set according to the bit b until the protocol terminates (steps 21–28). Finally, $\mathcal{A}_{\mathrm{prv}}$ outputs 1 if one of the following conditions hold: either $b = 1$ and $\mathcal{R}$ accepts $\mathcal{T}_{\mathtt{ID}}$, whose transcript has partially been computed by $\mathcal{A}_{\mathrm{prv}}$ with the real state (i.e., the output of Result is 1), or $b = 0$, and $\mathcal{R}$ rejected $\mathcal{T}_{\mathtt{ID}}$ since a part of the transcript has been generated using a faked state (i.e., the output of Result is 0).

Recall that Definition 6 requires the existence of a blinder $\mathcal{B}_D$ such that: $\mathbf{Adv}_{\mathcal{A}_{\mathrm{prv}}}^{\mathrm{prv}} = \left| \Pr\left[\mathbf{Exp}_{\mathcal{A}_{\mathrm{prv}}}^{\mathrm{prv}\text{-}0} = 1\right] - \Pr\left[\mathbf{Exp}_{\mathcal{A}_{\mathrm{prv}}}^{\mathrm{prv}\text{-}1} = 1\right]\right| = \epsilon(l)$ for a negligible function ϵ. If such $\mathcal{B}_D$ exists, then $\mathcal{B}_D$ must be able to do the following: first, $\mathcal{B}_D$ simulates both $\mathcal{R}$ and $\mathcal{T}_{\mathrm{ID}}$, then after $\mathcal{B}_D$ gets the state $S^{\mathcal{T}_{\mathrm{ID}}}$ of $\mathcal{T}_{\mathrm{ID}}$ from the Corrupt query, playing only at the reader side ($\mathcal{T}_{\mathrm{ID}}$ is simulated by $\mathcal{A}_{\mathrm{prv}}$ running the tag algorithm using either the real or a faked tag state), $\mathcal{B}_D$ can answer the Result query as $\mathcal{R}$ would do. Thus, $\mathcal{B}_D$ is able to recognize whether the messages received from $\mathcal{A}_{\mathrm{prv}}$ (simulating $\mathcal{T}_{\mathrm{ID}}$) are computed with the real state of $\mathcal{T}_{\mathrm{ID}}$ or not. One can think of $\mathcal{B}_D$ as a two-phase algorithm. In the first phase $\mathcal{B}_D$ simulates the protocol execution between $\mathcal{R}$ and a tag $vtag$. Then, in the second phase, upon receiving the state $S^{\mathcal{T}_{\mathrm{ID}}}$ of $vtag$, playing as the reader, $\mathcal{B}_D$ can distinguish if the tag messages received are computed according to the state of the tag simulated in first phase or not.

Now we show that if $\mathcal{B}_D$ exists, then $\mathcal{B}_D$ can be used to construct a narrow-forward adversary that distinguishes between any blinder $\mathcal{B}$ and the real oracles with non-negligible probability. Hence, the existence of $\mathcal{B}_D$ contradicts narrow-forward privacy and thus in turn destructive privacy. The idea of a narrow-forward adversary $\mathcal{A}_{\mathrm{prv}}^{\mathcal{B}_D}$ is to run $\mathcal{B}_D$ as subroutine showing to $\mathcal{B}_D$ a view that is identical to the ones that it gets when playing with $\mathcal{A}_{\mathrm{prv}}$ in Algorithm 7. The goal of $\mathcal{A}_{\mathrm{prv}}^{\mathcal{B}_D}$ is to exploit the capabilities of $\mathcal{B}_D$ to distinguish whether the output of the SendTag oracle is generated by the real oracle using the real tag state or by a blinder $\mathcal{B}$ for narrow-forward privacy having no information on the real tag state. Formally, $\mathcal{A}_{\mathrm{prv}}^{\mathcal{B}_D}$ is defined in Algorithm 8 and works as follows: first, $\mathcal{A}_{\mathrm{prv}}^{\mathcal{B}_D}$ creates two legitimate tags $\mathcal{T}_{\mathrm{ID}}$, $\mathcal{T}_{\mathrm{ID}}'$ (steps 1–2) and makes one of them accessible as $vtag$ (step 3). These three steps are also internally shown to $\mathcal{B}_D$. Then, $\mathcal{A}_{\mathrm{prv}}^{\mathcal{B}_D}$ internally asks $\mathcal{B}_D$ to start a new instance π of the Ident protocol with $vtag$ (step 4) and obtains the first protocol message m_1 generated by $\mathcal{B}_D$ (step 5). Then $\mathcal{A}_{\mathrm{prv}}^{\mathcal{B}_D}$ randomly chooses a protocol round $j_{\mathcal{R}}$ (step 6) and makes $\mathcal{B}_D$ to simulate the first $j_{\mathcal{R}}$ rounds of the protocol, up to the point after $\mathcal{B}_D$ has sent the reader message $m_{2j_{\mathcal{R}}-1}$ (steps 7–11). Then, $\mathcal{A}_{\mathrm{prv}}^{\mathcal{B}_D}$ queries the SendTag oracle with the message $m_{2j_{\mathcal{R}}-1}$ obtained by $\mathcal{B}_D$ (step 12). Next, $\mathcal{A}_{\mathrm{prv}}^{\mathcal{B}_D}$ makes $vtag$ inaccessible by querying the Free oracle (step 13) and makes accessible a randomly chosen tag $vtag'$ by querying the Draw oracle (step 14). Note that this step corresponds to the random selection of bit b in Algorithm 7. We stress that steps 12–15 are *not* shown to $\mathcal{B}_D$. Now $\mathcal{A}_{\mathrm{prv}}^{\mathcal{B}_D}$ queries the Corrupt oracle and obtains the state $S^{\mathcal{T}_{\mathrm{ID}}}$ of $vtag'$ (step 15). This query and $S^{\mathcal{T}_{\mathrm{ID}}}$ are also shown to $\mathcal{B}_D$ (step 16). Then $\mathcal{A}_{\mathrm{prv}}^{\mathcal{B}_D}$ sends to $\mathcal{B}_D$ the message obtained by the SendTag oracle in step 12, which has either been computed by the real SendTag oracle or the blinder $\mathcal{B}$ (step 17). Hereby, $\mathcal{B}_D$ expects to receive a message that has been computed according to the state $S^{\mathcal{T}_{\mathrm{ID}}}$ obtained by Corrupt. Now the second phase starts, where $\mathcal{A}_{\mathrm{prv}}^{\mathcal{B}_D}$ simulates the messages of $vtag'$ using $S^{\mathcal{T}_{\mathrm{ID}}}$ and the messages sent by $\mathcal{B}_D$, which is playing as a reader (steps 18–24), until the protocol terminates, as expected by $\mathcal{B}_D$. Now, for the hypothesis, $\mathcal{B}_D$ can distinguish whether the messages it receives are (i) computed according to the state of the tag simulated in the first phase (thus $\Gamma[vtag] = \Gamma[vtag']$) and in this case Result will output

Alg. 8. Narrow-forward adversary $\mathcal{A}_{\text{prv}}^{\mathcal{B}_D}$

```
 1: CreateTag(ID)                                              ▷ shown to 𝓑_D
 2: CreateTag(ID′)                                             ▷ shown to 𝓑_D
 3: vtag ← Draw(Pr[ID] = ½, Pr[ID′] = ½)                       ▷ shown to 𝓑_D
 4: π ← Launch( )                                              ▷ simulated by 𝓑_D
 5: m₁ ← SendReader(−, π)                                      ▷ simulated by 𝓑_D
 6: j_ℛ ∈_R {1, …, q_ℛ}
 7: i ← 1
 8: while i < j_ℛ do m₂ᵢ ← SendTag(m_{2i−1}, vtag)             ▷ simulated by 𝓑_D
 9:     m_{2i+1} ← SendReader(m₂ᵢ, π)                          ▷ simulated by 𝓑_D
10:     i ← i + 1
11: end while
12: m_{2j_ℛ} ← SendTag(m_{2j_ℛ−1}, vtag)                      ▷ computed by vtag
13: Free (vtag)
14: vtag′ ← Draw(Pr[ID] = ½, Pr[ID′] = ½)
15: S^{𝒯_ID} ← Corrupt(vtag′)
16: Show S^{𝒯_ID} ← Corrupt(vtag) to 𝓑_D
17: m_{2j_ℛ+1} ← SendReader(m_{2j_ℛ}, π)                      ▷ simulated by 𝓑_D
18: i ← j_ℛ + 1
19: while i < q_ℛ do
20:     if i ≤ q_𝒯 then m₂ᵢ ← 𝒯_ID^{(i)}(S^{𝒯_ID}, m_{2i−1}) ▷ computed by 𝒜_prv^{𝓑_D}
21:     end if
22:     m_{2i+1} ← SendReader(m₂ᵢ, π)                         ▷ simulated by 𝓑_D
23:     i ← i + 1
24: end while
25: b ← Result(π)                                             ▷ simulated by 𝓑_D
26: return (Γ[vtag] = Γ[vtag′] ∧ b) ∨ (Γ[vtag] ≠ Γ[vtag′] ∧ b̄)
```

1, or (ii) with a different state (thus $\Gamma[vtag] \neq \Gamma[vtag']$) and in this case Result will output 0. Now we show that $\mathbf{Adv}_{\mathcal{A}_{\text{prv}}^{\mathcal{B}_D}}^{\text{prv}}$ is non-negligible if $\mathcal{B}_D$ exists.

First, consider the case where $\mathcal{A}_{\text{prv}}^{\mathcal{B}_D}$ interacts with real oracles. If $\Gamma[vtag] = \Gamma[vtag']$, then due to the existence of $\mathcal{B}_D$ we have that Result returns 1 with overwhelming probability, which makes $\mathcal{A}_{\text{prv}}^{\mathcal{B}_D}$ to return 1 with the same probability. Note that even though $\mathcal{B}_D$ learns the state $S^{\mathcal{T}_{ID}}$ of $vtag$ only after obtaining message $m_{2j_\mathcal{R}}$ that has been computed from this state, by the stateless property of the scheme and thus by Lemma 2, there is no noticeable difference between the state of $vtag$ before and after the computation of $m_{2j_\mathcal{R}}$. In case $\Gamma[vtag] \neq \Gamma[vtag']$, we have that the first message $m_{2j_\mathcal{R}-1}$ received by $\mathcal{B}_D$ has been computed according to the state of $vtag$ and all subsequent messages are computed according to the state of $vtag'$. This deviates from what $\mathcal{B}_D$ expects and thus $\mathcal{B}_D$ could erroneously answer the Result query with 1. Let us denote with p the probability that $\mathcal{B}_D$ with input $S^{\mathcal{T}_{ID}}$ answers the Result query with 0 upon receiving a message computed with a random state followed by messages computed with $S^{\mathcal{T}_{ID}}$. Then we have $\Pr\left[\mathbf{Exp}_{\mathcal{A}_{\text{prv}}^{\mathcal{B}_D}}^{\text{prv-0}} = 1\right] = \frac{1}{2} \cdot (1 - \epsilon(l)) + \frac{1}{2} \cdot p \leq \frac{(1+p)}{2}$.

Now, consider the case where $\mathcal{A}_{\mathrm{prv}}$ interacts with $\mathcal{B}$. Here we have that in both cases ($\Gamma[vtag] = \Gamma[vtag']$ and $\Gamma[vtag] \neq \Gamma[vtag']$) the output of the SendTag oracle computed by $\mathcal{B}$ for the forward adversary is computed with a random state that with overwhelming probability is different from the state of $vtag$ and $vtag'$. Thus, in both cases $\mathcal{B}_D$ with input the state $S^{\mathcal{T}_{\mathrm{ID}}}$ receives the first message $m_{2j_{\mathcal{R}}-1}$ computed according to a state that is different from $S^{\mathcal{T}_{\mathrm{ID}}}$. Hence we have $\Pr\left[\mathbf{Exp}^{\mathrm{prv\text{-}0}}_{\mathcal{A}^{\mathcal{B}_D}_{\mathrm{prv}}} = 1\right] = \frac{1}{2} \cdot (1-p) + \frac{1}{2} \cdot p = \frac{(1-p)}{2} + \frac{p}{2} = \frac{1}{2}$. and it follows that $\mathbf{Adv}^{\mathrm{prv}}_{\mathcal{A}^{\mathcal{B}_D}_{\mathrm{prv}}} \leq \left|\frac{(1+p)}{2} - \frac{1}{2}\right| = \frac{p}{2}$. Note that if p is non-negligible, so is the advantage of $\mathcal{A}^{\mathcal{B}_D}_{\mathrm{prv}}$ and the proof is finished. If instead p is negligible, then $\mathcal{B}_D$ has non-negligible probability of answering 1 to a Result query when no message originates from a valid state. (In the above experiment, this case happens when $j_{\mathcal{R}}$ corresponds to the last round of the protocol.) Obviously a reader that always expects messages being computed according to a legitimate state would output 0 to a Result query in such an experiment, and this would contradict the fact that (even a variation of) $\mathcal{B}_D$ is successful against this variation of $\mathcal{A}_{\mathrm{prv}}$.

The last issue to address is the more general case where a reader admits wrong messages from a tag, still responding with 1 to a Result query when some messages are computed using a legitimate state. However, since the procedure of the reader is public, the above proof can be generalized to any reader strategy. Indeed, $\mathcal{A}_{\mathrm{prv}}$ must replace some correctly computed messages with messages computed with a random state such that the replacement of the valid messages exposes the failure of $\mathcal{B}_D$. This is achieved by asking $\mathcal{A}_{\mathrm{prv}}$ to compute each tag-side message either using a legitimate or an illegitimate tag state with probability q that comes from the description of the reader procedure for the Result query, so that the output of this query is noticeably perturbed by the replacement of a correctly computed message by a wrongly computed one.   □

7   Conclusion

In this paper, we revisited the security and privacy model for RFID systems proposed by Paise and Vaudenay (PV-Model) [30]. This model is very interesting since it covers many aspects of previous works and proposes a unified RFID security and privacy framework. We showed several impossibility results that show that the formalization given in the PV-Model is too restrictive and fails in modelling real-life scenarios, where interesting privacy notions and reader authentication are intuitively achievable. A partial and shorter version of this work appeared in [1].

Acknowledgments. We thank Paolo D'Arco for several useful discussions about RFID privacy notions. This work has been supported in part by the European Commission through the FP7 programme under contract 216676 ECRYPT II, 238811 UNIQUE, and 215270 FRONTS, in part by the Ateneo Italo-Tedesco under Program Vigoni and by the MIUR Project PRIN 2008 "PEPPER: Privacy E Protezione di dati PERsonali" (prot. 2008SY2PH4).

References

1. Armknecht, F., Sadeghi, A.R., Visconti, I., Wachsmann, C.: On RFID privacy with mutual authentication and tag corruption. In: Zhou, J. (ed.) ACNS 2010. LNCS, vol. 6123, pp. 493–510. Springer, Heidelberg (2010)
2. Atmel Corporation: Innovative IDIC solutions (2007),
   http://www.atmel.com/dyn/resources/prod_documents/doc4602.pdf
3. Avoine, G.: Adversarial model for radio frequency identification. ePrint, Report 2005/049 (2005)
4. Avoine, G., Lauradoux, C., Martin, T.: When compromised readers meet RFID. In: The 5th Workshop on RFID Security (RFIDSec) (2009)
5. Bellare, M., Fischlin, M., Goldwasser, S., Micali, S.: Identification protocols secure against reset attacks. In: Pfitzmann, B. (ed.) EUROCRYPT 2001. LNCS, vol. 2045, pp. 495–511. Springer, Heidelberg (2001)
6. Blundo, C., Persiano, G., Sadeghi, A.R., Visconti, I.: Improved security notions and protocols for non-transferable identification. In: Jajodia, S., Lopez, J. (eds.) ESORICS 2008. LNCS, vol. 5283, pp. 364–378. Springer, Heidelberg (2008)
7. Bringer, J., Chabanne, H., Icart, T.: Efficient zero-knowledge identification schemes which respect privacy. In: Proceedings of ASIACCS 2009, pp. 195–205. ACM Press, New York (2009)
8. Burmester, M., van Le, T., de Medeiros, B.: Universally composable and forward-secure RFID authentication and authenticated key exchange. In: Proc. of ASIACCS, pp. 242–252. ACM Press, New York (2007)
9. Canetti, R., Goldreich, O., Goldwasser, S., Micali, S.: Resettable zero-knowledge (extended abstract). In: STOC, pp. 235–244 (2000)
10. D'Arco, P., Scafuro, A., Visconti, I.: Revisiting DoS Attacks and Privacy in RFID-Enabled Networks. In: Dolev, S. (ed.) ALGOSENSORS 2009. LNCS, vol. 5804, pp. 76–87. Springer, Heidelberg (2009)
11. D'Arco, P., Scafuro, A., Visconti, I.: Semi-destructive privacy in DoS-enabled RFID systems. In: The 5th Workshop on RFID Security (RFIDSec) (2009)
12. Deng, R.H., Li, Y., Yao, A.C., Yung, M., Zhao, Y.: A new framework for RFID privacy. ePrint, Report 2010/059 (2010)
13. Deng, Y., Lin, D.: Instance-dependent verifiable random functions and their application to simultaneous resettability. In: Naor, M. (ed.) EUROCRYPT 2007. LNCS, vol. 4515, pp. 148–168. Springer, Heidelberg (2007)
14. EPCglobal Inc.. (April 2008), http://www.epcglobalinc.org/
15. Garcia, F., de Koning Gans, G., Muijrers, R., van Rossum, P., Verdult, R., Wichers Schreur, R., Jacobs, B.: Dismantling MIFARE Classic. In: Jajodia, S., Lopez, J. (eds.) ESORICS 2008. LNCS, vol. 5283, pp. 97–114. Springer, Heidelberg (2008)
16. Garcia, F.D., van Rossum, P.: Modeling privacy for off-line RFID systems. In: The 5th Workshop on RFID Security (RFIDSec) (2009)
17. Goyal, V., Sahai, A.: Resettably secure computation. In: EUROCRYPT, pp. 54–71 (2009)
18. Hutter, M., Schmidt, J.M., Plos, T.: RFID and its vulnerability to faults. In: Oswald, E., Rohatgi, P. (eds.) CHES 2008. LNCS, vol. 5154, pp. 363–379. Springer, Heidelberg (2008)
19. I.C.A. Organization: Machine Readable Travel Documents, Doc 9303, Part 1 Machine Readable Passports, 5th edn (2003)
20. Juels, A.: RFID security and privacy: A research survey. Journal of Selected Areas in Communication 24(2), 381–395 (2006)

21. Juels, A., Weis, S.A.: Defining strong privacy for RFID. ePrint, Report 2006/137 (2006)
22. Kasper, T., Oswald, D., Paar, C.: New methods for cost-effective side-channel attacks on cryptographic RFIDs. In: The 5th Workshop on RFID Security (RFIDSec) (2009)
23. Kirschenbaum, I., Wool, A.: How to build a low-cost, extended-range RFID skimmer. ePrint, Report 2006/054 (2006)
24. Mangard, S., Oswald, E., Popp, T.: Power Analysis Attacks Revealing the Secrets of Smart Cards. Springer, Heidelberg (2007)
25. Ng, C.Y., Susilo, W., Mu, Y., Safavi-Naini, R.: New privacy results on synchronized RFID authentication protocols against tag tracing. In: Backes, M., Ning, P. (eds.) ESORICS 2009. LNCS, vol. 5789, pp. 321–336. Springer, Heidelberg (2009)
26. Ng, C.Y., Susilo, W., Mu, Y., Safavi-Naini, R.: RFID privacy models revisited. In: Jajodia, S., Lopez, J. (eds.) ESORICS 2008. LNCS, vol. 5283, pp. 251–256. Springer, Heidelberg (2008)
27. Nithyanand, R., Tsudik, G., Uzun, E.: Readers behaving badly: Reader revocation in PKI-based RFID systems. ePrint, Report 2009/465 (2009)
28. NXP Semiconductors: MIFARE (May 2007), http://mifare.net/
29. NXP Semiconductors: MIFARE smartcard ICs (April 2010), http://www.mifare.net/products/smartcardics/
30. Paise, R.I., Vaudenay, S.: Mutual authentication in RFID: Security and privacy. In: Proc. of ASIACCS, pp. 292–299. ACM Press, New York (2008)
31. Sadeghi, A.R., Visconti, I., Wachsmann, C.: User privacy in transport systems based on RFID e-tickets. In: International Workshop on Privacy in Location-Based Applications (PiLBA) (2008)
32. Sadeghi, A.R., Visconti, I., Wachsmann, C.: Anonymizer-enabled security and privacy for RFID. In: Garay, J.A., Miyaji, A., Otsuka, A. (eds.) CANS 2009. LNCS, vol. 5888, pp. 134–153. Springer, Heidelberg (2009)
33. Sadeghi, A.R., Visconti, I., Wachsmann, C.: Efficient RFID security and privacy with anonymizers. In: The 5th Workshop on RFID Security (RFIDSec) (2009)
34. Sadeghi, A.R., Visconti, I., Wachsmann, C.: Location privacy in RFID applications. In: Bettini, C., Jajodia, S., Samarati, P., Wang, X.S. (eds.) Privacy in Location-Based Applications. LNCS, vol. 5599, pp. 127–150. Springer, Heidelberg (2009)
35. Sadeghi, A.R., Visconti, I., Wachsmann, C.: Enhancing RFID Security and Privacy by Physically Unclonable Functions. Springer, Heidelberg (2010)
36. Sadeghi, A.R., Visconti, I., Wachsmann, C.: PUF-enhanced RFID security and privacy. In: Workshop on Secure Component and System Identification (SECSI) (2010)
37. Vaudenay, S.: On privacy models for RFID. In: Kurosawa, K. (ed.) ASIACRYPT 2007. LNCS, vol. 4833, pp. 68–87. Springer, Heidelberg (2007)
38. Weis, S.A., Sarma, S.E., Rivest, R.L., Engels, D.W.: Security and privacy aspects of low-cost radio frequency identification systems. In: Hutter, D., Müller, G., Stephan, W., Ullmann, M. (eds.) Security in Pervasive Computing. LNCS, vol. 2802, pp. 50–59. Springer, Heidelberg (2004)

Implementation of Multivariate Quadratic Quasigroup for Wireless Sensor Network

Ricardo José Menezes Maia, Paulo Sérgio Licciardi Messeder Barreto, and
Bruno Trevizan de Oliveira

Escola Politécnica da Universidade de São Paulo,
Depatarmento de Engenharia Elétrica - Sistemas Digitais,
São Paulo, Brasil
ricardo.jmm@usp.br,
pbarreto@larc.usp.br,
btrevizan@larc.usp.br

Abstract. Wireless sensor networks (WSN) consists of sensor nodes
with limited energy, processing, communication and memory. Security
in WSN is becoming critical with the emergence of applications that require mechanisms for authenticity, integrity and confidentiality. Due to
resource constraints in WSN, matching public key cryptosystems (PKC)
for these networks is an open research problem. Recently a new PKC based on quasigroups multivariate quadratic. Experiments performed show
that MQQ performed in less time than existing major PKC, so that
some articles claim that has MQQ speed of a typical symmetric block
cipher. Considering features promising to take a new path in the difficult
task of providing wireless sensor networks in public key cryptosystems.
This paper implements in nesC a new class of public key algorithm called Multivariate Quadratic Quasigroup. This implementation focuses on
modules for encryption and decryption of 160-bit MQQ, the modules
have been implemented on platforms TelosB and MICAz. We measured
execution time and space occupied in the ROM and RAM of the sensors.

Keywords: Multivariate Quadratic Quasigroup, Implementation of
Modules Encryption and Decryption, Wireless Sensor Network.

1   Introduction

Advances in miniaturization and wireless communications provide the development of a new paradigm, where we highlight wireless sensor networks (WSN).
WSN's are composed of small devices equipped with the processing unit, sensing
and communication, called sensor nodes [15].

The sensors extract and transmit environmental data to one or more exit
points of the network, called nodes sinks. Later, the data sent by the sensors
will be stored and then processed. The installation of such sensors can be at predefined or not in the target area. These resources are extremely limited supply
of energy, processing power, memory, storage and communication systems with
low bandwidth [15,20].

M.L. Gavrilova et al. (Eds.): Trans. on Comput. Sci. XI, LNCS 6480, pp. 64–78, 2010.

WSN's can be used in various applications such as traffic control, monitoring of environmental variables, to detect the presence of hazardous materials, detection of enemy movements (military applications), identification and registration of persons in large environments (airports) , monitoring of human health, monitor the moisture levels in agricultural areas to carry out irrigation selectively detect intruders in border areas among others [3,16].

Certain applications require WSN's security requirements such as confidentiality, integrity and authenticity. Among these applications we can mention monitoring vital signs of patients. In an application for monitoring of environmental variables researchers can require the integrity of the data monitored. Applications for home automation authenticity is paramount, to enable the sensors are monitored only by the owners. In industrial applications the requirement of the three conditions can be crucial to prevent espionage or that other companies can gain competitive advantage [3,16].

WSN's using wireless communication, being more vulnerable to attacks, since this mode of communication, the mode of transmission is broadcast. By using broadcast, the network becomes more susceptible to the action of intruders, which can easily listen to, intercept and alter data traveling on the network [16].

Given the resource constraints in a WSN, there is a profound impact on the adoption of protocols and algorithms for communication and security. Therefore, a key issue is to satisfy the application requirements for a secure, considering the existing constraints on these networks [21,16].

The current solution to the problem of establishing security in WSN premises is around symmetric cryptosystems, even considering the increased security afforded by public key cryptosystems [18,21].

Consider symmetric schemes have drawbacks to security, such as one single key to encrypt and decrypt can jeopardize the entire system, if the key is exposed. In this case it is good to note that applications in WSN the sensors are usually exposed in the study environment, with a possible intruder violates any sensor to obtain the private key is also used by other sensors [18,16].

Although PKC's principle to permit greater security than symmetric schemes, restrictions imposed by WSN makes the deployment of PKC on WSN an open problem [21,18].

The difficulty of using PKC in WSN due to its speed a thousand times smaller than PKC with respect to symmetric algorithms. This occurs because the security of PKC be based on difficult mathematical problems as two discrete logarithm problem and factorization of integers [9].

Recently a new scheme of public keys, called Multivariate Quadratic Quasigroup (MQQ) based on multivariate polynomials and quadratic transformations of strings quasigroups. Experiments conducted show MQQ several orders of magnitude faster than the most popular public key algorithms like RSA, DH or ECC [9,7,1].

Experiments show that displays MQQ same level of security as RSA, where the MQQ with keys of 160 bits get the same level of security with RSA keys of

1024 bits. MQQ was very quick both to encrypt and to decrypt such speed makes authors of articles quoting MQQ speed of a typical block cipher symmetric [9,7].

Considering the growing need for increased security for WSN, the limited resources of this type of network and finally characteristics of promising MQQ for platforms with limited resources may consider the MQQ a new way to provide PKC for WSN . Despite the promising results of MQQ on other platforms, there is no previous work related MQQ the WSN.

The results in [9] and validated [7] are promising, after all these documents show that MQQ is faster than the more traditional public key cryptosystems such as RSA and ECC.

This MQQ performance with respect to traditional PKC (RSA and ECC) calls attention at all until the moment the performance of ECC was superior to other traditional PKC, such as RSA [18,21,19]. MQQ gets the same level of security than traditional PKCs, consuming much less computational resources [19,8,2].

Moreover MQQ displays the same level of security that other PKC requiring minor keys. The two characteristics mentioned draw attention, but the fact of MQQ need the basic operations XOR and AND between bits in the encryption and decryption processes are essential given the limited hardware of a WSN.

So MQQ becomes a promise not only for WSN, but for limited processing platforms as a whole. Since MQQ offers the security of conventional PKC's consuming much less computational resources [9,7].

Considering the fact WSN provide PKC to be an open research problem and the emergence of a new public key scheme called Multivariate Quadratic Quasi-group, which was proposed by Gligoroski, et al., [9]. Importantly, the fact that the author of MQQ has the speed of a typical symmetric cipher [9,7] opens new perspectives in way WSN provide PKC. Another fact that calls attention to MQQ use in sensors is that this approach has operations in need of basic instructions such as AND and XOR bit. This article describes the results of the implementation and execution of the modules for encryption and decryption of MQQ platforms TelosB [6] and MICAz [5].

The objective of this work is to analyze the performance of the encryption and decryption modules MQQ a platform for WSN. Was not made a comparison of MQQ with other PKC and was not made changes to make the MQQ safer.

Organization of the paper is the following: In Section 2 is a brief explanation of concepts necessary for understanding the MQQ. Details on the platform used in the experiments are found in section 3. Details of the algorithms for encryption and decryption and considerations made to facilitate its implementation can be found in sections 4 and 5. The results and analysis are shown in section 6. The final considerations are in the section 7.

2   Concepts

In this section will be made brief remarks about concepts relevant to MQQ. Further details about these concepts can be found in [7,10].

Definition 1. A quasigroup is a groupoid that satisfies the following law.

$$(\forall u, v \in Q)(\exists! x, y \in Q)(u * x = v, y * u = v).\tag{1}$$

Based on 1 we conclude that for every $a, b \in Q$ there is a single $x \in Q$ such that $a * x = b$. Then $x = a\backslash_* b$ where $\backslash_*$ is a binary operation on Q and the groupoid $(Q, \backslash_*)$ is also a quasigroup. The algebra $(Q, *, \backslash_*)$ satisfies the equation 2.

$$x\backslash_*(x * y) = y, x * (x\backslash_* y) = y\tag{2}$$

Assuming an alphabet (a finite set) Q and Q^+ the set of all nonempty words (finite set of strings) formed by elements of Q. In this study both the notations $Q^+ : a_1 a_2...a_n$ and $(a_1, a_2, ..., a_n)$ can be used, where $a_i \in Q$. Consider $*$ the quasigroup operation on the set Q. For each $l \in Q$ were defined two functions $e_{l,*}, d_{l,*} : Q^+ \rightarrow Q^+$.

Definition 2. Consider $ai \in Q, M = a_1 a_2...a_n$. Then
$e_{l,*}(M) = b_1 b_2...b_n \Longleftrightarrow$
$b_1 = l * a_1, b_2 = b_1 * a_2, ..., b_n = b_{n-1} * a_n,$
$d_{l,*}(M) = c_1 c_2...c_n \Longleftrightarrow$
$c_1 = l * a_1, c_2 = a_1 * a_2, ..., c_n = a_{n-1} * a_n,$
i.e., $b_{i+1} = b_i * a_{i+1}$ and $c_{i+1} = a_i * a_{i+1}$ for   each $i = 0, 1, ..., n - 1$, such that $b_0 = a_0 = l$.

The functions $e_{l,*}$ and $d_{l,*}$ transformations are called e and d of Q^+ based on operation $*$ with the head l respectively.

Theorem 1. If $Q, *$ is a finite quasigroup, then $e_{l,*}$ e $d_{l,-*}$ are mutually inverse permutations of Q^+, i.e.,

$$d_{l,\backslash_*}(e_{l,*}(M)) = M = e_{l,*}(d_{l,\backslash_*}(M))$$

for each $l \in Q$ and for each string $M \in Q^+$.

Definition 3. A quasigroup $Q, *$ of order 2^d is called Multivariate Quadratic QuasiGroup of type $Quad_{d-k}Lin_k$ if exactly $d - k$ polynomials f_i are of degree 2 (quadratic) and k of them are of degree 1 (linear), where $0 \leq k < d$.

Multivariate Quadratic Quasigroup (MQQ) is a special class of quasigroups [7,10].

3   Implementation Multivariate Quadratic Quasigroup

In this work the focus will be given in the algorithms to encrypt and decrypt, where details about the algorithms and MQQ key generation, encryption and decryption are found in [7,10].

The purpose of this work is to implement the encryption and decryption of MQQ on a node of wireless sensor networks, platforms using the crossbow's TelosB, 16-bit RISC processor with 10 KBytes of RAM and 48 KBytes of ROM for program MICAz and 4 KBytes of RAM and 128 KBytes of ROM for program.

The encryption and decryption modules were built into the operating system TinyOS 2.0.2 [13], having been written in C (ANSI) and then being called into nesC code have not been made aiming at optimizations perform in a specific sensor platform. In the implementations we used the nesC component Timer to check the execution time of each module, having been used for precision TMilli and getNow command to get current time [4]. The time was determined from the difference of the result of two calls, before and after calls of methods to encrypt and decrypt.

The development environment used was the linux distribution XubunTOS, being used as simulators to TOSSIM [14,12] and Avrora testing and validating the implementation. Later, the sources of the encryption and decryption were implemented in TelosB platforms and MICAz.

In the experiments we chose the platform of WSN's crossbow TelosB [6], which is an open platform designed to allow experiments and laboratory studies in the scientific community. It has features such as programming via USB, radio antenna integrated IEEE 802.15.4, data transmission rate of 250 kbps. It has the TI MSP430 microcontroller manufactured with Texas Instruments 8 MHz, 16-bit RISC architecture, 10 KBytes of RAM and 48 KBytes of ROM for program. In addition, optional features such as integrated sensors of temperature, humidity, light. Besides supporting the operating system TinyOS. Another WSN platform used was the crossbow MICAz [5], which has baud rate of 250 kbps and the rate of transmission of radio module can vary from 2.4 to 2.48 GHz has 4 KBytes of RAM and 128 KBytes of ROM for program, and supports TinyOS.

The implementation was done using MQQ with keys of 160 bits, 160 bits with second MQQ [7] has the same level of security as RSA 1024 bits.

Will be collected memory space occupied by MQQ in RAM and ROM on the platforms TelosB and MICAz. Analyze the space on memory is vital to assess whether MQQ can be a viable proposition for WSN, after all space is needed in the sensor memory to run useful applications. Importantly, the programmable flash memory in the two existing platforms will be called a ROM.

To facilitate the experiments, the data structures that represent the eight quasigroups and not natural and inverse matrices are represented by multidimensional vectors. The values of the eight quasigroups and the natural and inverse matrices are fixed in ROM of the sensors.

4   Encryption

The algorithm for encryption is the multiplication of a set of n multivariate polynomials $\mathbf{P} = \{P_i(x_1, ..., x_n) | i = 1, ..., n\}$ on a vector $x = (x1, x2, ..., x_n)$, i.e., $y = P(x)$ [9,7].

In this implementation each polynomial P_i is interpreted with its coefficients $c_i \in \{0, 1\}$,

Example: Consider $P_i = c_0 + c_1 \times x_1 + c_2 \times x_2 + c_3 \times (x_1 \times x_2)$, $P_i = 0110 \equiv$ $P_i = x_1 + x_2$

To implement the encryption MQQ P has been represented as a matrix and x as a vector, where $y = P(x)$ was represented with the result of multiplying P

for x, where operations are represented by the sum an XOR operation between bits and the multiplication of two variables is represented by an AND operation among bits.

Example: Consider $x = \{x_1 = 1, x_2 = 0, x_3 = 1\}$ and the polynomial P_1

$P_1(x) = (x_1 \times x_2 + x_1 \times x_3)$

$P_1(x) = (x_1 \text{ AND } x_2 \text{ XOR } x_1 \text{ AND } x_3)$

$P_1(x) = 1 \text{ AND } 0 \text{ XOR } 1 \text{ AND } 1 = 0 \text{ XOR } 1$

$P_1(x) = 1$

We can represent $y = P(x)$ as follows below [7,10]:

$P_i(x_1, ..., x_n) = a_{i,0,0} + \sum_{j=1}^{n} a_{i,j,0} x_j + \sum_{j=1}^{n-1} \sum_{k=j+1}^{n} a_{i,j,k} x(k-j) x_k,$

Then $a_{i,0,0} \in \{0,1\}$.

Then $y = P(x) \equiv y = A \times X$.

In this case A is the public key of MQQ and represents the coefficients in each polynomial, i.e., term will not exist in polynomial coefficient 0. By owning coefficients for n polynomials and the permutation the n terms $x_1, ..., x_n$ of a polynomial, then the matrix A has the dimension of $A_{n \times (1+n+\frac{n \times (n-1)}{2})}$. The matrix A consists of a vector of n positions, where for each position matrix A is a vector with 160 positions locations, where each element of the array can get a bit. Since the defined position vector of polynomials (bit 1) means that a term belongs to the determined polynomials.

This means that for an implementation of MQQ with keys of 160 bit public key and the matrix A with size would be $A_{160 \times 12881}$ which would give a size of $160 \times 12881 = 2060960$ bits, turning in kilobytes $(2060960 \div 8) \div 1024 = 251.58$ KBytes.

X is a vector of size 12 881 $X_{12881 \times 1}$, being obtained by permutation of $x = (x_1, ..., x_n)$. Once the encryption process can be described as:

$y(y_1, ..., y_n) = A_{160 \times 12881} \times X_{12881 \times 1}$

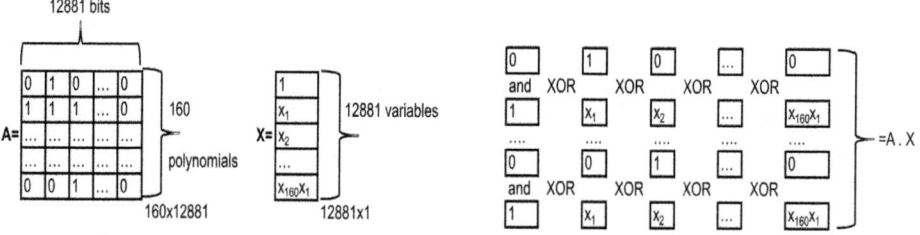

Fig. 1. Representation of multiplication of the public key with the vector X

According to [7] $y = A \times X$ represents the encryption process, according to the figure 1.

But it is important to make some considerations about the size of the keys and instructions used in encryption. The first consideration relates to the instructions used in the encryption that are summarized AND and XOR bit instructions simple to be implemented in a sensor node.

Considering the size of the public key of 251.58 KBytes prevents the allocation of these keys in memory of a sensor, either RAM or even ROM.

This paper presents a proposal to shrink the public key based on the fact that a term may be redundant in several polynomials [9]. Once you use a vector of length 12 881 bits, if one of the bits of A is 0 means that the term does not exists in the polynomials.

A bit of A is a term that means there are some polynomials, having been created an auxiliary 160-bit vector what is the 160 reports in which the polynomial term is present. Assist in the vector when the bit is 1 means that the term is a given polynomial, as depicted in figure 2.

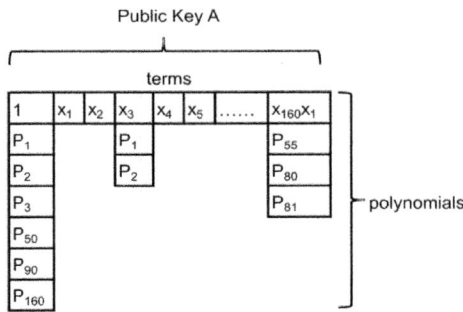

Fig. 2. Proposed new representation for public key used MQQ

In this solution the 12881 bits represented by the permutation of the terms $(x_1, ..., x_n)$ are fixed in ROM of the sensor, however the 160-bit vector auxiliary are dynamically allocated case in which to inform the polynomial term is present.

The idea is to have a vector with all permutations of terms. Example x1, x2, ..., x1x3, ... x1x160. So the technique used to reduce the size of the public key A is to advise that particular term belongs simultaneously to several polynomials.

In figure 2 in the term 1 is the polynomials P1, P2, P3, P50, P90 and P160. The term x3 is the polynomials P1 and P2. While the term x160x1 is found in polynomials P55, P80 and P81. In polynomials the item $x3$ belongs to the polynomials P1 and P2.

Example:
$P1 = 1 + x3 + x1x160$
$P2 = 1 + x3$
$P3 = 1$
$P50 = 1$
$P55 = x160x1$
$P80 = x160x1$
$P81 = x160x1$
$P90 = 1$
$P160 = 1$

In the new approach the matrix A is represented in a vector of length $A[12881/16]$ getting $A[806]$ a 16-bit vector, and the auxiliary structure has size $Auxiliar[160/16]$ which is equivalent to a vector $Auxiliar[10]$ of 16 bits. The advantage this approach is to reduce the size required to store the public key, and in some cases the optimal public key to decrease to 251.58 KBytes $(12881 \div 8) \div 1024 = 1.57$ KBytes. So in that scenario is to implement MQQ of encryption in wireless sensor networks is the greatest difficulty in finding mechanisms to reduce the size public key and may use other forms of representation as sparse matrices.

5 Decryption

The definition of the decryption algorithm MQQ is described in the table 1 [9].

Table 1. Algorithm to decrypt and sign

Algorithm to decrypt / sign with private key $T, S, *_1, ..., *_8$
Input: Vector $y = y_1, ..., y_n$.
Output: Vector $x = (x_1, ..., x_n)$ such that $P(x) = y$
1. $y' = T^{-1}(y)$.
2. $W = y'_1, y'_2, y'_3, y'_4, y'_5, y'_6, y'_{11}, y'_{16}, y'_{21}, y'_{26}, y'_{31}, y'_{36}, y'_{41}$.
3. $Z = Z_1, Z_2, Z_3, Z_4, Z_5, Z_6, Z_7, Z_8, Z_9, Z_{10}, Z_{11}, Z_{12}, Z_{13} = Dob^{-1}(W)$.
4. $y'_1 \longleftarrow Z_1, y'_2 \longleftarrow Z_2, y'_3 \longleftarrow Z_3, y'_4 \longleftarrow Z_4, y'_5 \longleftarrow Z_5, y'_6 \longleftarrow Z_6,$ $y'_{11} \longleftarrow Z_7, y'_{16} \longleftarrow Z_8, y'_{21} \longleftarrow Z_9, y'_{26} \longleftarrow Z_{10}, y'_{31} \longleftarrow Z_{12}, y'_{41} \longleftarrow Z_{13}$.
5. $y' = Y_1...Y_k$ where Y_i are vectors of dimension 5.
6. Being $*_i$, $i = 1, ..., 8$, obtaining $x' = X_1...X_k$, so that, $X_1 = Y_1, X_2 = X_{1\backslash 1}Y_2, X_3 = X_{2\backslash 2}Y_3$ and $X_i = X_{i-1\backslash 3+((i+2) \bmod 6)}Y_i$
7. $x = S^{-1}(x')$

The algorithm used to decrypt a private key comprised two nonsingular matrices T and S and eight quasigroups $*_1, *_2, *_3, *_4, *_5, *_6, *_7, *_8$. The matrices $T_{n \times n}$ and $S_{n \times n}$, in the case of this implementation dimensions are $T_{160 \times 160}$ and $S_{160 \times 160}$. So to store the matrices T and S will be used a structure that stores $2 \times (160 \times 160) = 51200$ bits, leaving $(51200 \div 8) \div 1024 = 6.25$ KBytes.

Each of the eight quasigroups $*_1, ..., *_8$ is a matrix of size 32 x 32 = 1024, where one of 1024 positions of a virtual group has 5 bits.

Once the size of one of the eight quasigroups is $32 \times 32 \times 5 = 5120 = 5120$ bits, so to store the eight quasigroup will need a space for $8 \times (32^2 \times 5) = 40$ 960 bits which would $(40960 \div 8) \div 1024 = 5$ KBytes.

In all the private key needs to be stored 11.25 KBytes and be made feasible the implementation of the encryption a private key has been placed in ROM of the sensor. But it is important to consider that within each quasigroup is polynomial terms of opening up space for redundant implementation that optimizes the space occupied by quasigroups.

The entrance to the decryption algorithm consists of a vector y of 160 bits representing the encrypted message. While the output of the algorithm is represented by a vector x of 160 bits representing the message decrypted. Since y and x represented in a vector of 16 bits with 10 positions.

Step 1 of table 1 is the inverse of matrix multiplication $T^{-1}_{160 \times 160}$ with the vector entry $y_{160 \times 1}$. In this implementation the inverse matrix T^{-1} is fixed in ROM of the sensor, being represented as an array of 16 bits with size $T^{-1}_{160 \times 16}$. The operations of multiplication and sum consisting of ANDs and XOR between the bits of T^{-1} and y, the result being assigned to the vector y'.

Step 7 of table 1 is the same procedure of step 1, so that instead of $y' = T^{-1}(y)$ would be $x = S^{-1}(x')$.

To implement the sensor arrays in MQQ T^{-1} and S^{-1} will be fixed in ROM of the sensor.

The second step aims to achieve 13 bits of the vector $y' = T^{-1}(y)$, which are arranged in 13-bit vector W. The vector W receives each bit $y'_1, y'_2, y'_3, y'_4, y'_5, y'_6, y'_{11}, y'_{16}, y'_{21}, y'_{26}, y'_{31}, y'_{36}, y'_{41}$ vector y'.

The third step is to research the inverse matrix of Dob^{-1}, taking as input parameter the vector of W 13 bits.O result of this research in the matrix Dob^{-1} is assigned to the vector Z 13 bits. The size Dob^{-1} is $13 \times 2^{13} = 106496$, totaling 13 KBytes.

The fourth step is to assign each of the bit vector Z again to the vector y'.

In the fifth step the bits of y' 5 will be organized and arranged in 5 bits in the vector $Y = Y_1, ..., Y_{160/5} = Y_1, ..., Y_{32}$, and Y be a vector of dimension 32x5.

The sixth step results in the vector $x' = X_1, ..., X_{32}$ of dimension 32 x 5, where each X_i has 5 bits resulting from research groups in the quasigroups $*1, ..., *8$. For an element of quasigroup is given a month 10-bit value representing the address of an element of quasigroup.

In decrypting the data stored in the ROM of the sensor were T^{-1}, S^{-1}, Dob^{-1} and quasigroups $*1, ..., *8$ giving a total of static structures to be stored $(2 \times 160^2) + (13 \times 2^{13}) + (8 \times 32^2 \times 5) = 198656$ bits, i.e., 24.25 KBytes.

Whereas in the quasigroups may be redundant terms were not made in this implementation optimizations to reduce the space occupied by quasigroups.

6   Results

The results were obtained from the implementation of the modules for encryption and decryption on the platforms TelosB and MICAz, the parameters were measured execution time and memory used. We obtained 10 samples of encryption and decryption modules on platforms TelosB and MICAz.

The space occupied by the modules for encryption and decryption are shown in table 2.

According to Table 2 the space occupied by the RAM and ROM in the process of encrypting and decrypting in MICAz are larger than other values, for two reasons. The first is due to the fact that the amount of static arrays to be stored in the memory of the sensor in decryption to be greater and should be considered

Table 2. Space occupied by the modules to decrypt and decrypt

Platform	TelosB		MICAz	
Algorithm	Encryption	Decryption	Encryption	Decryption
RAM (bytes)	26	28	1658	31024
ROM (bytes)	3436	34582	2778	33748
RAM (KBytes)	0.025	0.027	1.61	30.29
ROM (KBytes)	3.35	33.77	2.71	32.95

that no optimization was done in order to reduce the size of quasigroups that have redundant terms. The second fact is due to implementation, because the declaration of the policy matrix was placed *const* and when this policy TelosB find it automatically allocates these structures in the ROM which does not occur with the MICAz.

In table 3 is the percentage occupied by MQQ on TelosB and MICAz where we can determine the percentage of occupancy in the memory (RAM and ROM) the procedures for encryption and decryption of MQQ platforms TelosB and MICAz.

Table 3. Percentage of memory space occupied by the procedure encrypt and decrypt

Platform	TelosB		MICAz	
MQQ	Encryption	Decryption	Encryption	Decryption
RAM %	0,25	0,27	40,25	757,25
ROM %	6,97	70,35	2,11	25,74

Figures 3 and figure 4 is the relationship between the space occupied in memory by MQQ and size of memory available on the platforms TelosB and MICAz.

Fig. 3. Memory Used by the MQQ on TelosB

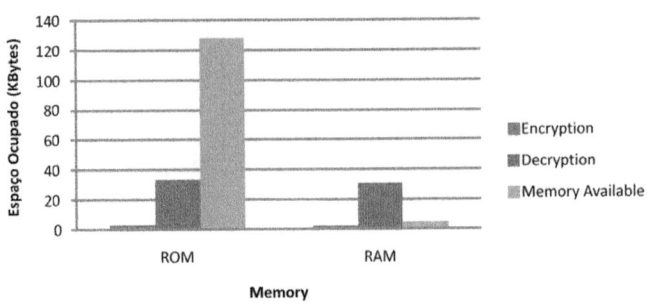

Fig. 4. Memory Used by the MQQ on MICAz

We can see that the platform TelosB consumption of RAM is minimal. The consumption of the ROM decryption TelosB is greatest in connection with encryption, where the largest concentration in the ROM is due to the use of policy *const*, thereby placing data in ROM. The decryption requires more memory space due to the need of data structures represented by the quasigroups, the inverse matrices T, S and reverse Dobbertin. Whereas the objective of this work is a fair comparison of MQQ platforms tested, so no specific optimization was performed for MICAz in order to put the static structures in the ROM. Importantly, this study does not address the optimization algorithm proposed by [9], but considering that the quasigroups may have repeated elements can reduce the space required to store these quasigroups.

On the platform TelosB MQQ is promising after the higher consumption of memory is in ROM, in addition there are prospects of reducing the size occupied by data structures used in the encryption and decryption MQQ. The platform also MICAz is promising, after putting the data structures in the ROM space is not critical in MICAz platform.

In the experiment with the decryption MICAz with the space occupied by decrypting extrapolated the available memory, so could not execute and thus obtain the execution time MQQ. But as has been reviewed with the problem MICAz can be circumvented by placing the data structures in ROM and improved by optimizing the space occupied by data structures MQQ.

It is important to emphasize that it is possible to reduce the size of the data structures needed by MQQ, because the polynomials usually have repeated terms. Soon there is room for further research in order to reduce the size of the data structures used by MQQ.

Reduce the space occupied by MQQ is essential for real applications requiring security in WSN, after all the little available memory on the sensor should coexist both the actual application of WSN as the cryptographic algorithms.

Samples with the execution times of the modules for encryption and decryption are shown in table 4.

Table 4. Runtime modules and decrypt decrypt

Platform	TelosB		MICAz
Algorithm	Encryption	Decryption	Encryption
sample 1 (milliseconds)	825	117	445
sample 2 (milliseconds)	822	116	445
sample 3 (milliseconds)	826	117	445
sample 4 (milliseconds)	826	116	445
sample 5 (milliseconds)	823	117	445
sample 6 (milliseconds)	827	117	445
sample 7 (milliseconds)	827	116	445
sample 8 (milliseconds)	825	117	445
sample 9 (milliseconds)	825	116	445
sample 10 (milliseconds)	825	117	445
Average (milliseconds)	825.1	116.6	445

As can be seen in table 5 decryption module not implemented in MICAz, missing modify the decryption to allocate the necessary static structures in the ROM. The structures were declared as *const static* and when TelosB find these policies allocates the data in the ROM the same is not true of MICAz.

Despite not having been registered in the decryption of the MICAz MQQ, since the size of the decryption MICAz exceeded the available space in RAM. In figure 5 can have a relationship of runtime on both platforms tested. As can be seen from time to encrypt TelosB is approximately seven times greater than the time to decrypt further finds that the time to encrypt the platform MICAz is almost double what the platform TelosB.

Considering the relationship of performance on TelosB, we can estimate the time to decrypt the MICAz would be approximately 63.00 milliseconds, it is important to note that in the decryption MICAz is only an estimate based on implementation of the TelosB MQQ.

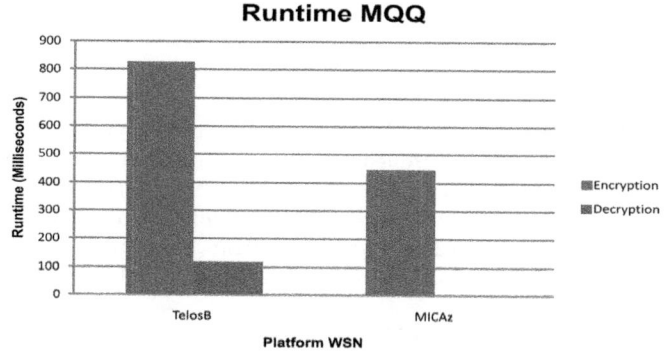

Fig. 5. Runtime MQQ

7   Conclusions

This paper proposes a new approach to solve the problem of providing PKC in WSN to enable the implementation of 160-bit MQQ platforms TelosB and MICAz. Is important to consider this innovative work in the proposal to bring this new PKC for WSN based multivariate quadratic quasigroups.

The main contribution of this paper is to propose a new approach to providing wireless sensor networks with public key cryptosystems, using the algorithm based on quasigroups multivariate quadratic. Also proposed is a way to accommodate the data structures necessary for the process of encryption MQQ in the original works to encrypt data structures require more space than the memory available on the platforms used TelosB and MICAz. In this approach it is considered that in terms of the polynomials involved MQQ may be repeated.

The platform TelosB had a larger footprint in ROM because every time a structure declared with *const* is found, the data are stored in ROM.

In any case it is necessary to perform optimizations to reduce the space occupied by the eight quasigroups in the decryption module and the public key. The reduction of the space occupied by these structures is possible because there is redundant elements in quasigroups and the public key.

The times obtained in the experiments with the sensors show that it is feasible to use MQQ in wireless sensor networks, but is important to report that in [17] the MQQ with up to 160 variables was broken. In [11] perceives that if a suitable replacement for the Dobbertin transformation is found, MQQ can possibly be made strong enough to resist pure Gröbner attack for correct choices of quasigroups size and number of variables. Therefore the Dobbertin transformation is weakness in the MQQ [11]. Although MQQ have been broken keys of 160 bits, the principle of cryptographic on MQQ was not broken. Despite MQQ be promising is essential to find a suitable replacement for the Dobbertin transformation [11].

The MQQ exhibited favorable performance in platforms TelosB and MICAz, but there are aspects of MQQ that still need to be analyzed in WSN, such as key generation algorithm of MQQ, energy consumption of MQQ algorithms, performance comparison between ECC and MQQ.

References

1. Ahlawat, R., Gupta, K., Pal, S.K.: From mq to mqq cryptography: Weaknesses new solutions. In: Western European Workshop on Research in Cryptology (2009)
2. Bogdanov, A., Eisenbarth, T., Rupp, A., Wolf, C.: Time-area optimized public-key engines: Mq-cryptosystems as replacement for elliptic curves? Cryptology ePrint Archive, Report 2008/349 (2008), http://eprint.iacr.org/
3. CERTICOM. Securing sensor networks - getting it right from the start, with public-key (2006),
   http://certicomcenterofexcellence.com/
   pdf/white_paper_sensor_networks_login.pdf

4. Gay, D., Sharp, C., Turon, M.: Timers (2007),
   http://www.tinyos.net/tinyos-2.x/doc/html/tep102.html.
5. Inc CrossBow Technology. Micaz wireless measurement system (2008),
   http://www.xbow.com/Products/Product_pdf_files/
   Wireless_pdf/MICAz_Datasheet.pdf
6. Inc CrossBow Technology. Telosb mote platform (2008),
   http://www.xbow.com/Products/Product_pdf_files/
   Wireless_pdf/TelosB_Datasheet.pdf
7. El-Hadedy, M., Gligoroski, D., Knapskog, S.J.: High performance implementation
   of a public key block cipher - mqq, for fpga platforms. In: RECONFIG 2008:
   Proceedings of the 2008 International Conference on Reconfigurable Computing
   and FPGAs, Washington, DC, USA, pp. 427–432. IEEE Computer Society, Los
   Alamitos (2008)
8. Gaubatz, G., Kaps, J.P., Öztürk, E., Sunar, B.: State of the art in ultra-low power
   public key cryptography for wireless sensor networks. In: 2nd IEEE International
   Workshop on Pervasive Computing and Communication Security (PerSec 2005),
   Kauai Island, pp. 146–150 (2005)
9. Gligoroski, D., Markovski, S., Knapskog, S.J.: Public key block cipher based on
   multivariate quadratic quasigroups. Cryptology ePrint Archive, Report 2008/320
   (2008), http://eprint.iacr.org/
10. Gligoroski, D., Markovski, S., Knapskog, S.J.: Multivariate quadratic trapdoor
    functions based on multivariate quadratic quasigroups. In: MATH 2008: Procee-
    dings of the American Conference on Applied Mathematics, Stevens Point, Wis-
    consin, USA, pp. 44–49. World Scientific and Engineering Academy and Society
    (WSEAS) (2008)
11. Perret, L., Gligoroski, D., Faugère, J.-C., Odegard, R.: Analysis of the mqq pu-
    blic key cryptosystem. In: The Ninth International Conference on Cryptology
    And Network Security (CANS 2010), Kuala Lumpur (Malaysia). LNCS, Springer,
    Heidelberg (2010)
12. Levis, P.: Tossim: Accurate and scalable simulation of entire tinyos applications.
    In: Proceedings of the First ACM Conference on Embedded Networked Sensor
    Systems (SenSys) (2003),
    http://webs.cs.berkeley.edu/papers/tossim-sensys03.pdf
13. Levis, P.: Tinyos programming. TinyOS Programming (2006),
    http://csl.stanford.edu/~pal/pubs/tinyos-programming.pdf
14. Levis, P., Lee, N.: Tossim: A simulator for tinyos networks. TOSSIM: A Simulator
    for TinyOS Networks (2003),
    http://www.eecs.berkeley.edu/~pal/pubs/nido.pdf
15. Liu, D., Ning, P.: Security for Wireless Sensor Networks. Springer, Heidelberg
    (2007)
16. Lopez, J., Roman, R., Alcaraz, C.: Analysis of security threats, requirements, tech-
    nologies and standards in wireless sensor networks, pp. 289–338 (2009)
17. Mohamed, M.S., Ding, J., Buchmann, J., Werner, F.: Algebraic attack on the mqq
    public key cryptosystem. In: Garay, J.A., Miyaji, A., Otsuka, A. (eds.) CANS 2009.
    LNCS, vol. 5888, pp. 392–401. Springer, Heidelberg (2009)
18. Oliveira, L.B., Scott, M., López, J., Dahab, R.: TinyPBC: Pairings for authentica-
    ted identity-based non-interactive key distribution in sensor networks. In: 5th Inter-
    national Conference on Networked Sensing Systems (INSS 2008), Kanazawa/Japan
    (2008) (to appear)

19. Pal, S.K., Sumitra.: Development of efficient algorithms for quasigroup generation encryption. In: Proc. IEEE International Advance Computing Conference IACC 2009, March 6–7, pp. 940–945 (2009)
20. Palafox, L.E., Garcia-Macias, J.A.: XXXIV - Security in Wireless Sensor Networks. In: Handbook of Research on Wireless Security (2008)
21. Szczechowiak, P., Oliveira, L.B., Scott, M., Collier, M., Dahab, R.: NanoECC: Testing the limits of elliptic curve cryptography in sensor networks. In: Verdone, R. (ed.) EWSN 2008. LNCS, vol. 4913, pp. 305–320. Springer, Heidelberg (2008)

Hardware Architectures for Elliptic Curve Cryptoprocessors Using Polynomial and Gaussian Normal Basis over GF(2^{233})

Vladimir Tujillo-Olaya and Jaime Velasco-Medina

Universidad del Valle, Bionanoelectronics Group
Cali, Colombia
{vlatruo,jvelasco}@univalle.edu.co

Abstract. This work presents efficient hardware architectures for elliptic curves cryptoprocessors using polynomial and gaussian normal basis. The scalar point multiplication is implemented using random curves over GF(2^{233}) and the Lopez-Dahab algorithm. In this case, the GF(2^m) multiplication is implemented in hardware using three algorithms for polynomial basis (PB) and three for gaussian normal basis (GNB). The cryptoprocessors based on PB with D=32 and GNB with D=30 use 76 μs and 60 μs for scalar multiplication and 26697 and 18567 ALUTs, respectively. The compilation and synthesis results show that the GNB cryptoprocessor presents a better performance than PB cryptoprocessor. However, the last one is less complex and more scalable from the design point of view.

Keywords: Polynomial basis, normal basis, elliptic curves.

1 Introduction

In order to protect or exchange confidential information, cryptography plays an important role in the security of the information. Therefore, it is necessary to implement efficient cryptosystems, which can support applications economically feasible. In this context, public key cryptography based on elliptic curves is widely used in applications like: private key exchange and digital signatures [1]. Additionally, the Elliptic Curve Cryptography (ECC) can be used in applications where the computational resources are limited such as smart cards, cellular telephones and wireless systems which are gradually replacing many traditional communication systems [2]. The ECC systems are included in the NIST and ANSI standards, and the principal advantage over other systems of public key like RSA is the size of the parameters, which are very small, however the ECC systems provide the same level of computational security.

New techniques like parallelized algorithms and efficient cryptographic algorithms are used for communications security. However, the performance of cryptographic methods is crucial for real world applications. Due to hardware technologies can present a physically communication security, they can be used for implementing cryptographic algorithms with high performance and low cost. The public-key cryptography

M.L. Gavrilova et al. (Eds.): Trans. on Comput. Sci. XI, LNCS 6480, pp. 79–103, 2010.

plays an important role in protecting the information. In this context, elliptic curve cryptography (ECC) is emerging as a very good alternative to RSA, the conventional public-key system used on the Internet today. The main advantage of ECC over other public-key systems is that ECC presents higher security per key bit compared with RSA. That means ECC offers equivalent security with smaller key sizes and uses fewer computational resources. It is generally accepted that a 233-bit ECC key provides comparable levels of computational security as a 2048-bit RSA key [3]. Due to its computational advantages, ECC is particularly well suited for mobile and wireless applications where the computational platforms are constrained in the amount of available computational resources and battery power.

The ECC is included in several international standards such as ANSI X9.62 [4], IEEE P1363 [5] and NIST [6]. Recently, the National Agency of Security (NSA) recommended the use of ECC to protect sensitive US Government information.

The hardware implementation of ECC can be divided into three levels: finite field arithmetic, elliptic group operation and scalar (or point) multiplication. Therefore, in order to achieve efficient hardware implementations it is important to reach the best algorithm optimization for each level. However, it is important to mention that the most expensive operation applied in ECC systems is the "scalar multiplication" of a large natural number with a point on an elliptic curve [7]. In this case, the performance of an elliptic curve cryptoprocessor depends on the multiplication over $GF(2^m)$. So, the finite field multiplier is the most important functional block of the cryptoprocessor.

2   Related Work

Several algorithms for finite field operations and point multiplication have been proposed and efficiently implemented in hardware [8-14].

B. Ansari and M. Anwar in [8] presented a high-performance architecture of elliptic curve scalar multiplication based on the Montgomery ladder method over finite field $GF(2^m)$, and using a pseudopipelined word-serial finite field multiplier that works in parallel with other finite field blocks.

C.H. Kim, S. Kwon and C.P. Hong in [9] proposed an architecture based on the López–Dahab elliptic curve point multiplication and used gaussian normal basis for $GF(2^{163})$. In that work, three 55-bit word level multipliers were employed to parallelize Lopez-Dahab algorithm.

K. Järvinen and J. Skyttä in [10] presented an implementation of point multiplication on Koblitz curves with parallel finite field multipliers. In that work, polynomial and gaussian normal basis are used to represent finite field elements.

W. Chelton and M. Benaissa in [11] proposed the design of a high-speed pipelined application-specific instruction set processor (ASIP) for ECC over $GF(2^{163})$. In that work, different levels of pipelining were applied to the data path to find an optimal pipeline depth and the Mastrovito bit parallel multiplier was chosen because it was suitable for pipelining.

M. Juliato, G. Araujo, J. López and R. Dahab in [12] implemented finite field operations on NIOS II processor. That work evaluated finite field operations using gaussian normal basis over $GF(2^{163})$.

S. Antao, R. Chaves, and L. Sousa in [13] presented a very compact and flexible processor. In that case, the processor supported ECC systems using polynomial basis representation over $GF(2^{163})$.

B. Muthukumar, Dr. S. Jeevanantharr in [14] presented an elliptic curve cryptography coprocessor. In that work, an FPGA-based modular multiplier architecture over $GF(2^{233})$ is proposed using both shifted canonical basis and type II optimal normal basis.

3 Mathematical Background

3.1 Elliptic Curves Arithmetic over GF(2^m)

The *non-supersingular* curves are usually chosen to elliptic curve cryptosystems, and an elliptic curve E over the binary field $GF(2^m)$, is defined by Equation (1),

$$y^2 + xy = x^3 + ax + b .$$

(1)

where a and $b \in GF(2^m)$, $b \neq 0$. It is well known that the set of points $P = (x, y)$, where $x, y \in GF(2^m)$, that satisfy the equation, together with the point ∞, called the *point at infinity*, form an additive abelian group $E_{a,b}$ with ∞ serving as its identity. Next, the group laws for *non-supersingular* curve are described:

1. *Identity.* $P + \infty = \infty + P$ for all $P \in E_{a,b}$
2. *Negative.* If $P = (x, y) \in E_{a,b}$ then $(x, y) + (x, x + y) = \infty$. The point $(x, x + y)$ is denoted by P and called the negative of P. Also, $\infty = \infty$.
3. *Point addition.* Let $P = (x_1, y_1) \in E_{a,b}$ and $Q = (x_2, y_2) \in E_{a,b}$, where $P \neq \pm Q$. Then the addition $P + Q = (x_3, y_3)$, where

$$x_3 = \lambda^2 + \lambda + x_1 + x_2 + a .$$

(2)

$$y_3 = \lambda(x_1 + x_3) + x_3 + y_1 .$$

(3)

with $\lambda = \dfrac{y_1 + y_2}{x_1 + x_2}$.

(4)

4. *Point doubling.* Let $P = (x_1, y_1) \in E_{a,b}$ where $P \neq -P$. Then the point doubling $2P = (x_3, y_3)$, where

$$x_3 = \lambda^2 + \lambda + a .$$

(5)

$$y_3 = x_1^2 + \lambda x_3 + x_3 .$$

(6)

with $\lambda = x_1 + \dfrac{y_1}{x_1}$.

(7)

3.2 Representation of Elements of Binary Fields

The binary field $GF(2^m)$ or *characteristic two finite field* contains 2^m elements and can be view as a vector space over GF(2) with dimension m. All field elements can be represented uniquely as binary vectors of dimension m. There is a variety of ways to represent elements in a binary finite field, depending on the choice of a basis for representation. Polynomial basis and normal basis are commonly used and supported by NIST and other standards.

Polynomial Basis. *Finite fields of order 2^m* are called binary fields. One way to construct $GF(2^m)$ is to use a *polynomial basis* representation: The elements of $GF(2^m)$ are the binary polynomials of degree at most $m - 1$:

$$GF(2^m) = \left\{ a_{m-1}x^{m-1} + \cdots + a_2 x^2 + a_1 x + 1 : a_i \in \{0,1\}, 0 \le i \le m-1 \right\}. \tag{8}$$

Let $P(x) = x^m + p_{m-1}x^{m-1} + \cdots + p_2 x^2 + p_1 x + 1$ (where $p_i \in GF(2)$) be an irreducible polynomial of degree m over $GF(2)$. Irreducibility of $p(x)$ means that $p(x)$ cannot be factored as a product of binary polynomials with degree less than m.

The field element is usually denoted by the bit string $(a_{m-1}a_{m-2}...a_1 a_0)$ of length m, thus the elements of $GF(2^m)$ can be represented by the set of all binary strings of length m. The multiplicative identity element '1' is represented by the bit string $(00...01)$ while the additive identity element is represented by the bit string of all 0's.

Field operations: the following arithmetic operations are defined on the elements of $GF(2^m)$ when using a polynomial basis representation with reduction polynomial $p(x)$:

- *Addition*: If we define the elements $a, b \in GF(2^m)$ to be the polynomials
  $A(x) = \sum_{i=0}^{m-1} a_i x^i$, $B(x) = \sum_{i=0}^{m-1} b_i x^i$ respectively, then their sum is written

$$C(x) = A(x) + B(x) = \sum_{i=0}^{m-1} (a_i + b_i)_i x^i. \tag{9}$$

  Where, $a=(a_{m-1}a_{m-2}...a_1 a_0)$ and $b=(b_{m-1}b_{m-2}...b_1 b_0)$ are elements of $GF(2^m)$, then $a + b = c = (c_{m-1}c_{m-2}...c_1 c_0)$ where the bit additions in Equation (9) $(a_i + b_i)$ are performed modulo 2.

- *Multiplication*: the finite field multiplication of two field elements, where
  $A(x) = \sum_{i=0}^{m-1} a_i x^i$, $B(x) = \sum_{i=0}^{m-1} b_i x^i$ and $C(x) = \sum_{i=0}^{m-1} c_i x^i$, can be carried out by

multiplying $A(x)$ and $B(x)$ and performing reduction modulo $p(x)$ or alternatively by interleaving multiplication and reduction, then the multiplication is shown as follows:

$$(b(x)a_{m-1}x^{m-1} + ... + b(x)a_2x^2 + b(x)a_1x + b(x)a_0)mod\ p(x)\ . \tag{10}$$

$$C(x)= \sum_{i=0}^{m-1} b(x)a_i x^i\ mod\ p(x)\ . \tag{11}$$

Inversion: if a is a non zero element in $GF(2^m)$, the multiplicative inverse $a^{-1}(x)$ of element $a(x)$ in the finite field $GF(2^m)$ is defined as the element that satisfies the multiplication $a(x).a^{-1}(x) = 1\ mod\ f(x)$. Where $f(x)$ is an irreducible polynomial. Commonly, methods for finite field inversion over $GF(2^m)$ are mainly based on Fermat's theorem and on Euclid's algorithm.

Normal Basis. ANSI X9.62 describes detailed specifications of ECC protocols and allows *Gaussian Normal Basis* be used to represent finite field elements [15]. An element in the $GF(2^m)$ has the computational advantage that squaring can be done very efficiently. However, multiplying distinct elements can be cumbersome. In this case, there are multiplication algorithms that make both simpler and more efficient this operation.

A normal basis for $GF(2^m)$ is as follows:

$$\{ \beta,\beta^2,\beta^{2^2},...,\beta^{2^{m-1}} \},\ \text{where } \beta \in GF(2^m)$$

Where, any element $\alpha \in GF(2^m)$ can be written as follows:

$$a = \sum_{i=0}^{m-1} a_i \beta^{2^i}\ ,\ \text{where } a_i \in \{0,1\}. \tag{12}$$

The *type T* of a GNB is a positive integer, and allows measuring the complexity of the multiplication operation with respect to that basis. Generally, the type T of smaller value allows for a more efficient multiplication. For a given m and T, the field $GF(2^m)$ can have at most one GNB of type T.

A GNB exists whenever m is not divisible by 8. Let m be a positive integer and let T be a positive integer. Then the type T of a GNB for $GF(2^m)$ exists if and only if $p = Tm+1$ is prime.

If $\{ \beta,\beta^2,\beta^{2^2},...,\beta^{2^{m-1}} \}$ is a GNB in $GF(2^m)$, then the element $a = \sum_{i=0}^{m-1} a_i \beta^{2^i}$ is represented by the binary string $(a_0 a_1 a_2 .... a_{m-1})$, where $a_i \in \{0,1\}$.

In this case, the multiplicative identity element is represented by the bit string of all 1s. The additive identity element is represented by the bit string of all 0s. An important result for the GNB arithmetic is the Fermat's Theorem. For all $\beta \in GF(2^m)$ so that:

$$\beta^{2^m} = \beta\ . \tag{13}$$

This theorem is important to carry out the squaring of an element in $GF(2^m)$.

Field operations: the following arithmetic operations are defined on the elements of $GF(2^m)$, when using a normal basis representation. The following arithmetic operations are defined on the elements of $GF(2^m)$, when using a GNB of type T:

- *Addition*: If $a = (a_0 a_1 a_2 ..... a_{m-1})$ and $b = (b_0 b_1 b_2 .. b_{m-1})$ are elements of $GF(2^m)$, then $a + b = c = (c_0 c_1 c_2 ... c_{m-1})$ where $c_i = (a_i + b_i) \bmod 2$.

- *Squaring*: Let $a = (a_0 a_1 a_2 .... a_{m-1}) \in GF(2^m)$, then $a^2 = \left(\sum_{i=0}^{m-1} a_i \beta^{2^i} \right)^2$

  $= \sum_{i=0}^{m-1} a_i \beta^{2^{i+1}} = \sum_{i=0}^{m-1} a_{i-1} \beta^{2^i}$ due to Fermat's Theorem; $\beta^{2^m} = \beta$, then

  $a^2 = (a_{m-1} a_0 a_1 a_2 ..... a_{m-2})$, in this case, squaring is a simple rotation of the vector representation.

- *Multiplication*: in order to perform multiplication, first, it is necessary to construct a function $F(U,V)$ on inputs $U = (\ U_0\ U_1\ ...\ U_{m-1}\ )$ and $V = (\ V_0\ V_1\ ...\ V_{m-1}\ )$ as follows:

$$F(U,V) = \sum_{k=0}^{p-2} U_{j(k+1)} V_{j(p-k)} \cdot \tag{14}$$

From Equation (14) the sub indexes $j(k+1)$ and $j(p-k)$ can be computed as shown in algorithm 1.

4 Hardware Architectures for Finite Field Arithmetic

Finite field arithmetic has increased the attention of the researchers due to cryptographic applications and error correction codes. It is well known that $GF(2^m)$ is easier to implement in hardware than other finite fields (GF(p)). This section presents the developed hardware architectures for finite field arithmetic in this work.

4.1 Algorithms for Polynomial Basis Multiplication over GF(2^m)

Finite field multiplier always plays an important role determining the performance of the cryptoprocessors due to the elliptic curve point multiplication involves intensive finite field multiplications.

Bit-serial Multiplication Algorithms. Two algorithms, right-to-left and left-to-right, are generally used to derive the least significant or most significant bit for bit-serial multipliers (LSB or MSB) [16-20].

LSB and MSB first multipliers are polynomial basis multipliers and compute the $GF(2^m)$ multiplication in m cycles. The product is obtained by the addition of partial-products, and the reduction is interleaved with the addition steps and performed by additions of the irreducible polynomial. Fig. 1 shows the MSB first polynomial basis multiplication algorithm [19].

Linear feedback shift registers are used to perform reductions $B=x.B(x)$ mod $p(x)$ or $C=Cx+a.B(x)$ mod $p(x)$. However, the register of the operand B can be saved in the MSB first algorithm compared with the LSB first in which both the contents of B and C need to be updated on each iteration.

Algorithm 1: MSB first polynomial basis multiplication algorithm

    Input: A, B $\in$ GF(2^m)   ***Output:*** C=AB mod $p(x)$

        0.   *C(x)=0*
        1.   ***For*** $k = m\text{-}1$ ***to*** 0 ***do***
        2. $C=C.x + A_{m\text{-}1}B$ *mod* $p(x)$
        3.   *A=A<<1*
        4.   ***End for***

Fig. 1. LSB first polynomial basis multiplication algorithm

The hardware implementation of the MSB first multiplication algorithm proposed in [19] is shown in Fig. 2, and uses m cells and calculates the multiplication in m cycles.

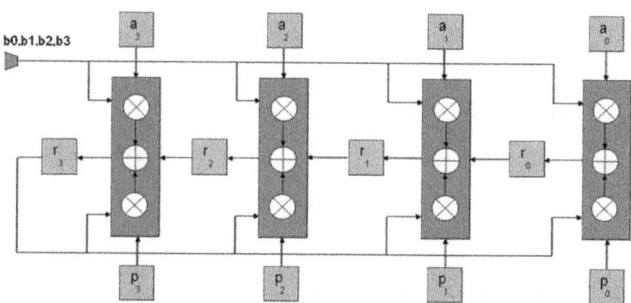

Fig. 2. Block diagram for MSB first based multiplier in GF(2^4)

Digit-Serial Multiplication Algorithm. The digit-serial multiplier is a parallel version of the bit-serial multiplier [18-20]. Digit-serial multiplier can compute several bits of the product in each clock cycle. Let D denote the digit size, then it takes m/D clock cycles to complete one multiplication in GF(2^m). The digital-serial left to right multiplication algorithm is shown in Fig. 3.

Algorithm 2: Digit-serial multiplication algorithm

Input: A,B $\in$ GF(2^m)   ***Output:*** C=AB mod $p(x)$

0. $C(x)=0, a' =A$
1.   ***For*** k = m/D-1 ***to*** 0 ***do***
2. $d_{D-1}=a'_{n-1}.x^{D-1} B \ mod \ p(x)$
3. $d_{D-2}=a'_{n-2}.x^{D-2} B \ mod \ p(x)$
4.   ... *(computations of dj can be done in parallel)*
5. $d_1=a'_{n-D+1}.xB \ mod \ p(x)$
6. $d_0=a'_{n-D}.xB \ mod \ p(x)$
7. $C=(x^{D}.C \ mod \ p(x))+ \sum^{D-1}d_j$
8. $d = a' <<D$
9.   ***End for***

Fig. 3. Digital serial multiplication algorithm

The advantage of the digit-serial multiplier over serial multipliers is that it can increase the speed of multiplication, but the digit-serial multiplier requires to use a reduction module. The hardware implementation of the digit-serial multiplication algorithm presented in [18] is similar to the MSB first multiplier and is shown in Fig. 4. It contains LFSRs for computing $c(x)=xDc(x)+s(x)$ and XOR-AND arrays for computing $s(x)$.

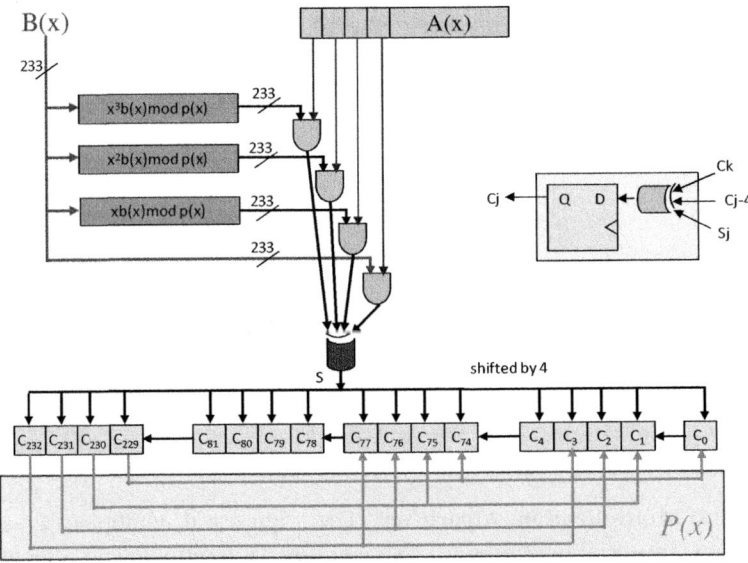

Fig. 4. Block diagram for Digit-serial multiplier in GF(2^{233})

Hardware Architecture for Serial Multiplier using PCA cell. In [16], H. Li and C. N Zhang presented a low complexity Programmable Cellular Automata (PCA) based versatile modular multiplier in GF(2^m) and this is shown in Fig. 5. In this case, the PCA rules are shown in Table 1. Where, Cm is configured as the coefficients of $B(x)$,

Cr is configured as the coefficients of *P(x)*, *Xs* is configured as coefficients of *A(x)*, and *Xl* and *Xr* are partial results of neighborhood PCA. The architecture of PCA cell is shown in Fig. 6.

Table 1. PCA rules

Cm	Cr	Xso
0	0	Xl
0	1	Xl+Xr
1	0	Xl+Xs
1	1	Xl+Xr+Xs

Algorithm 3: PCA based modular multiplication algorithm

Input: $A(x), B(x), p(x)$ **Output:** $C=AB \bmod p(x)$

1. Reset PCA
2. Configure coefficients of $B(x)$ as *Cm*, and coefficients of $P(x)$ as *Cr*
3. Run PCA *m* clock cycles

Fig. 5. Multiplication algorithm based on PCA

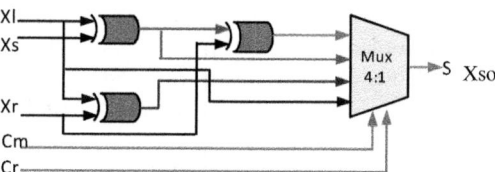

Fig. 6. PCA cell

An array of PCA cells is used to implement the finite field multiplier over $GF(2^4)$ which is shown in Fig. 7. The PCA array is suitable for both parallel and serial multiplier implementations.

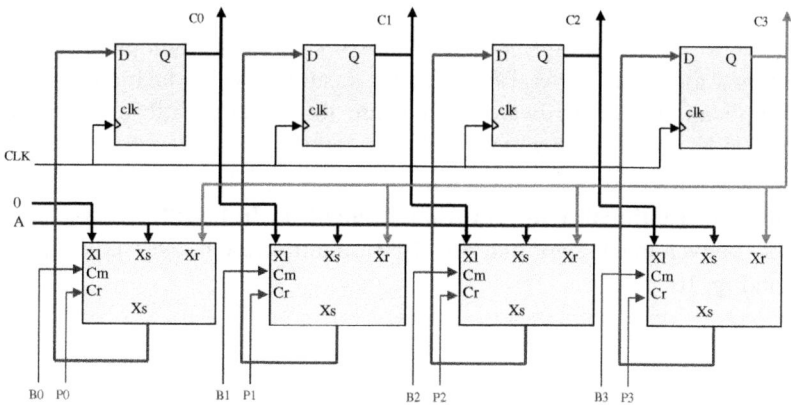

Fig. 7. Block diagram of serial multiplier using PCA cell over $GF(2^4)$

4.2 Algorithms for Gaussian Normal Basis Multiplication over GF(2^m)

Conventional Multiplication Algorithm. In [15], NIST presents a conventional algorithm for the GNB multiplication over GF(2^m), which is shown in Fig. 8.a. In this case, an algorithm is used to generate the $J(k)$ subindexes for the $F(U,V)$ array, which is shown in Fig. 8.b, and some other parameters must be taken into account, for example, the parameters recommended by NIST for GF(2^{233}) are:

$m = 233$,   number of bits
$T = 2$,   number recommended by NIST for GF(2^{233})
$p = 467$,   prime number $p = Tm + 1$
$U = 466$,   number that satisfies the relation U^2 mod $p = 1$

<table>
<tr><td>

Algorithm 4: conventional
algorithm
Input: a, b $\in$ GF(2^m)
Output: c = a.b $\in$ GF(2^m)

1.   U $\leftarrow$ a = (a_0, a_1, ...a_{m-1})
2.   V $\leftarrow$ b = (b_0, b_1, ...b_{m-1})
3.   ***For*** k = 0 ***to*** m-1 ***do***
4. $c_{(k)}$ = F(U,V)
5.   U = left rotation of U
6.   V = left rotation of V
7.   ***End***
8.   c $\leftarrow$ = (c_0, c_1,... c_{m-1}) = a.b
     where:

$$F(U,V) = \sum_{k=1}^{p-2} U_{J(k+1)} V_{J(k)}$$

</td><td>

Algorithm 5: j(k) generation for
f(u, v)
Input: m, T, U, p
Output: J(1), J(2), ..., .J(p-1)

1.   w $\leftarrow$ 1
2.   ***For*** j=0 ***to*** T-1 ***do***
3.   n $\leftarrow$ w
4.   ***For*** i=0 ***to*** m-1 ***do***
5.   j(n) $\leftarrow$ i
6.   n $\leftarrow$ 2n mod p
7.   ***End***
8.   w $\leftarrow$ UW mod p
9.   ***End***

</td></tr>
</table>

Fig. 8. a) Conventional algorithm for the GNB multiplication over GF(2^m) b) $J(k)$ generation for $F(U, V)$

In order to achieve a higher performance for hardware multipliers based on the conventional algorithm, several $F(U,V)$ arrays can be used, which allow to speed up the multiplication. The hardware architecture for the conventional GNB multiplier using 4 $F(U,V)$ arrays is shown in Fig. 9.

Modified Conventional Multiplication Algorithm. In [21][22] Lopez presented a modified conventional algorithm for GNB multiplication over GF(2^m), which is shown in Fig. 10.

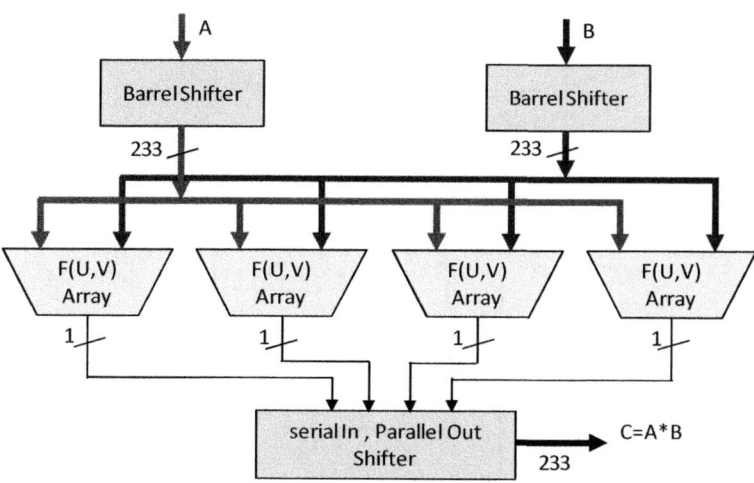

Fig. 9. Hardware architecture for GF(2^{233}) multiplier based on conventional algorithm using 4 *F(U, V)*

Algorithm 6: modified algorithm: GNB multiplication over GF(2^m)

Input: A, B in GF(2^{233})
Output: C = A.B

1. T = B^2 (Rot_right(B,1)) C = 0
2. *For* i = m-1 *to* 0 *do*
   2.1 C = C^2 (C = Rot_right(C,1))
   2.2 if a_i = 1 then C = C xor mbeta(T)
   2.3 T = T^2 (T = Rot_right (T,1))
   End
3. *Return* C = A. B

T(p1) = [0, $T_{p1}(1)$, $T_{p1}(2)$, $T_{p1}(3)$, ... ,$T_{p1}(162)$]
T(p2) = [0, $T_{p2}(1)$, $T_{p2}(2)$, $T_{p2}(3)$, ... ,$T_{p2}(162)$]
T(p3) = [T_1, $T_{p3}(1)$, $T_{p3}(2)$, $T_{p3}(3)$, ... ,$T_{p3}(162)$]
T(p4) = [0, $T_{p4}(1)$, $T_{p4}(2)$, $T_{p4}(3)$, ... ,$T_{p4}(162)$]
Then

Fig. 10. Modified conventional algorithm for the GNB multiplication over GF(2^m)

The hardware architecture for the GNB multiplier based on modified conventional algorithm is shown in Fig. 11.

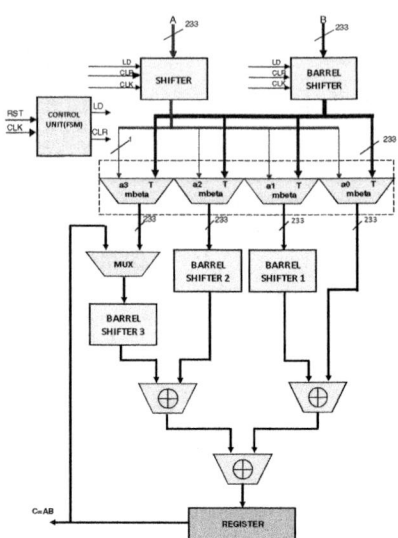

Fig. 11. Hardware architecture for $GF(2^{233})$ multiplier based on modified conventional algorithm

4.3 Squarer Using Polynomial Basis over $GF(2^m)$

Let $p(x)$ be the irreducible polynomial over $GF(2^m)$ [17]. Let:

$$A = \sum_{i=0}^{m-1} a_i x^i \ . \tag{15}$$

Be a polynomial representation of an arbitrary element of $GF(2^m)$. The squaring operation of A is represented by Equation (16)

$$C = \sum_{i=0}^{m-1} c_i x^i = A^2 \bmod p(x) = a'_0 + a'_1 x^2 + a'_1 x^4 + ... + a'_{\left\lceil \frac{m}{2} \right\rceil} x^{2\left\lceil \frac{m}{2} \right\rceil} + ... + a'_{m-1} x^{2m-2} \bmod p(x) \ . \tag{16}$$

In order to reduce the element A^2, the Equations (17-21) are presented. In this case, If $p(x) = x^m + x^k + 1$, with $1 < k < m/2$, and k is even and m is odd then:

$$c_i = \begin{cases} a'_i + a'_{2m-k+1}, & i = 0, 2, ..., k-2, & (17) \\ a'_{m+i}, & i = 1, 3, ..., k-1, & (18) \\ a'_i + a'_{2m-2k+i}, & i = k, k+2, ..., 2k-2, & (19) \\ a'_{m+i} + a'_{m-k+i}, & i = k+1, k+3, ..., m-2, & (20) \\ a'_i & i = 2k, 2k+2, ..., m-1 & (21) \end{cases}$$

The equations for the bit-parallel implementation of the squaring operation over $GF(2^7)$ are : $c_0 = a_0$; $c_1 = a_4$ xor a_6; $c_2 = a_1$; $c_3 = a_5$; $c_4 = a_2$ xor a_4 xor a_6; $c_5 = a_6$; $c_6 = a_3$ xor a_5.

When the irreducible polynomial is $p(x) = x^7 + x^3 + 1$.

Then, considering both m and k are odd, the squaring operation can be solved using the Equations (22 - 26).

$$c_i = \begin{cases} a'_i & i = 0,2,...,k-1, & (22) \\ a'_{m+i} + a'_{2m-k+i}, & i = 1,3,...,k-2, & (23) \\ a'_i + a'_{m-k+i} + a'_{2m-2k+i}, & i = k+1, k+3,...,2k-2, & (24) \\ a'_{m+i}, & i = k, k+2,...,m-2, & (25) \\ a'_i + a'_{m-k+i} & i = 2k, 2k+2,...,m-1 & (26) \end{cases}$$

4.4 Squarer Using Gaussian Normal Basis over GF(2^m)

Let $a = (a_0 a_1 a_2 .... a_{m-1}) \in$ GF(2^m), then $a^2 = \left(\sum_{i=0}^{m-1} a_i \beta^{2^i} \right)^2 = \sum_{i=0}^{m-1} a_i \beta^{2^{i+1}} = \sum_{i=0}^{m-1} a_{i-1} \beta^{2^i}$

taking into account the Fermat's Theorem; $\beta^{2^m} = \beta$, then $a^2 = (a_{m-1} a_0 a_1 a_2 ..... a_{m-2})$. In this case, the squaring operation requires a simple cyclic shift by one bit of the vector representation and is free in terms of both timing and area.

4.5 Polynomial Basis Inversion over GF(2^m)

In [22] is presented an alternative approach to exploit Fermat's little theorem by computing the multiplicative inverse using several multiplications. However, the Extended Euclidean Algorithm (EEA) is an alternative for computing inverses in polynomial representations. The EEA is shown in Fig. 12.

Algorithm 7: polynomial basis inversion algorithm using EEA

Input: a in GF(2^m)
Output: $a^{-1} \bmod P(x)$

1. $u=a$, $v= P(x)$, $g_1=1$, $g_2=0$
2. **While** $u \neq 1$ **do**
3. $j = deg(u)-deg(v)$
4.   **if** j < 0 **then**
5. $u \leftrightarrow v$, $g_1 \leftrightarrow g_2$, $j=-j$
6.   **end if**
7. $u = u + x^j v$, $g_1 = g_1 + x^j g_2$
8. **End While**
9. **Return**(g_1)

Fig. 12. Polynomial Basis Inversion Algorithm using EEA

The hardware implementation of polynomial inversion using EEA is shown in Fig. 13. In this case, A is the input element and $P(x)$ is the irreducible polynomial. If the MSB of the registers *Shifter-U* and *n-bits-shifter-V* are both '1', the value stored on register *Shifter-U* is replaced by *Shifter-U* ⊕ *n-bits-shifter-V* and the value stored on register *Shifter-G1* is replaced by *Shifter-G1* ⊕ *n-bits-shifter-G2*; otherwise the appropriate pair is shifted to the left until the MSB of the register *Shifter-U* or the *n-bits-shifter-V* is '1'. Also, *Encoder-blocks* are used to find the degree of *U* and *V*, and the *Subs-block* is used to calculate the *j* value.

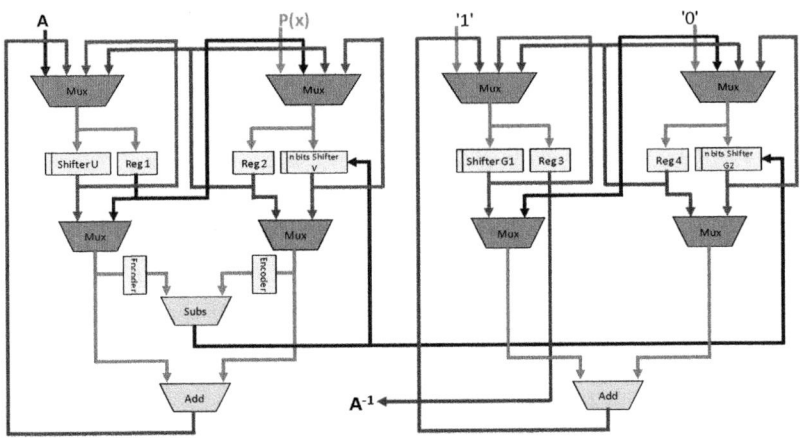

Fig. 13. Hardware implementation for Polynomial Basis Inversion over GF(2^m)

4.6 Normal Basis Inversion over GF(2^m)

In [23], Itoh and Tsujii proposed a method for computing inversion over GF(2^m), which minimizes the number of field multiplications. Fig. 14. shows the inversion algorithm proposed by Itoh and Tsujii. In this case, the recursive formula for a^{-1} is:

$$
a^{2^{m}-1} = \begin{cases} (a^{2^{\frac{m-1}{2}}-1})^{2^{\frac{m-1}{2}}} \cdot a^{2^{\frac{m-1}{2}}+1} & , m \ odd. \\ a \cdot \left(a^{2^{m-2}-1}\right)^{y} & , m \ even \end{cases} \quad . \tag{27}
$$

This algorithm can be applied to achieve the inversion operation in polynomial or normal basis representation. In this case, squaring operation can be easily computed.

Algorithm 8: Inversión Algorithm in $GF(2^m)$
                by *Itoh-Tsujii.*

Input: $\beta \neq 0$, $\beta \in GF(2^m)$.
Output: β^{-1}

1. let $m - 1 = b_r,..., b_1b_0 = 1b_{r-1},..., b_1b_0$
2. $n = \beta$, $k = 1$
3. *For* $i = r-1$ *downto* 0 *do*:
4. $\mu = n$
5.   *For* $j = 0$ *to* k *do*:
6. $\mu = \mu^2$
7. $n = \mu.n$, $k = 2.k$
8.   *if* $b_i = 1$ *then*
9. $n = n^2.\beta$, $k = k + 1$
10. *Output* n^2

Fig. 14. Inversion Algorithm proposed by Itoh-Tsujii

Table 2 shows the sequence of multiplications and squarings for inversion over $GF(2^{233})$, in this case, $m-1 = 232 = 11101000_2$ and $n = \beta$ is the input element.

Table 2. Sequence of multiplications and squarings for inversion over $GF(2^{233})$

i	b_i	operations
6	1	$n = n^2\beta$ $n = n^2\beta$
5	1	$n = n^{2^3}n$ $n = n^2\beta$
4	0	$n = n^{2^7}n$
3	1	$n = n^{2^{14}}n$ $n = n^2\beta$
2	0	$n = n^{2^{29}}n$
1	0	$n = n^{2^{58}}n$
0	0	$n = n^{2^{116}}n$ $n = n^2 = \beta^{-1}$

As it can be seen from Table 2 an inversion operation requires 10 $GF(2^m)$ multiplications and several squaring, independent of the finite field representation. Inversion based on multiplication operation does not increase significantly the complexity of the hardware design, but can severely impact the performance. Then, most hardware designers try to avoid the inversions during intermediate computations. Fig. 15 shows the hardware implementation for multiplication and inversion operations over $GF(2^m)$.

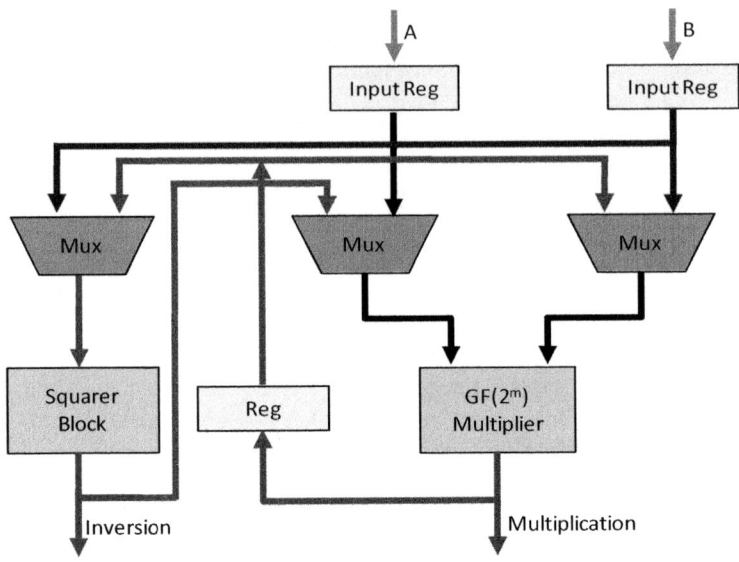

Fig. 15. Hardware implementation for GNB Multiplication-Inversion over GF(2^m)

5   Hardware Architectures for Elliptic Curve Cryptoprocessors

This work uses the López and Dahab point multiplication algorithm presented in [24], which does not has any extra storage requirements and the same operations (doubling and addition points) are performed in each iteration of the main loop, then it potentially increases the resistance to timing attacks. In terms of finite field multiplication, the approximate cost of computing kP using López and Dahab algorithm is $6m + 20$, which is an efficient implementation of Montgomery's ladder method for computing kP on non-supersingular elliptic curves over $GF(2^m)$.

5.1   Hardware Architectures for Elliptic Curves Cryptoprocessors Using Polynomial Basis

The cryptoprocessor architecture using polynomial basis has two register files, two FSMs, two digit-serial multipliers, two squaring blocks, two addition blocks and one inversion block, which allow for the calculation of addition, squaring, multiplication and inversion arithmetic over GF(2^{233}). In this case, the first FSM controls the I/O registers, generates the control sequences for the scalar multiplication, processes the key and initializes the cryptoprocessor. The second FSM carries out the point multiplication kP. The cryptoprocessor allows parallel processing by considering the duplication of functional blocks and its architecture is shown in Fig. 16.

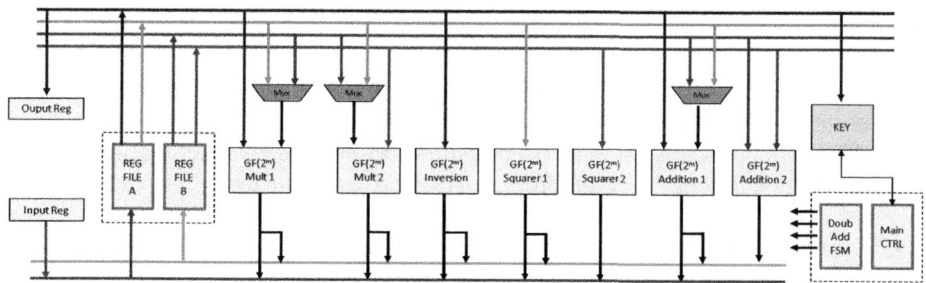

Fig. 16. Elliptic curve cryptoprocessor using polynomial basis

In this design a high flexibility is achieved due to the finite field arithmetic is implemented using "generic" parameters for the functional blocks design. That is, the blocks are parameterized using VHDL description, which allows achieving the modularity of the architecture and a very good trade-off between performance and area.

5.2 Hardware Architectures for the Elliptic Curves Cryptoprocessor Using Gaussian Normal Basis

The cryptoprocessor architecture using gaussian normal basis uses two register files, two FSMs, one GNB multiplier, one inversor-multiplier block, two addition blocks and several hardwired squaring blocks, which allow to calculate the addition, multiplication and inversion arithmetic over $GF(2^{233})$. In this case, the first FSM controls the I/O registers, generates the control sequences for the scalar multiplication, processes the key and initializes the cryptoprocessor. The second FSM carries out the point multiplication kP. The cryptoprocessor is shown in Fig. 17.

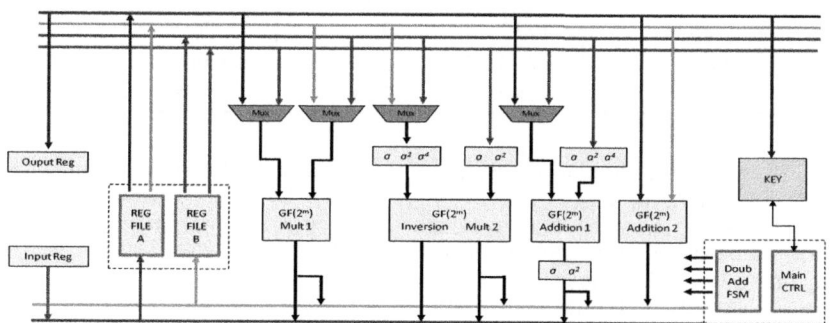

Fig. 17. Elliptic curve cryptoprocessor using Gaussian normal basis

As it can be observed from Fig. 16 and Fig. 17 there are hardware differences between the cryptoprocessor based on polynomial and gaussian normal basis. These differences are due to the basis representation in order to obtain the best performance for each processor. For example, two squaring blocks are implemented in PB cryptoprocessor while the squaring blocks from the GNB cryptoprocessor are hardwired.

6 Experimental Results

The cryptoprocessors were designed using structural VHDL description and synthesized by Quartus II v.9.0. on the FPGA EP3SE50F780C2. Some of the developed VHDL models are parameterized in order to synthesize different architectures and the functionality of the cryptoprocessors is verified using the NIST parameters.

6.1 Experimental Results for $GF(2^{233})$ Polynomial Basis Multipliers

Tables 3 and 4 show the synthesis results for serial and digit-serial multipliers, respectively. It can be observed from Table 3 and 4, the multipliers based on MSB first, PCA and digit-serial algorithms present a good performance using small area, which is very suitable for elliptic curve cryptoprocessor design. However, the serial multipliers need *233* clock cycles for the multiplication operation while digit-serial multipliers need 233/D clock cycles.

Table 3. Synthesis results for $GF(2^{233})$ serial multipliers

Serial algorithm	ALUTs	Total registers	F_{MAX}(MHz)	Tend (µs)
MSB	246	233	251.21	0.92
PCA	252	233	251.21	0.92

Table 4. Synthesis results for $GF(2^{233})$ digit-serial multipliers

Digit-serial algorithm	ALUTs	Total registers	F_{MAX}(MHz)	Tend (µs)
D=2	465	233	235.5	0.493
D=4	657	233	221.7	0.262
D=8	1735	233	206.3	0.140
D=16	3268	233	193.2	0.075
D=32	6789	233	180.8	0.041

Fig. 18. shows the simulation results for the MSB multiplier, considering a clock of 100MHz (10ns). In this case, $C(x)=A(x)B(x) \ mod \ (x^{233}+ x^{74} + 1)$, where:
A(x)= 1B34FD6E213A880DB4CCD1B83009BA66A1F25FE4AFA4C3B618BCE7BFD95
B(x)= 1DCB2A1003EF56D83CB9E0CA501F90E288927F964B43752A43FFF88F675
Then, after 233 clock cycles the multiplication result is:
C(x)= 149D2ABF4C87453BDBAA7067C8DC9A223DE6742F15AE77FA827DB472E3

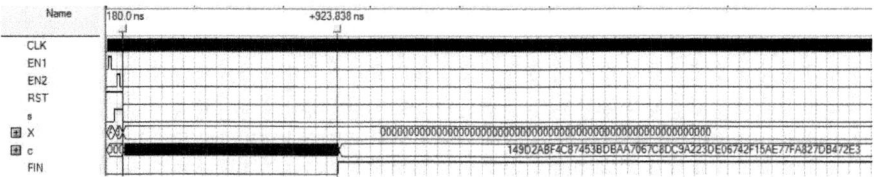

Fig. 18. Simulation results for the polynomial multiplication based MSB first multiplier

6.2   Experimental Results for GF(2^{233}) Gaussian Normal Basis Multipliers

Tables 5 and 6 show the synthesis results for hardware implementations for the conventional and modified algorithm, respectively.

Table 5. Synthesis results for the GF(2^{233}) multiplier based on conventional algorithm

F(U,V) m=233	ALUTs	Total registers	F_{MAX}(MHz)	Tend (μs)
2	595	708	202.2	0.576
4	1180	1295	181.8	0.323
8	2331	2587	166.6	0.179
16	4585	4054	142.8	0.106
32	8335	5689	133.3	0.058

Table 6. Synthesis results for the GF(2^{233}) multiplier based on modified algorithm

mbeta(T) m=233	ALUTs	Total registers	F_{MAX}(MHz)	Tend (μs)
3	713	707	223.6	0.523
5	1322	1260	190.4	0.309
9	2026	2467	172.9	0.171
30	5498	4056	149.7	0.057

From Tables 5 and 6, it is possible to observe that multipliers based on the modified conventional algorithm present a good performance using smaller area than the multipliers based on conventional algorithm. This is because of mapping the logic F(U,V) from conventional algorithm requires several levels of XOR gates while mapping the logic of mbeta(T) from modified algorithm uses only 1 level of XOR gate. Therefore, performance of multiplier based on modified algorithm is better than multiplier based on conventional algorithm.

Fig. 19 shows an example for the simulation results with the GNB multiplication by using the modified conventional algorithm using 5 mBeta blocks. In this case, C=AB, where:

A=18B863524B3CDFEFB94F2784E0B116FAAC54404BC9162A363BAB84A14C5
B=04925DF77BD8B8FF1A5FF519417822BFEDF2BBD752644292C98C7AF6E02
C=06E783B4C5CE979C6AB1709DB668BF3889E9A1C189787C2868D7321F516

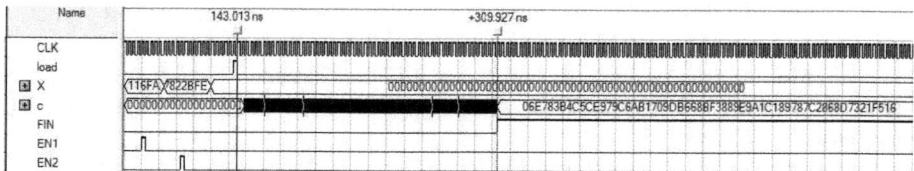

Fig. 19. Simulation results for the GNB multiplication based on modified conventional algorithm

6.3 Experimental Results for GF(2^{233}) Squaring

The synthesis results for GF(2^{233}) squaring using polynomial basis and GNB are shown in Tables 7.

Table 7. Synthesis results for GF(2^{233}) squaring

Basis	ALUTs	Tend (ns)
Polynomial	153	11.085
GNB	0	2.8

Polynomial basis uses more ALUTs and time than gaussian normal basis to perform the squaring operation. However, the polynomial basis squaring can be performed at the same time that multiplication operation.

6.4 Experimental Results for GF(2^{233}) Inversion

The synthesis results for the GF(2^{233}) inversion based on the Extended Euclidean Algorithm for polynomial basis are shown in Table 8. In this case, the hardware architecture uses 1435 ALUTs and 955 registers.

Table 8. Synthesis results for the GF(2^{233}) inversion based on polynomial basis

Inversion	ALUTs	Total registers	Fmax(MHz)	Tend (us)
EEA	1435	955	109.31	4.68

Fig. 20 shows the simulation results for the polynomial basis inversion and considering a clock of 100Mhz. In this case, the input data is:
A(x) = 0000000000000000018000000000000000000000000000000000000000003
A^{-1}(x)= 1FFFFFFFFFFFFFFFFFFFFFFFFFFFFFFFFFFFFFFFFFFFFFFFFFFFFFFFFFFFFF

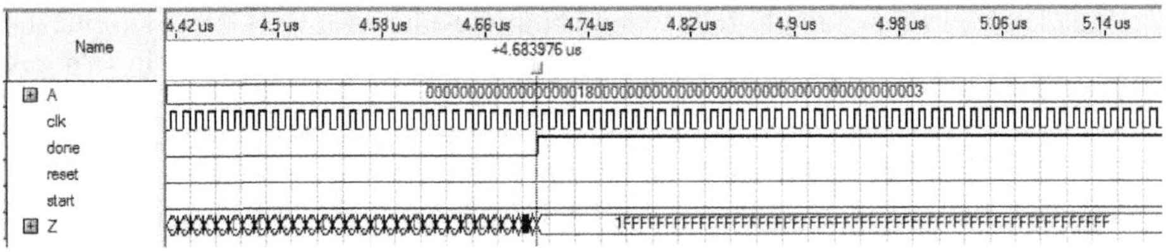

Fig. 20. Simulation results for the polynomial basis inversion based on EEA

From Fig. 20, it is possible to observe that if the Extended Euclidean algorithm is used, an inversion takes $2m$ clock cycles. In this case, the computation time for EEA is 4.68us.

The synthesis results for the GNB inversion based on Itoh-Tsujii algorithm are shown in Table 9. In this case, the hardware architecture was implemented by using the modified conventional multiplier considering different mbeta values.

Table 9. Synthesis results for the Gaussian normal basis inversion based on ITA

Inversion	ALUTs	Total registers	Fmax(MHz)	Tend (us)
mBeta=3	986	1225	218.7	6.25
mBeta=5	2150	2480	185.5	3.26
mBeta=9	3688	3564	168.5	1.95
mBeta=30	7005	5956	138.6	0.587

Fig. 21 shows the simulation results for the GNB inversion. In this case, the input data is:

A(x) = 924BBEEF7B171FE34BFEA3282F457FDBE577AEA4C885259318F5EDC04
A^{-1}(x)= 11A36620F8481692B2B5D66641A847D986968281183CD331BCF59743EF

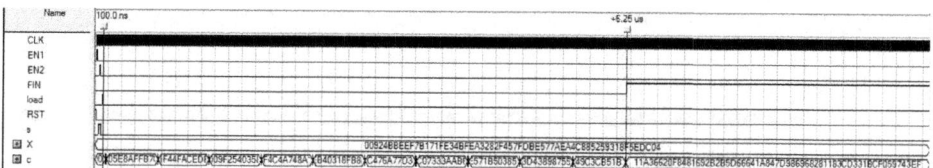

Fig. 21. Simulation results for the Gaussian normal basis inversion based on ITA

From Fig. 21, it can be observed that the processing time for inversion operation is almost 10 times the multiplication time. In this case, the computation time of the GNB inversion is 6.25us by using 3 mBeta modified conventional multiplier.

6.5 Experimental Results for Elliptic Curve Cryptoprocessors over GF(2^{233})

In order to perform the synthesis and simulation of the cryptoprocessors, the following parameters for a pseudo-random elliptic curve are used:

The form of the pseudo-random curve is: $E: y^2 + xy = x^3 + x^2 + b$,
The type and irreducible polynomial for GF(2^{233}) are: $T = 2$, $p(x) = x^{233} + x^{74} + 1$
The base point order given in decimal form is:
r=6901746346790563787434755862277025555839812737345013555379383634485463
The parameter b and the base point for polynomial basis are:
   b = 066647ede6c332c7f8c0923bb58213b333b20e9ce4281fe115f7d8f90ad
   G_x = 0fac9dfcbac8313bb2139f1bb755fef65bc391f8b36f8f8eb7371fd558b
   G_y = 1006a08a41903350678e58528bebf8a0beff867a7ca36716f7e01f81052
The parameter b and the base point for gaussian normal basis are:
   b = 1a003e0962d4f9a8e407c904a9538163adb825212600c7752ad52233279
   G_x = 18b863524b3cdfefb94f2784e0b116faac54404bc9162a363bab84a14c5
   G_y = 04925df77bd8b8ff1a5ff519417822bfedf2bbd752644292c98c7af6e02

The synthesis results for elliptic curve cryptoprocessor using polynomial basis are shown in Table 10.

Table 10. Synthesis results for GF(2^{233}) elliptic curve cryptoprocessor using polynomial basis

Multiplication Algorithm	ALUTs	Total registers	F_{MAX}(MHz)	Tend(ms)
MSB first	8486	5241	104.42	1.61
Digit serial-D=2	9634	5475	89.31	0.94
Digit serial-D=4	11934	5849	86.03	0.51
Digit serial-D=8	19869	6321	81.03	0.27
Digit serial-D=16	22459	6856	76.68	0.153
Digit serial-D=32	26697	7012	71.32	0.076

The simulation results for GF(2^{233}) cryptoprocessor using polynomial basis and the MSB first multiplier are shown in Fig. 22. In this case, the private key is:
k=18B863524B3CDFEFB94F2784E0B116FAAC54404BC9162A363BAB84A14C5

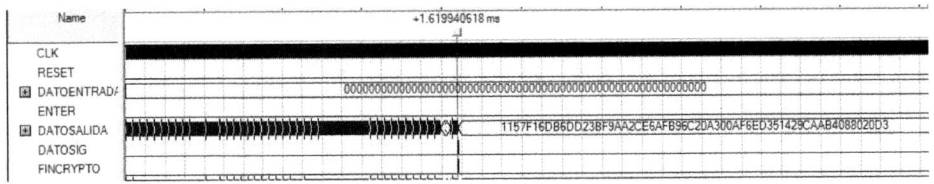

Fig. 22. Simulation results for GF(2^{233}) cryptoprocessor using polynomial basis

The synthesis results for the cryptoprocessor using GNB are shown in Table 11.

Table 11. Synthesis results for GF(2^{233}) cryptoprocessor using GNB

Multiplication Algorithm	ALUTs	Total registers	F_{MAX}(MHz)	Tend(ms)
3 mBeta	7835	4335	123.76	0.689
5 mBeta	8079	5426	118.57	0.496
9 mBeta	8587	6512	110.98	0.204
30 mBeta	18567	8469	97.51	0.060

Fig. 23 shows the simulation results for GF(2^{233}) cryptoprocessor using GNB and the modified conventional multiplier with 5 mBeta blocks. In this case, the private key is:
k=18B863524B3CDFEFB94F2784E0B116FAAC54404BC9162A363BAB84A14C5

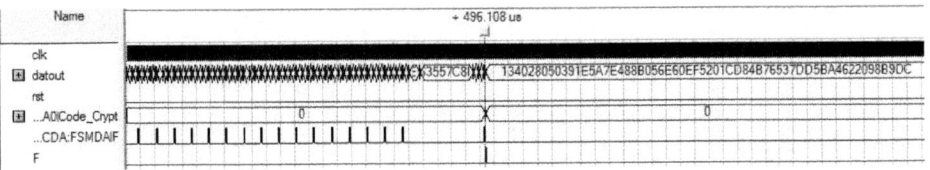

Fig. 23. Result for the elliptic curve point multiplication using GNB

6.6   Comparison Results

Although, it is not correct to compare hardware architectures implemented on different platforms and tools, the obtained results in this work are compared with recent works presented in Table 12.

Table 12. Performance comparison results

Design	m		D	Clk (MHz)	kP us	FPGA	Hardware Resource
[21]	163	PB	41	100	41	XC2V2000	LUT 6095, FF 2398
[22]	163	GNB	3*55	143	10	XC4VLX80	24,363 SLICES
[23]	k-233	PB	30	205.72	10.34	EP2S180F1020C3	ALUTS 31567, FF 11369
[24]	163	PB	163	153.9	19.55	XC4VLX200	16209 SLICES
[25]	163	GNB	1	120	1586	EP1S10F780C6ES	Les 3246
[26]	163	PB	-	150	1345	XC4VSX35	1095 SLICES
[27]	233	ONB	117	80	2280	XC3S1000	-
This work	233	PB	32	71.32	76	EP3SE50F780C2	ALUTS 26697, FF 7012
This work	233	GNB	30	97.51	60	EP3SE50F780C2	ALUTS 18567, FF 8469

Additionally, the comparison is very difficult due to there are other technical considerations such as different FPGA devices, point multiplication algorithms, finite field representations, size of the field, etc. Then, the performance of the designed cryptoprocessors in this work cannot be compared with the presented ones in the literature.

However, a brief comparison is described. In [21], only one 41-bit finite field multiplier with pipeline to reduce the critical path is used. A polynomial basis representation over $GF(2^{163})$ and a fixed irreducible polynomial are assumed. In turn, this work uses two polynomial basis multipliers over $GF(2^{233})$ for any irreducible polynomial. In [22], a parallel architecture using three 55-bit GNB multipliers is presented. In turn, this work uses two 30-bit GNB multipliers. In [23], an FPGA-based implementation of point multiplication on Koblitz curve (K-233) using four polynomial basis multipliers is presented. Also, that work implemented normal basis on k-163 and k-283. In turn, this work implements point scalar multiplication on generic curves over $GF(2^{233})$. In [24], a bit-parallel multiplier pipelined into 7 stages is implemented over $GF(2^{163})$. In [25], a hardware/software implementation of finite field arithmetic using GNB is presented. In that case, a 32-bit NIOS II is used. In [26] an elliptic curve cryptoprocessor is implemented using a Karatsuba-Offman finite field multiplier which uses few area. In [27], an elliptic curve cryptography coprocessor is presented. In that work a multiplication requires 117 clock cycles in $GF(2^{233})$. In turn, this work presents two cryptoprocessors using GNB and PB representations with higher performance than those ones presented in [25][26][27].

7 Conclusions

This work presents two elliptic curve cryptoprocessors suitable for the computation of point multiplication over $GF(2^m)$ using GNB and polynomial basis. In this case, efficient hardware architectures are designed for finite field multiplication, in order to select the best implementation for the cryptoprocessor design. These multiplier architectures incorporate bit-serial and digit-serial algorithms.

Also, some optimization design considerations were carried out. First, the algorithms were implemented using finite state machine instead of a stored-program machine due to the simplicity of the group operations when using Lopez-Dahab algorithm. Second, parallel processing was used by using two multipliers in the $GF(2^m)$ arithmetic unit. Finally, the digit size (D) for the $GF(2^m)$ multiplication based on GNB or polynomial basis determinates the performance of these.

Taking into account the experimental results, it is possible to conclude that serial polynomial basis $GF(2^m)$ multipliers present better performance than GNB, but the performance of the EAA algorithm for the polynomial basis $GF(2^m)$ inversion is not as good as the ITA for GNB inversion. Also, squaring operation based on GNB present better performance than squaring operation based on polynomial basis, due to squaring based on GNB is a simple rotation and it does not need hardware to be implemented.

Due to the cryptoprocessors were designed using the same tools, FPGA, finite field m size and hardware description language, the GNB cryptoprocessor presents a higher performance than the polynomial basis cryptoprocessor. However, the scalability is an advantage of polynomial basis.

Finally, the designed cryptoprocessors present a high performance, use small area and provide a good time-area trade-off. Therefore, these can be used for embedded applications such as smart cards, cellular telephones and IP cores for SoC.

References

1. Koblitz, N., Vastone, S., Menezes, A.: The State of Elliptic Curve Cryptography. Designs, Codes and Cryptography 19(2/3), 173–193 (2000)
2. Sklavos, N., Zhang, X.: Wireless Security & Cryptography: Specifications and Implementations. CRC-Press, A Taylor and Francis Group (2007) ISBN: 084938771X
3. Certicom research, The Elliptic Curve Cryptosystem, Certicom (April 1997)
4. ANSI X9.62-1999. The Elliptic Curve Digital Signature Algorithm. Technical report, ANSI (1999)
5. IEEE. P1363: Editorial Contribution to Standard for Public Key Cryptography, http://grouper.ieee.org/groups/1363/
6. FIPS 186-2, Digital Signature Standard (DSS), http://csrc.nist.gov/publications/ps/ps186-2/ps186-2-change1.pdf
7. Blake, I., Seroussi, G., Smart, N.: Elliptic Curves in Cryptography. Cambridge University Press, Cambridge (1999)
8. Ansari, B., Anwar, M.: High-Performance Architecture of Elliptic Curve Scalar Multiplication. IEEE Trans. on Computers 57(11), 1443–1452 (2008)

9. Kim, C.H., Kwon, S., Hong, C.P.: FPGA implementation of high performance elliptic curve cryptographic processor over $GF(2^{163})$. Journal of Systems Architecture 54(10), 893–900 (2008)
10. Järvinen, K., Skyttä, J.: Fast point multiplication on Koblitz curves: Parallelization method and implementations. Journal of Microprocessors and Microsys- tems (2009)
11. Chelton, W., Benaissa, M.: Fast elliptic curve cryptography on FPGA. IEEE Transactions on Very Large Scale Integration (VLSI) Systems 16(2) (February 2008)
12. Juliato, M., Araujo, G., López, J., Dahab, R.: A Custom Instruction Approach for Hardware and Software Implementations of Finite Field Arithmetic over F2163 using Gaussian Normal Bases. The Journal of VLSI Signal Processing Systems (2005)
13. Antao, S., Chaves, R., Sousa, L.: Compact and Flexible Microcoded Elliptic Curve Processor for Reconfigurable Devices. In: 17th IEEE Symposium on Field Programmable Custom Computing Machines (2009)
14. Muthukumar, B., Jeevanantharr, D.S.: Design of an Efficient Elliptic Curve Cryptography. In: First International Conference on Advanced Computing, pp. 34–37 (December 2009)
15. National Institute of Standards and Technology, Digital Signature Standard, FIPS Publication 186-2 (February 2000), http://csrc.nist.gov/fips
16. Li, H., Zhang, C.N.: Efficient cellular automata versatile multiplier for $GF(2^m)$. Journal of information science and engineering 18, 479–488,
    http://www.iis.sinica.edu.tw/JISE/2002/200207_01.pdf
17. Wu, H.: bit-parallel finite field multiplier and squarer using polynomial basis. IEEE transactions on computers 51(7),
    http://www.ieeexplore.ieee.org/iel5/12/21897/01017695.pdf?ar number=1017695
18. Hütter1, M., Großschädl2, J., Kamendje, G.-A.: A Versatile and Scalable Digit-Serial/Parallel Multiplier Architecture for Finite Fields $GF(2^m)$. In: Proceedings of the 4th International Conference on Information Technology: Coding and Computing (ITCC 2003), pp. 670–692 (2003)
19. Song, L., Parhi, K.K.: Efficient Finite Field Serial/Parallel Multiplication. In: Proceedings of International Conference on Application Specific Systems, Architectures and Processors - ASAP 1996, pp. 72–82 (1996)
20. Karatsuba, A., Ofman, Y.: Multiplication of multidigit numbers on automata. Sov. Phys.-Dokl (Engl. transl.) 7(7), 595–596 (1963)
21. Dahab, R., Hankerson, D., Long, M., Lopez, J., Menezes, A.: Software multiplication using Gaussian normal basis. IEEE Transactions on Computers Archive 55(8), 974–984 (2006)
22. Fong, K., Hankerson, D., Lopez, J., Menezes, A.: Field Inversion and Point Halving Revisited. IEEE Transactions on Computers 53, 1047–1059 (2004)
23. Itoh, T., Tsujii, S.: A fast algorithm for computing multiplicative inverses in $GF(2^m)$ using normal bases. Information and Computing 78(3), 171–177 (1988),
    http://www.eprint.iacr.org/2006/035.pdf
24. Lopez, J., Dahab, R.: Fast multiplication on elliptic curves over $GF(2^n)$ without precomputation. In: Koç, Ç.K., Paar, C. (eds.) CHES 1999. LNCS, vol. 1717, pp. 316–327. Springer, Heidelberg (1999)

GPU Accelerated Cryptography as an OS Service

Owen Harrison and John Waldron

Computer Architecture Group, Department of Computer Science, Trinity College
Dublin, Dublin 2, Ireland

Abstract. Graphics processing units (GPUs) have become popular devices for accelerating general purpose computing. In recent years there has been a surge in research involving the use of GPUs as cryptographic accelerators. Research has shown that contemporary GPU architectures can achieve higher throughput in the context of both symmetric and asymmetric key cryptography than a traditional CPU. Despite the existence of these new approaches, there remains no way for OS kernel services or userspace applications to make use of these implementations in a practical manner. To overcome this shortcoming, this paper investigates the integration of GPU accelerated cryptographic algorithms with an established service virtualisation layer within the Linux kernel, the OCF-Linux framework. This paper demonstrates that it is feasible to use a centralised kernel service to provide a standardised abstraction to GPU accelerated cryptographic functions for both kernelspace and userspace components.

1 Introduction

Symmetric-key algorithms such as AES, DES, ARIA; symmetric-key modes of operations; and asymmetric-key algorithms such as RSA, DSA and those based on ECC have recently been explored in the context of GPU acceleration [1–10]. It has been demonstrated that the GPU can act as an effective accelerator of symmetric-key algorithms using sufficiently large buffers and of asymmetric-key algorithms using a sufficient number of concurrent primitives. Despite the existence of these new approaches, there remains no way for OS kernel services or userspace applications to make use of these implementations in a practical manner. The use of these implementations require interaction with GPU specific interfaces such as the CUDA API, which is inconvenient for application developers and unavailable to kernel services. With the increasing number of GPU accelerated cryptographic algorithms, there is a need to provide an efficient and standardised operating system wide interface to these implementations. To overcome this shortcoming, this paper investigates the integration of GPU accelerated cryptographic algorithms with an established service virtualisation layer within the Linux kernel. The OpenBSD Cryptographic Framework (OCF) provides the basis for such a virtualisation layer.

M.L. Gavrilova et al. (Eds.): Trans. on Comput. Sci. XI, LNCS 6480, pp. 104–130, 2010.

The original OCF was developed for OpenBSD and has since been ported to FreeBSD [11], NetBSD and Linux [12]. It was created to provide uniform access to cryptographic accelerator functionality by hiding hardware specific details behind a standardised API. It provides access to this functionality for kernelspace services as well as normal userspace applications and APIs. For our investigation we use the Linux port of the OCF and the 2.6.26 Linux kernel. Although we do not directly use the native linux-crypto (Crypto API) project [13], which has in-built support for some crypto-cards, we note that the OCF acts as a wrapper for this library. We did not use linux-crypto for this work due to its current lack of support for asymmetric algorithms and the fledgling status of its userspace interface, however the main contributions in this paper are also relevant to this project.

The main contributions of this paper are: the effective integration of the GPU within the OCF model; the observation that the GPU interface is userspace only and the mechanisms introduced to allow it to be part of a kernel service; the introduction of a new memory management system within the OCF to allow efficient handling of memory transfers between multiple address spaces; and also an implementation of a general purpose multi-request batching scheme for asymmetric-key requests with regard to the GPU. Note that the source code of the core implementation functions presented within this paper are available online [14].

The motivation for this work is to provide a standard method of access to the latest GPU crypto acceleration work to all components within an operating system, with minimal loss of performance. This will allow application, kernel and driver developers to transparently include the GPU as part of their cryptographic solutions. We also observe that the GPU has a requirement of high work loads to achieve its peak performance. By using a centralised framework, which is used for all system-wide cryptographic needs, we increase the likelihood of high occupancy on the GPU and thus its potential to act as an effective crypto-accelerator.

2   Background and Related Work

2.1   OCF Background

Figure 1 shows a high level view of the OCF framework. The core component of the framework, the main "Crypto" layer, provides two APIs - the producer API for use by crypto-card device drivers and the consumer API for use by other kernel subsystems. An ioctl interface, which uses the /dev/crypto device file, provides a mechanism through which normal userspace applications can issue cryptographic requests. This interface is provided by the "Cryptodev" layer and uses the consumer API to pass on userspace requests to the Crypto layer. Device drivers can register their support for various cryptographic algorithms with the Crypto layer. Cryptographic requests received directly by the Crypto layer or sent via the Cryptodev layer are matched with capable devices and issued to the corresponding device driver. The device driver ID is recorded within the request, which is returned to the requesting application or kernel component along with the results of the processed request. Further requests can be issued to the same device within the OCF by maintaining the driver ID within the request or if left unset the OCF will again select a suitable device dynamically.

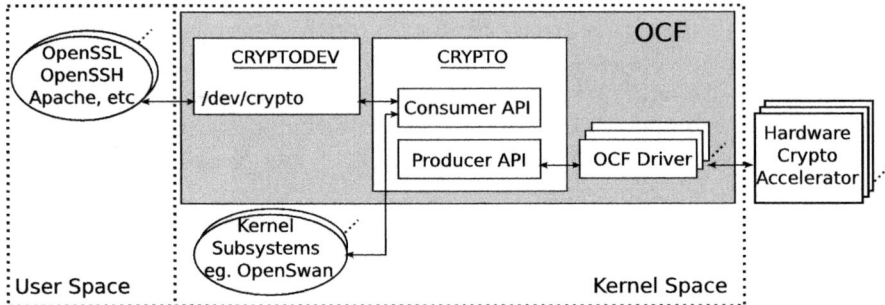

Fig. 1. Original OCF Architecture

2.2 GPU Background

The GPU used in our implementations is the Nvidia GeForce 8800GTX, which was the first DirectX 10 [15] compliant GPU released by Nvidia. It is Nvidia's first processor that supports the CUDA API [16]. The 8800GTX consists of 16 SIMD processors, called Streaming Multiprocessors (SM), each of which contain 8 ALUs. A single instruction is issued to an SM every 4 clock cycles, which is executed by all 8 ALUs. This creates an effective SIMD width of 32 operands for an SM. The code that runs on the GPU is referred to as a kernel. Via the CUDA API, the programmer can specify the number of threads that are required for execution on the GPU during a kernel call. These threads are grouped into programmer defined numbers of CUDA blocks, where each block of threads is guaranteed to run on a single SM. The number of threads per block is also programmer defined. Programmers should allocate threads in groups of 32, called a CUDA warp, to match the effective SIMD width mentioned above. A point of note relevant for this paper regards thread divergence. If any thread execution path diverges from the execution path of other threads within a CUDA warp, all the divergent code paths must be executed serially on the SM. An important note regarding GPU performance is its level of occupancy. This refers to the number of threads available for execution at any one time, and is important for hiding memory latency. It is desirable to have as high a level of occupancy as possible.

2.3 Related Work

The only previous attempt to provide a form of uniform access to GPU cryptographic acceleration involved AES via an OpenSSL engine by Rosenberg [17]. This implementation was applicable to userspace applications only and reported a 0 to 3% improvement over the CPU. There has been much recent research into using the GPU to accelerate various cryptographic algorithms as mentioned in Section 1. A sampling of the results achieved include 2,414 280-bit elliptic-curve point multiplications per second for a general 280-bit modulus on an Nvidia GeForce 8800GTS [1]; 1024-bit RSA decryption running on an Nvidia GeForce 8800GTX with a peak throughput of 0.18 ms/op [5]; and processing of AES in CTR mode on a GeForce 8800GTX at peak rates of 15,423 Mbps without including data transfer to the GPU and 6,914 Mbps when data transfers are included [4]. Harrison [18] presents a detailed account of related work in the field of GPU cryptographic acceleration for the interested reader.

3   Integration of GPU and OCF

3.1   Overview

The OCF provides a standardised method for the integration of any cryptographic accelerator device driver using its producer API. This API allows a device driver to register itself and its supported algorithms with the OCF, making it a target for processing cryptographic requests. The device driver is responsible for registering four callback functions with the OCF, which are used for the set up and tear down of symmetric algorithm sessions and also for the processing of symmetric and asymmetric requests. We have created a GPU cryptographic driver that fulfils the producer API requirements. The driver currently supports AES and modular exponentiation with CRT, suitable for RSA-1024. Supporting these two algorithms allows an analysis of the main issues arising from GPU integration with the OCF for both symmetric and asymmetric functions.

The algorithms supported by the GPU driver are implemented using Nvidia's CUDA API. The CUDA interface is provided via a userspace runtime library and as such requires its usage to be from userspace processes. Unfortunately Nvidia do not provide a driver that allows the direct control of their cards from within the kernel. This restriction forces all interactions with their cards to originate from userspace processes, making the provision of CUDA services from within the kernel a challenge. To overcome this restriction, we have split our GPU driver into two parts, a kernelspace driver and a userspace daemon. Regarding the possibility of Nvidia releasing direct access interface to their drivers, it is unlikely given that Nvidia have a history of only releasing binary versions of their drivers. This has continued unchanged even though it has hampered the GPGPU movement since its inception. Also, CUDA userspace libraries perform a number of runtime code translations for converting CUDA assembly into optimised GPU ready code, it is likely that this task is more suitable to staying within userspace.

Figure 2 shows a high level overview of the GPU driver integration into the OCF. It illustrates the separation of the GPU driver into the kernelspace part, *Gpucrypt*; and the userspace part, *Gpucryptd*. Gpucryptd follows the normal daemon convention, and as such runs as a high privilege background OS process. Gpucryptd is responsible for receiving cryptographic requests from Gpucrypt

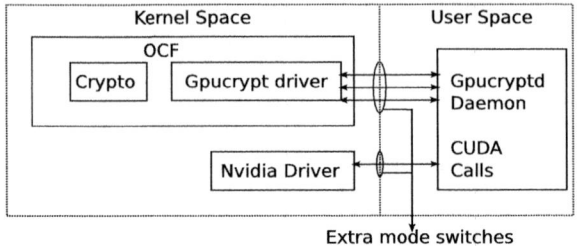

Fig. 2. OCF and GPU: High Level View - Different Address Space Problem

and processing them using Nvidia's userspace runtime API. A major disadvantage of this separation is the use of extra address spaces within the processing pipeline making data transfer more complex. Extra address spaces can result in a critical bottleneck in performance when processing requests unless memory is carefully managed. We explore this issue in full later. Another disadvantage of the driver separation is the introduction of two extra OS mode switch points within the processing pipeline. This becomes more of an overhead when the number of cryptographic requests increase, particularly for small request buffer sizes. Unfortunately there is no way to avoid these mode switches, however since the GPU only suits cryptographic acceleration with large workloads and high arithmetic intensity we will see that this overhead has limited effect.

3.2   Kernelspace Motivation

Considering the added complexing of separating the communication with the GPU between user and kernelspace, one could posit that it isn't worth the effort of creating a kernelspace GPU driver and that all communication with the GPU should be directly via the CUDA userspace library. Also, one could posit that a similar abstraction layer could be provided by including the CUDA related code within the standard cryptographic libraries, such as OpenSSL. However, there are a number of advantages to using a kernelspace GPU driver, these include:

– Kernelspace processes cannot use userspace libraries directly. Thus if we wish that kernelspace processes, which currently can use cryptographic hardware via the OCF, be able to use the GPU for cryptographic acceleration, then we must provide a bridging mechanism from kernelspace to userspace.
– As will be shown, concurrency is important when achieving effective throughput rates using a GPU for cryptographic performance. Thus, it is important to ensure that cryptographic requests generated by disparate applications can be pooled to increase the occupancy on the GPU. If applications process their cryptographic requests via a userspace cryptographic library that in turn uses the userspace CUDA library, there is no potential for pooling the requests. The requests will exist within separate address spaces. When a kernelspace GPU driver is used, the driver can pool the requests from multiple processes, thus increasing GPU occupancy and the overall system

cryptographic throughput. This is particularly pertinent in the context of asymmetric-key requests.

- The startup costs of requesting work to be done on the GPU is high. The CUDA libraries must create a context and perform code translation and optimisation. This cost can be amortised if a single process sends a lot of requests to the GPU, reusing the context and avoiding repeated code translation overheads. However, the startup cost can make up a significant proportion of the total GPU process time, if the requesting process is short lived, or doesn't require repeated processing of cryptographic requests. Using a central GPU driver, this problem can be eliminated as the startup costs can be incurred a single time at OCF startup. The userspace component of the GPU driver can create a CUDA context and cryptographic requests to have the CUDA library perform the necessary code translations. The userspace component would be expected to have a lifetime similar to the OS lifetime.

- The provision of a kernelspace driver allows the integration of GPU cryptographic processing within frameworks such as the OCF and also the native linux-crypto (Crypto API) layer mentioned previously. Userspace cryptographic libraries use these OS provided layers to avoid dealing directly with the complexities of the underlying hardware. Thus, including the GPU driver at the OS layer, allows cryptographic libraries to include GPU support without change if they already support such OS frameworks, as OpenSSL currently does.

3.3   Memory Management

When using devices that handle high volumes of data transfer it is common practice to ask the driver to allocate the memory used in these transfers. This has the advantage that the driver knows what type of memory (contiguous/non-contiguous, zone location) suits the corresponding device for DMA transfers. It is also common practice that allocated memory is shared between the driver and the calling process, either by driver allocation (`mmap()` kernel function) or by mapping userspace pages (`get_user_pages()` kernel function). If memory is not shared then userspace processes must undergo a copy of memory between user address space and kernel address space using the `copy_from_user()` and `copy_to_user()` kernel functions. Using an abstraction framework like the OCF, or the linux-crypto project, removes the direct line of communication required for standard requests for driver memory allocation, either by userspace processes or kernelspace subsystems.

Integration of the GPU with such a framework emphasises this deficiency due to two factors. First, the GPU requires large volumes of data for symmetric algorithms to reach its performance potential [4]. The larger the volumes of data, the worse the memory copy overhead. The OCF Cryptodev layer implements a copy from and to userspace policy for data transfer. Figure 3 shows the Cryptodev layer's performance with and without these copies as the buffer sizes increase. To explain the drop in performance of the Cryptodev layer, we note that it uses the `copy_from_user()` and `copy_to_user()` kernel functions to

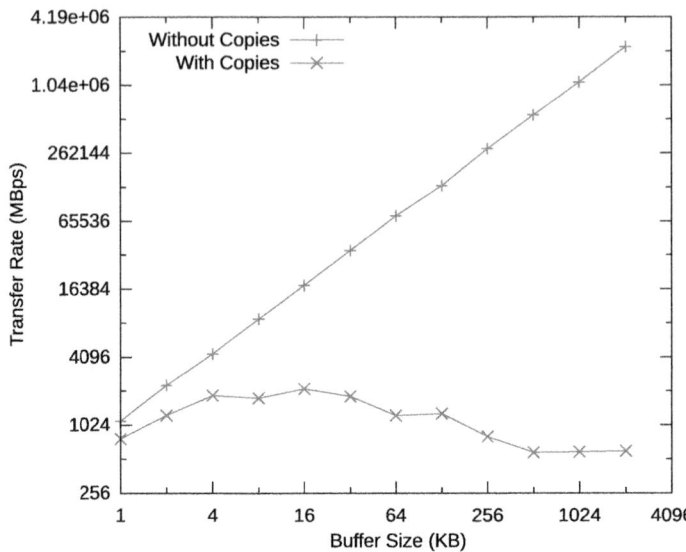

Fig. 3. Illustration of the Cryptodev Layer Memory Management Overhead

transfer memory between userspace and kernelspace. We test the performance of copy_from_user(), shown in Figure 4, and can see that the memory copy loses efficiency as the buffer sizes increase. Second, one cannot give memory to the Nvidia driver and request it to be used for DMA acceleration, the memory must be requested from the driver. Thus, even using a direct I/O approach (as in the new linux-crypto userspace API), where userspace pages are mapped in by the kernel on request, we must still perform a memory copy into and out of GPU DMA memory. Thus for any device that has DMA memory restrictions, it can be beneficial to have a mechanism for allowing the framework's drivers to manage their own memory.

We have added a new memory management system to the OCF that allows a consumer component (userspace application or kernelspace component) to directly use memory that is managed by the OCF drivers. This system allows the GPU driver to reduce the number of memory copies required during request processing to zero. Each memory allocation is recorded centrally by the OCF as a new memory mapping, which stores the driver's address (map_ptr) and the consumer component's address (app_ptr) of the allocated memory. map_ptr is the address the driver uses to refer to the allocated memory, which would normally be a kernelspace address, however in the case of the GPU it will belong to the Gpucryptd daemon address space. The app_ptr is the address the consumer component uses to refer to the memory and will belong to a userspace processes' address space if the OCF was called via the Cryptodev interface, or otherwise belong to the kernel address space.

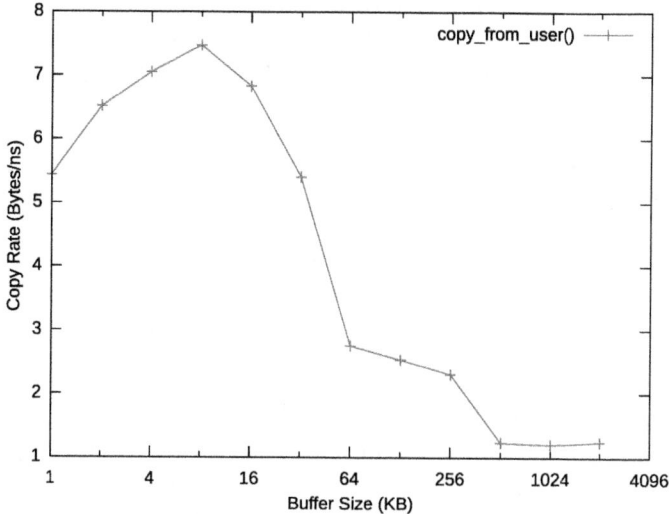

Fig. 4. Performance of the `copy_from_user()` Function

Figure 5 illustrates our implementation of the memory mappings for all consumer components. We create a separate mapping space, indexed by the `current` thread's thread group ID, to store all mappings for each consumer component (the ID is zero in the case of kernel consumer components). This ensures allocated memory can only be accessed by the process that requested the allocation. Within each space the mappings are grouped according to the device that allocated the memory. If memory allocation fails due lack of device support for the new memory management system, where possible we still wish to avoid the memory copies performed by the Cryptodev layer. An allocation request can be tied specifically by the consumer to a particular underlying device. In this case, if the device fails to allocate memory, then this failure is reported to the consumer. Otherwise, when a memory allocation fails, the Cryptodev layer allocates its own memory for sharing with the userspace consumer.

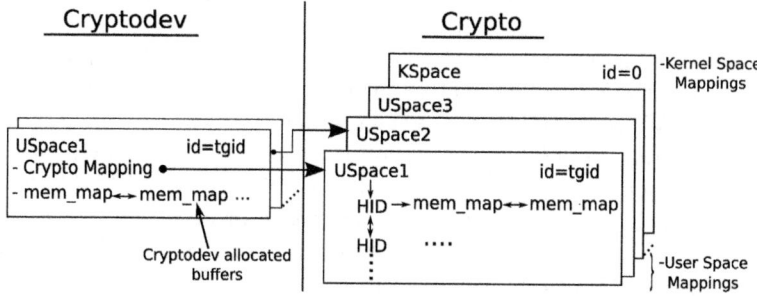

Fig. 5. Crypto and Cryptodev Layers:Memory Mappings Internal Structure

The Cryptodev layer maintains its own memory mappings for such allocations. This can be seen on the left hand side of Figure 5. Each userspace client process is represented by its own mapping space in the Cryptodev component. Each of these mapping spaces have a list of mappings for memory that is allocated by the Cryptodev layer directly. Also, each of these mapping spaces has a reference to a mapping space within the Crypto layer, representing all the mappings of memory allocations performed by the OCF device drivers on behalf of userspace client processes. This is shown on the right hand side of the figure. The "KSpace" area on the right hand side of the figure represents the mappings of the memory allocations performed by the OCF device drivers on behalf of kernelspace client processes.

The memory consumption overhead of this memory management approach is small and scales well. Compared to normal use of the OCF, the new memory management system allows the reduction in the total amount of system memory in use for a given request by sharing memory between kernel and userspace. Compared to a native approach, bypassing the OCF, there is the added memory consumption of the OCF and the extra pointers required to keep track of the memory translations for each request buffer associated with the new memory management system. These overheads are relatively low assuming the request buffer is of a reasonable size (e.g. $\geq 1\,$KB). If the buffers are small enough that memory overheads become a substantial overhead, the performance overhead of redirecting such small requests to hardware accelerators would become the primary system bottleneck. The following is a detailed account of how mappings are created, removed and used within the new memory management system. The new memory management API is listed in A.1.

3.3.1   Memory Map Creation

Userspace, allocation request: A userspace process makes a request for memory via a new ioctl added to the Cryptodev layer. This allocation request can suggest a specific device to carry out the allocation or allow the OCF to choose a device and report the used device back to the consumer. The standard mmap() system call is not used as it cannot support device specification in this way. The OCF relays the allocation request to the device driver, which performs the allocation and returns the underlying memory pages and map_ptr to the OCF. The OCF takes these pages and manually calls the internal kernel version of mmap(), which generates a new virtual memory area (VMA) for the userspace process. We use the returned pages from the device driver to map to this VMA, and return the VMA start address to the userspace process. In the case where no device can be found to fulfil the allocation request, the Cryptodev layer allocates its own kernel memory, and maps this to the calling process' VMA. The VMA start address is the pointer used by the userspace process and is the aforementioned app_ptr. The OCF uses the app_ptr and map_ptr to add a memory map to the appropriate "USpace" as shown in Figure 5.

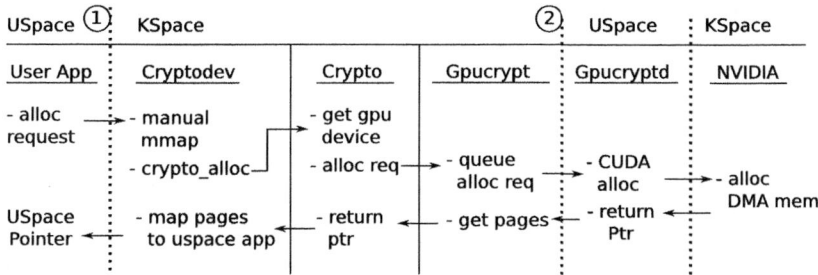

Fig. 6. New OCF Memory Allocation: Cryptodev to GPU

We illustrate an example of a userspace process allocation request to the GPU in Figure 6. We can see six cases of mode switch between user and kernel mode; four of them involve the added trip out of and back into the kernel due to userspace CUDA calls. Here we have highlighted mode switch 1 and 2. The above explains the mapping of kernel memory to userspace memory to eliminate a potential memory copy at mode switch 1. We discuss the GPU driver part of the new memory management system, which eliminates the copy at mode switch 2 in Section 3.4. The remaining mode switch is handled internally by the CUDA runtime.

Userspace, fork: On fork, a child process will have shared access to OCF allocated memory. Allocated memory returned by the OCF is set as always shared. Supporting private memory by implementing a copy-on-write procedure or executing a new allocation for each fork were deemed unnecessary considering that a child process can allocate its own memory. We share the memory between parent and child by overriding the VMA's vm_open() kernel function, which is called by the kernel when a new memory reference is created, such as on fork(). When this function is called, we create a new Cryptodev or Crypto memory mapping, within a new mapping space representing the process' new address space. Note that light-weight processes automatically share allocated memory as the VMAs of the processes are shared.

Kernelspace, allocation request: A process in kernel mode, or a kernel component, requests memory directly from the Crypto layer using the new crypto_alloc() function. This processes the request as above, selecting an appropriate device. However, instead of dealing with a userspace process' VMA, the OCF returns a kernelspace pointer, which references the new memory. The memory returned is not necessarily within the kernel address space, in the case of the GPU it belongs to the Gpucryptd address space. Thus, the OCF selectively performs a vmap() to map the memory into the kernel virtual address space if necessary. The kernelspace pointer returned is the app_ptr in this context, and is used to create a new memory map in the Crypto layer within the kernelspace mappings, see "KSpace" in Figure 5.

3.3.2   Memory Map Removal

Userspace: A userspace process can free OCF allocated memory by executing the munmap() Unix command or by terminating. On such an event the kernel calls the vm_close() kernel function for the allocated memory's VMA, which we have overwritten. If this process is the last to hold an open reference to the shared memory we issue a free command to the Crypto layer for device allocated memory or to the Cryptodev layer for Cryptodev allocated memory. After the memory is freed, the memory map is removed from the appropriate Crypto or Cryptodev memory map USpace.

Kernelspace: Kernelspace components issue a free directly to the Crypto layer via the new crypto_free() function. Unlike the Cryptodev free process above, the OCF must ensure that there exists a valid mapping for the kernelspace pointer for the specified device. If found, a free command is issued to the device driver and on return the memory mapping is removed from the kernel mapping space, KSpace.

3.3.3   Memory Map Translation

Userspace: All buffer pointers within cryptographic requests received via the Cryptodev layer interface are processed for potential address translation. First, the Crytpo layer is called to find a mapping that matches the userspace pointer (app_ptr) and the device specified within the request. This is done by finding the mapping space corresponding to the calling process using its thread group ID. Once the space is found we scan for a matching map within the corresponding device list of memory mappings. If a match is found, the device address (map_ptr) recovered replaces the userspace pointer within the cryptographic request and can be used directly by the device without any copies taking place. If no match is found, we repeat a similar process for any Cryptodev allocated memory. If still no match is found we default to the original OCF behaviour of using kmalloc() and the copy_from/to_user() kernel functions to copy the userspace buffers into the new kernelspace buffers.

Figure 7 gives a brief illustration of the interactions between the Crypto and Cryptodev layer during this translation. The left hand side of the figure shows the memory map process followed when a userspace process makes a cryptographic request. The Cryptodev component first searches the Crypto user spaces for a memory map that represents device allocated memory. Failing to find a map, the component then searches for mappings that represent memory allocated by the Cryptodev component. If a mapping is still not found, then the OCF defaults to its original functionality and allocates kernel memory to hold the contents of the request buffer. The original OCF behaviour should possibly be upgraded to use the get_user_pages() call, as in the linux-crypto project, to default to using direct I/O when no mapping is found thereby eliminating the use of the expensive copy_from/to_user() kernel functions. The right hand side of Figure 7 shows that requests made by kernel processes are handled by the Crypto layer, and

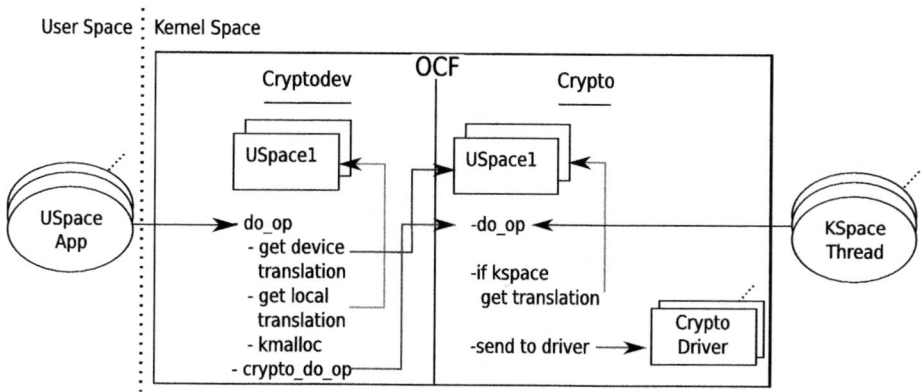

Fig. 7. Crypto and Cryptodev Layers: Memory Mapping Translation Process

only the "KSpace" mappings are searched, representing memory allocated by device drivers. All cryptographic requests are tagged internally to ensure that device drivers can detect if a request pointer is a native device address or a normal kernelspace address.

Kernelspace: Cryptographic requests received via the Crypto kernel interface undergo a similar procedure as above, except only the kernel mapping space is searched and only the Crypto memory mappings are searched. Considering that, as kernel mode processes are trusted, we provide the ability for these processes to translate the allocated memory before the request is sent to the Crypto layer. This allows the kernel processes to use native device driver pointers in their requests with tagging thus avoiding translation overhead.

Existing consumers: Care has been taken to ensure legacy consumers can continue to use the OCF with minimal impact to performance. Regarding cryptographic requests made via the Cryptodev interface, the new memory management system will only impose a small translation overhead for userspace applications not using OCF allocated memory. This overhead consists of the failure to retrieve a USpace for requests, which is very fast. For processes in kernel mode using the Crypto layer directly, the overhead depends on how many mappings are being used by other kernel components, as this determines the size of the translation search space.

3.4   GPU Driver and Daemon

Here we discuss in detail the GPU driver component within the OCF and in particular its separation into a kernel driver and a userspace daemon. As previously mentioned this separation is necessary due to the requirement of using a userspace API to communicate with Nvidia's GPUs. If Nvidia provided kernel

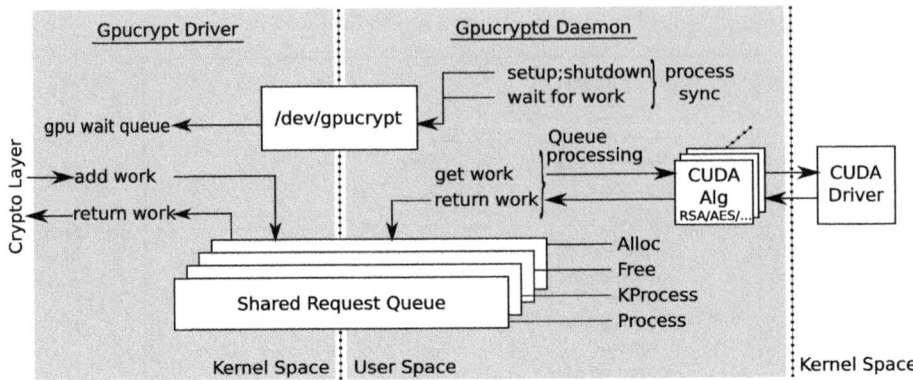

Fig. 8. GPU Driver Gpucrypt and GPU Daemon Gpucryptd

level access to its device drivers this separation could be avoided, as would the extra kernel to user mode switches. Figure 8 illustrates both the Gpucrypt driver and Gpucryptd daemon components and an overview of how they cooperate to fulfil the requests delivered by the Crypto layer. This is further discussed below.

/dev/gpucrypt: for the purpose of providing a communications channel between the two components we have created a new OS character device file called /dev/gpucrypt. On OCF startup, the Gpucrypt driver module is initialised and connects itself with the /dev/gpucrypt device file. The Gpucryptd component can subsequently open this device file and communicate with Gpucrypt via ioctls. These ioctls are used for initial handshake of Gpucryptd with Gpucrypt when the daemon sets up shared buffers for use in request processing. It also uses the interface to send a "ready for work" and "shutdown" signals. When the Gpucrypt driver receives these signals it correspondingly registers and unregisters with the OCF Crypto layer. The /dev/gpucrypt device is most intensively used to co-ordinate the processing of cryptographic and memory requests. When no work is available on the request queues, the Gpucryptd daemon calls the driver to passively wait for more work by putting itself to sleep. Thus, whenever work is received from the Crypto layer the driver calls wake on the daemon process' wait queue. Whenever work is finished and requires returning to the driver, the daemon uses an ioctl to signal that the work is finished, to remove the work from the queue and to call the Crypto layer for request return. The ioctls are listed and detailed in A.2.

Processing Requests: the Gpucrypt driver implements four shared request queues, one for each type of OCF request supported: symmetric, asymmetric, alloc and free requests. The advantage of using separate queues for each request type is that it simplifies queue management. It allows a straightforward grouping of cryptographic requests for batching purposes rather than dealing with a single queue of mixed requests. These queues are allocated by the

Gpucryptd daemon at start-up and memory mapped into the Gpucrypt driver, thus allowing efficient transmission of request data. When the Gpucrypt driver receives a requests from the OCF, it copies all the necessary instructions into the relevant queue. All pointers used in the requests at this stage have undergone address translation, and the addresses used within the queue are from the Gpucryptd daemon address space. Thus, the Gpucryptd daemon does not have to worry about address mapping, it can treat all pointers as native in a normal manner.

Cryptographic Requests: the Gpucrypt driver supports multi-threaded and asynchronous cryptographic requests, helping to increase the concurrency of requests on the process queues. Calls from the Crypto layer to process a symmetric or asymmetric request are returned immediately after the Gpucrypt driver has queued the request and signalled for the Gpucryptd process to awaken if necessary and process the request. All manipulations of the queues are thread safe. The only time a cryptographic request blocks is when the corresponding queue is full. The results of the processed requests are returned asynchronously when the Gpucryptd daemon issues an ioctl to instruct the driver that it is finished. This in turn calls the Crypto layer to inform it that the request is finished.

Memory Requests: as with standard memory allocation and free operations, we have implemented these as blocking requests. Apart from blocking the consumer thread, memory requests do not block any other request from being processed within the OCF. Figure 6, which served as an example of an OCF alloc request, can now be discussed in the context of Gpucrypt and Gpucryptd. To service an allocation request, the Gpucrypt driver first puts the allocation details on the shared alloc request queue. The Gpucryptd daemon processes this by executing the `cudaMallocHost()` function call, which allocates pinned DMA accelerated memory. The returned address is placed back on the shared request queue, which is then used by the Gpucrypt driver to access the underlying pages. On initialisation of the Gpucryptd daemon, it registers with the Gpucrypt driver its internal task kernel pointer. This is used to retrieve access to the daemon's underlying virtual memory areas and pages. A note should be made that the virtual memory area used to reference the CUDA allocated pages is flagged with `VM_IO`. Device driver programmers commonly use this flag to prevent memory from being included in core dumps, however it also has the effect of treating the memory area as backed by non system RAM. For I/O mapped memory it is necessary to restrict access to the underlying pages as they don't exist in RAM, however in our experience, CUDA only returns RAM backed memory. We must temporarily disable this flag in order to retrieve the underlying pages, though we take the precaution of acquiring the Gpucryptd's memory map semaphore during this period. Our experience is that this technique has successfully returned the underlying pages to the Crypto layer in all of our tests.

Request Order: maintaining separate shared queues has advantages as stated above, however it has a disadvantage of not automatically preserving the original request ordering between the differing types of requests. This can cause faults when memory requests are run out of order with respect to cryptographic requests. If we solve this problem using a single request queue, then batching is less effective as the queue requires processing in order. This can lead to memory requests unnecessarily splitting groups of cryptographic requests. The solution adopted for this problem was the use of read-write semaphores within the OCF. We have used a read-write semaphore for each mapping space (i.e. one per consumer process) within the OCF and found the solution to give minimal overhead. Each cryptographic request is responsible for acquiring a read-write semaphore for *reading* if a memory translation has occurred and releasing the semaphore on request completion. Each memory request must acquire a read-write semaphore for *writing*, which ensures the memory request is the only request for the consumer process within the OCF pipeline. This ensures that any translations that were valid at the start of the processing of request, remain so until the end. The use of read-write semaphores as opposed to normal semaphores allows the most common type of request, i.e. cryptographic requests, that share a mapping space to exist concurrently within the OCF pipeline. Also if no memory translation is used, e.g. legacy consumer processes, then no semaphores are used as in the original OCF.

Driver Removal: a driver can be removed at any time, and thus we must deal with the case of allocated memory when such an event occurs. Requests can be migrated to another device by the OCF and thus memory allocated for one device can be sent to another device. The Cryptodev layer sees this event as a failed translation and defaults to copying the memory from the userspace process, thus the requests will continue to proceed, however at a slower pace. To avoid this slow down the consumer process must monitor the requests for a change in device used and if a change occurs the OCF allocated memory should be freed and allocated again by the new device. If no OCF allocated memory is used then no action is required. Note that even though the device may have freed the memory, its pages are kept alive due to the consumer process' reference. Requests sent directly to the Crypto layer will also fail the translation stage and the memory will be treated as a standard kernelspace memory pointer. Again the kernel consumer thread must monitor requests for changes in the device used and reallocate memory when this occurs.

3.5   Security

We look at each of the changes made to the OCF in terms of their security implications. The new memory allocation functionality requires that all memory returned is automatically zeroed to protect from leaking information. The use of memory translation for userspace requests bypasses the need for the copy_to/from_user() kernel functions. These kernel functions perform important validation, ensuring that the addresses are part of the calling process' address space. We ensure that this validation is maintained by only searching for translations within a mapping space which is indexed by the thread group ID. This combined with the fact that the mappings within the space only contain userspace pointers, which are generated by the kernel on behalf of the process, ensure that any match found during translation are valid userspace addresses for that process. During translation we also check the size of the buffers specified within the cryptographic requests, to ensure no buffer overflow will occur. Regarding the Gpucrypt driver, it must be ensured that the /dev/gpucrypt file is accessible by the root user only. If this is not the case, then any userspace program may connect to the Gpucrypt driver and receive OCF cryptographic requests. Also, it should be noted that no official statement from Nvidia is available on the purpose of the VM_IO flag and if its temporary disabling can cause security problems, see Section 3.4. Even though extensive testing has shown the correct function of mapping pages as required, and the Gpucryptd's mmap semaphore is acquired during the flag's disabled period, the GPU drivers are proprietary and the source code is not available. This results in a difficultly in verifying the temporary disabling of the flag does not present a possible security problem. In general, the approach presented in this paper has undergone a basic security evaluation, however, it is not recommended that the implementation be used in a security sensitive context where it is possible that an intruder may have access to the hosting server. Before such a step can be made, a dedicated security study should be performed.

4   Concurrent Request Processing

4.1   Asymmetric Request Batching

The batching of requests allows for an increase in system wide throughput by permitting the combination of separate requests to be sent to and processed by the GPU. The types of requests that comprise a batch and the preprocessing that can be done to this batch can have significant effects on performance. General purpose symmetric-key batching on the GPU has been discussed in detail elsewhere [4], so we will not go into this further in this paper. However, to date there has been no treatment of general purpose batching of distinct asymmetric-key requests on the GPU. Considering the OCF presents the opportunity to batch requests for delivery to the GPU we investigate the possibility of asymmetric request batching here.

Single Request: Currently the OCF does not support a method of executing more than one asymmetric cryptographic operation within a single request. The framework does provides the ability to chain multiple requests with a link list. This gives the ability to send in a single call multiple requests to the Crypto or Cryptodev layer. We have made a small change to the OCF API to allow the request's input buffers to contain multiple instances of its input vectors. For example, in relation to modular exponentiation, this permits a request to contain multiple bases for each exponent/modulus pair (analogous to multiple messages per key). Although only giving a slight performance improvement, it simplifies the process for clients to send multiple requests with a single key. We are reluctant to make any modifications to existing API structures for reasons of compatibility with existing applications, thus this change is suggestive and can be safely omitted if compatibility is required.

General Purpose Request Batching: We have based our asymmetric-key implementation on the serial radix algorithm for CRT modular exponentiation suitable for RSA-1024 on a pre-existing implementation [5]. This involves spawning a new CUDA thread to handle the exponentiation of each base. As there can be multiple bases per request, and one thread per base, we require a mechanism that allows each thread to dynamically discover its request data. We must also take into consideration that the base, modulus and exponent for each operation is split into two, due to the CRT technique [19]. This involves using the prime factors P and Q of the modulus N to generate smaller pairs of bases, mod P and Q, and smaller pairs of exponents, mod $P-1$ and $Q-1$. We can then separately calculate the resultant smaller modular exponentiation within the residue number system $\{P, Q\}$ and recombine at the end to produce the final result.

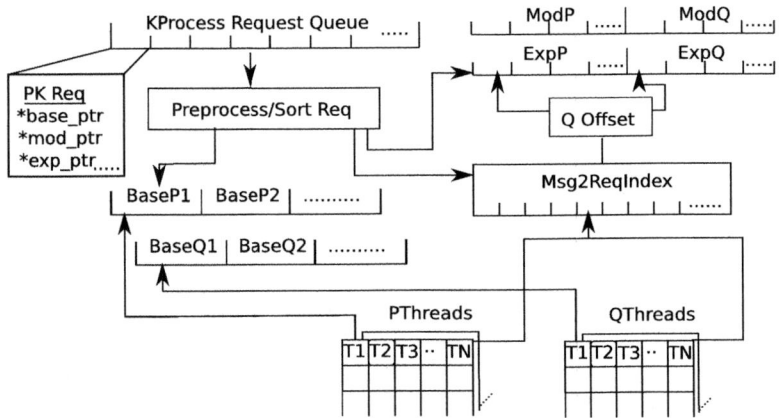

Fig. 9. Mechanism for Processing Multiple Distinct Asymmetric Key Requests

Figure 9 illustrates the mechanism used to direct the threads to their corresponding data. As described separately [5], we direct all even numbered CUDA blocks to P related data and all odd CUDA blocks to Q related data. The base data is configured in a manner so that each CUDA thread can simply scale their global thread ID to find the offset of their base data. During the preprocessing stage (discussed next), we generate a message to request index, labelled Msg2ReqIndex. This index is used to translate the message number, i.e. the base number within the full batch of requests, to the OCF request number. The request number is used to generate an offset into the modulus, exponent and related per request data. In the figure we can see that the modulus, exponent and related data is split into two groups. This allows a simple conditional addition of a single offset to the request offset to direct a thread to the P or Q related data depending on whether the CUDA block is odd or even.

Request Preprocessing: The Gpucryptd daemon can have access to multiple asymmetric requests at any one time. The GPU's processing performance of these requests can depend greatly on the order in which they appear within the GPU buffers. Concerning an efficient modular exponentiation implementation, the code path taken is largely dependent on the exponent. When the GPU executes modular exponentiations with different exponents within the same CUDA warp, we experience thread divergence and a cost is incurred. This is due to having to execute the separate code paths serially rather than concurrently, see Section 2.2. The more varied the exponents within a warp, the higher the warp cost, and thus the higher the CUDA block cost, up to a limit. More concretely, we use the well known Sliding Window [20] technique for exponentiation. This technique involves traversing the exponent as a binary string from left to right. Based on the value of a binary digit at a particular position within the exponent, the algorithm determines whether a square, a series of squares or a multiply is performed. If two threads within the same SM are executing an exponentiation using different exponents, then depending on the binary make-up of the exponent the thread code paths can diverge at different stages of the exponentiation process. Divergence of threads results in a loss of performance due to the inability of an SM to concurrently execute different instructions. In effect, the SM executes all code paths for all threads in a warp, disregarding the results of some operations appropriately. To measure the performance cost of the possible thread divergence we ran a series of tests with randomly generated exponents. From these tests we have measured the different warp costs for a GPU execution of a modular exponentiation in various divergent scenarios and the cost ranges between 1 and 2.5, where 1 is equal to the minimum run time of a non-divergent modular exponentiation. Ideally we would be able to efficiently take any array of varying sized and keyed requests and reorder them to derive the minimum total cost, or runtime.

We can draw a loose analogy between this problem and the perfect packing version of the 2-dimensional strip packing problem [21]. If we let the cost of each CUDA block become the height of an object, the width of the object is 1 and the width of the container is the number of available SMs, then we wish to minimise the height of the container holding all the objects. The analogy is not exact as we also have the added complexity that the height of each object, i.e. the CUDA block cost, can vary depending on how requests are ordered. To find the optimal solution to this is computationally impractical. However, we can use heuristics to arrive at a reasonable solution. If we first consider that the block cost increases whenever an exponent changes within the array of requests, we should sort all requests according to their exponent, thus creating a list of non-divergent groups of requests. We perform this sort by an approximation, using only the first integer of the exponent, giving a good accuracy/efficiency trade off. We label this approach as "1 pass".

We do not have control over the order in which the Nvidia driver chooses its CUDA blocks for execution when an SM becomes free, however it is reasonable to assume it follows a first fit approach, i.e. whenever an SM is free it takes the next lowest block by ID and assigns it to the SM. A reasonable close to optimal approach to solving the strip packing problem is to use the first fit descending heuristic. We follow this heuristic by sorting the non-divergent groups in increasing order of the number of operations within each group. This ensures that the most costly CUDA blocks occur in the lower block IDs. We call this approach "2 pass" as it involves the 1 pass sort above and an extra sort.

Both the 1 and 2 pass techniques are contrasted with no sorting (0 pass) in various scenarios in Figure 10. The scenarios are run outside of the OCF as they concern modular exponentiation batching on the GPU in general and not just in the context of the framework. The tests consisted of sending multiple requests to the GPU for concurrent execution. The size of each request within each test was randomly chosen in a guided manner. The "Large" tests restricted the sizes of the requests (the number of bases per request) to be high, typically 100-300; the "Small" tests contained only small requests, typically 1-10; and the "Mixed" tests contained a random mixture of large and small request sizes. Each test was run with a varying probability for each request to be followed by a request with the same exponent and modulus, i.e. the same key. This is labelled "Collision Probability", with 1 meaning all requests are using the same key and 1/512 meaning a 1 in 512 chance of two requests chosen at random from the test having the same key. This collision probability simulates a multithreaded environment sending requests to the OCF with differing numbers of system wide keys.

Figure 10 analyses the relative performance of the 0, 1 and 2 pass techniques within each scenario. It can be seen that the 0 pass approach underperforms in all scenarios. The 1 and 2 pass techniques mostly perform the same with a general slight overhead noticeable for the 2 pass approach. The 2 pass approach substantially outperforms the 1 pass approach when the collision probability is low and there is a mix of request sizes. The performance improvement of 2 pass at a collision probability of 1/512 is 24%. Small requests are more costly than large requests as the rate of change of the exponent is higher. These small messages when mixed randomly between lower cost large requests, form a layer of thinly distributed costly warps. It is beneficial to move these costly small requests into a small number of high cost blocks as the block costs converge on a relatively small overhead. This increases the number of low cost blocks that can run concurrently and finish while the high costs blocks complete. We recommend the use of this 2 pass approach due to its better performance in this scenario and relatively small overhead in the general case.

4.2   Request Pipelining

In the scenario where we have multiple cryptographic requests outstanding on the Gpucryptd daemon queue, we have the opportunity to split the processing and return of requests into two concurrent operations. The CUDA API allows for the execution of a kernel on the GPU asynchronously. The Gpucryptd daemon permits both asymmetric and symmetric CUDA modules (see Figure 8) to retrieve more requests from the queue without returning. The daemon also

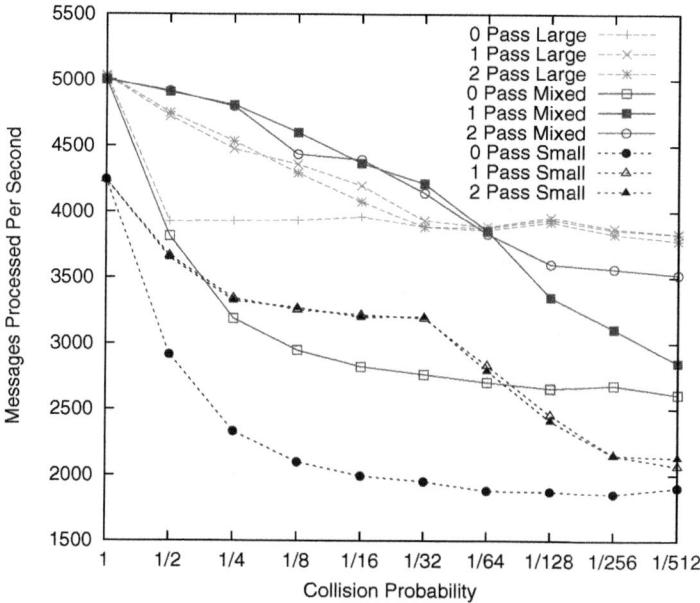

Fig. 10. Comparison of Pre-processing techniques for RSA-1024 Request Batching

supports callbacks to return completed requests. This allows the Gpycryptd daemon to delegate the flow control of request retrieval and request return to the cryptographic modules. Thus, when implementing a cryptographic algorithm for the GPU, it is straight forward to overlap the return of previously completed requests with the execution of the next requests. We present the effects of pipelining in Section 5.1.

5   Performance

5.1   Symmetric-Key Performance

To analyse the overhead of using the OCF for symmetric-key performance, we use an AES implementation based on an existing implementation [4]. Figure 11 shows the performance of AES when operating on different sized buffers with and without going through the OCF. The non-OCF version of the implementation, labelled as "Standalone", performs comparably [4]. We compare this standalone version to four other tests. Two tests were performed using normal userspace processes to initiate the requests and thus go via the Cryptodev layer of the OCF, labelled as "Cryptodev with/without MM". The remaining two tests were performed using a kernel thread which initiated the requests directly via the Crypto layer of the OCF, labelled "Crypto with/without MM". The "with MM" and

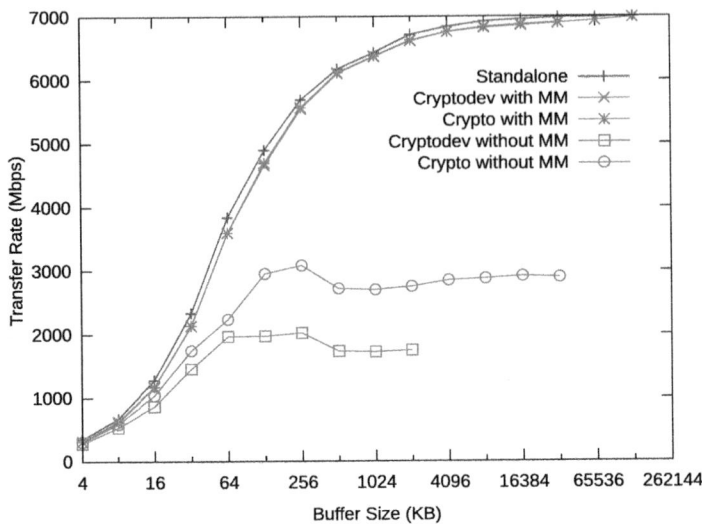

Fig. 11. Performance of GPU accelerated AES using the OCF

"without MM" tags, refer to variants of the tests whereby we either include our new memory management system or use the original OCF memory management respectively.

We can see that the two tests which use the OCF and the new memory management system, perform with a small overhead compared to the standalone version. Based on the Cryptodev interface, the average percentage overhead of using the OCF is 3.4%, with a range of 9.3% for the smallest request buffers through to 0.2% for the largest buffers. The spread in overhead percentage is due to the smaller request buffers requiring more calls through the OCF to perform the same amount of processing compared with larger request buffers. Although it cannot be seen here, there is a slight advantage to executing the cryptographic requests from the kernel as the Cryptodev layer overhead is removed.

The two tests, which are performed without the new memory management system, experience a substantial reduction in performance as the buffer sizes increase. This is due to having to perform extra memory copies for each transition between address spaces. The shape of the graphs can be understood when compared to Figure 4. The reason for "Crypto without MM" outperforming "Cryptodev without MM", is that the direct calls to the Crypto layer from kernelspace eliminates one of the address space transitions, thus reducing the number of memory copies performed. The reason for "Crypto without MM" not covering the full range of buffer sizes is that it hits the default maximum vmap() limit in the kernel. "Cryptodev without MM" is also limited in the buffer sizes used due to the original limit imposed by the OCF. These limits can be removed, though the results show little difference.

Figure 12 is used to investigate both multithread scalability and pipelining, as discussed in Section 4.2, for symmetric-key processing on the GPU. The "Single

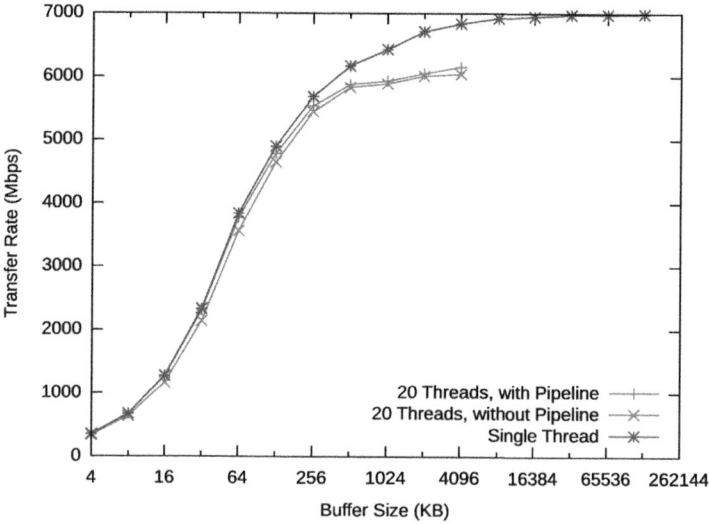

Fig. 12. Multithreaded performance of GPU accelerated AES using the OCF

Thread" test is the same as "Standalone" in Figure 11, involving multiple iterations of symmetric-key requests with varying buffer sizes. The multithreaded tests consist of executing the same amount of operations as the Single Thread test, using the same sized requests, however the requests are split across 20 threads. One of the multithreaded test runs using the previously mentioned request pipelining, and the other without. Note that the multiple threads referred to are the consumer threads making requests to the OCF, the Gpucryptd daemon itself remains a single thread. It can be seen at small buffer sizes that the multithreaded scenario using the pipeline slightly out performs the scenario without pipeline use. It achieves this improvement from asynchronously returning request results, thus hiding (or partially hiding) the return cost to the consumer. Both of the multithreaded tests taper prematurely in performance as the buffer size increases. This is presumed to be due to the multithread version of these tests requiring 20 times more active memory at any one time than the single thread version. Thus the performance degrades due to increased pressure on system memory.

5.2 Asymmetric-Key Performance

For our tests of asymmetric-key performance we used an implementation based on the modular exponentiation approach for RSA-1024 presented by Harrison and Waldron [5]. Figure 13 shows a comparison of running a standalone version of this implementation and using the implementation via the OCF Cryptodev layer from a userspace process. We can see here that there is no discernible difference in performance, in fact it is difficult to see there are two plotted graphs in the figure. This is due to the high arithmetic intensity inherent in the modular

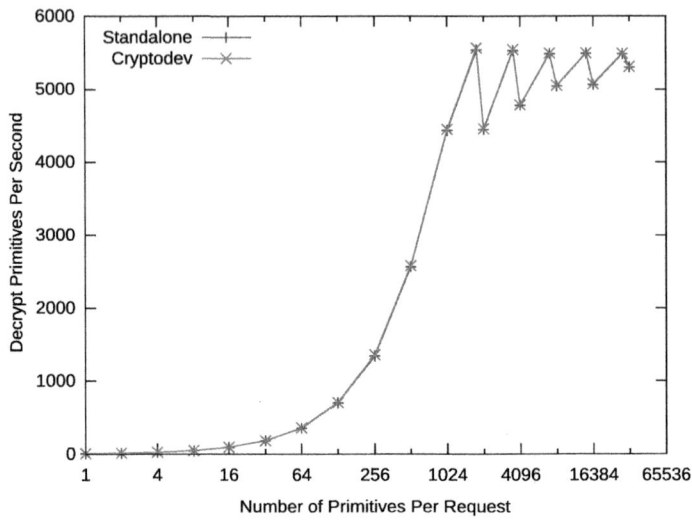

Fig. 13. Performance of GPU accelerated RSA-1024 using the OCF

exponentiation algorithm and thus the OCF overhead is relatively quite small. The average percentage overhead of using the OCF via the Cryptodev interface compared to the standalone version is 0.4%, with a range of 0.1% for the smallest number of messages per request to 0.6% for the largest. A related point is that we have performed these tests with and without the new memory management system and also with and without pipelining as in the symmetric-key tests above. The results were indistinguishable from the standalone version due to the small overhead associated with data transfer through the OCF compared to the work done on the GPU.

Figure 14 illustrates the behaviour of the OCF when processing multiple asymmetric-key requests with the same key concurrently. We achieve concurrency by using multiple threads via the Cryptodev interface. As it is a blocking interface there is no other way a userspace thread can achieve concurrency. We also test concurrency via direct kernel calls to the Crypto layer, which permits asynchronous request execution. This permits multiple outstanding requests within the Gpucrypt request queue at one time using a single kernel thread. The "Concurrency Level" label in the figure refers to either the number of threads (Cryptodev test) or the number of concurrent requests sent asynchronously (Crypto test). All tests, both multithreaded and single threaded, perform the same total number of asymmetric operations. Thus, as the concurrency increases the number of messages per request decreases, shown in brackets on the x-axis. We have highlighted the performance improvement when processing requests consisting of 112 primitives concurrently versus serially. This improvement is due to the use of batching as described in Section 4.1. From Figure 14 we can also see that the multithreaded tests lose performance as the concurrency level increases. The main reason for this performance degradation is the inability to

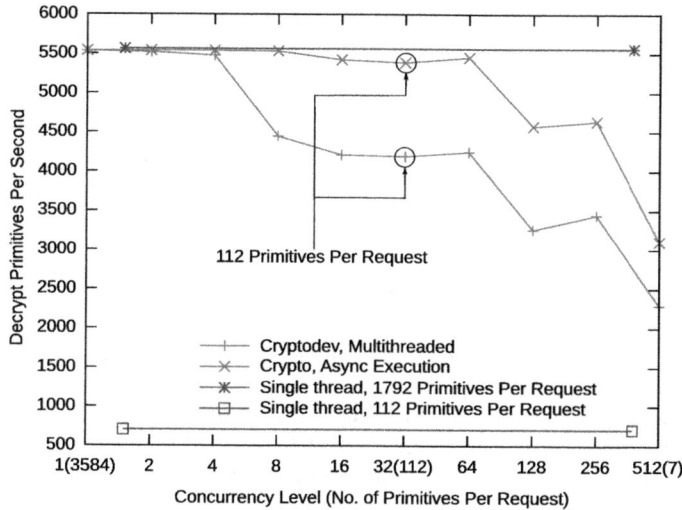

Fig. 14. Concurrency and GPU accelerated RSA-1024 using a single key within the OCF

maintain an occupied GPU. In the tests, as the concurrency increases the request sizes decrease, the OS has a harder time to deliver sufficient numbers of requests to the queue for batching due to process switching overheads. This relates to the reason the Crypto test outperforms the Cryptodev test. The OS does not have to reschedule processes as frequently to deliver the same amount of data to the GPU.

6   Conclusions

We have seen that the GPU can be effectively integrated into the OCF with careful design of a driver consisting of a kernelspace OCF driver and a userspace daemon. The paper shows that there is an average overhead of 3.4% when using the OCF for AES over a standalone implementation. In the context of RSA-1024 we see that there is a very low 0.3% average overhead when compared to a standalone version. A new memory management system within the OCF was shown to be critical in maintaining this performance for symmetric-key operations. Without its use we see a drop in performance of over 50% when using the OCF's kernelspace Crypto interface, and over a 70% drop via the OCF's userspace Cryptodev interface.

We presented a new general purpose mechanism for processing multiple asymmetric-key requests on the GPU and found that the preprocessing of mixed key requests is crucial to maintaining performance. We have also shown the effectiveness of integrating this mechanism as part of the OCF and its use within multithreaded and asynchronous scenarios. The most important factor regarding performance in these scenarios is the ability of the OS to schedule multiple threads efficiently so as to provide enough work for the GPU to reach peak performance. We have seen that GPU accelerated cryptographic functions can be made available in a uniform manner to all OS components, both in-kernel and userspace, via the OCF without excessive overhead.

References

1. Bernstein, D., Chen, T.-R., Cheng, C.-M., Lange, T., Yang, B.-Y.: ECM on Graphics Cards. In: International Conference on Advances in Cryptology - Eurocrypt, pp. 483–501 (April 2009)
2. Fleissner, S.: GPU-Accelerated Montgomery Exponentiation. In: International Conference on Computational Science ICCS, pp. 213–220 (May 2007)
3. Harrison, O., Waldron, J.: AES Encryption Implementation and Analysis on Commodity Graphics Processing Units. In: Paillier, P., Verbauwhede, I. (eds.) CHES 2007. LNCS, vol. 4727, pp. 209–226. Springer, Heidelberg (2007)
4. Harrison, O., Waldron, J.: Practical Symmetric Key Cryptography on Modern Graphics Hardware. In: USENIX Security Symposium, pp. 195–209 (July 2008)
5. Harrison, O., Waldron, J.: Efficient Acceleration of Asymmetric Cryptography on Graphics Hardware. In: Preneel, B. (ed.) AFRICACRYPT 2009. LNCS, vol. 5580, pp. 350–367. Springer, Heidelberg (2009)
6. Manavski, S.A.: CUDA Compatible GPU as an Efficient Hardware Accelerator for AES Cryptography. In: IEEE International Conference on Signal Processing and Communications, pp. 65–68 (November 2007)
7. Moss, A., Page, D., Smart, N.P.: Toward Acceleration of RSA Using 3D Graphics Hardware. In: IMA International Conference on Cryptography and Coding, pp. 364–383 (December 2007)
8. Szerwinski, R., Güneysu, T.: Exploiting the Power of GPUs for Asymmetric Cryptography. In: Oswald, E., Rohatgi, P. (eds.) CHES 2008. LNCS, vol. 5154, pp. 79–99. Springer, Heidelberg (2008)
9. Yang, J., Goodman, J.: Symmetric Key Cryptography on Modern Graphics Hardware. In: Kurosawa, K. (ed.) ASIACRYPT 2007. LNCS, vol. 4833, pp. 249–264. Springer, Heidelberg (2007)
10. Yeom, Y., Cho, Y., Yung, M.: High-Speed Implementations of Block Cipher ARIA Using Graphics Processing Units. In: International Conference on Multimedia and Ubiquitous Engineering, pp. 271–275 (April 2008)
11. Leffler, S.J.: Cryptographic device support for FreeBSD. In: Usenix, BSD Conference, pp. 69–78 (September 2003)
12. OCF-Linux Project, http://ocf-linux.sourceforge.net/
13. linux-crypto (Crypto API), http://mail.nl.linux.org/linux-crypto/
14. Harrison, O.: Source code, http://www.scss.tcd.ie/~harrisoo/code.html
15. Blythe, D.: The Direct 3D 10 System. ACM Transactions on Graphics 25(3), 724–734 (2006)
16. Nvidia Corporation, CUDA, http://developer.nvidia.com/object/cuda.html
17. Rosenberg, U.: Using Graphic Processing Unit in Block Cipher Calculations. Master's Thesis, University of Tartu (2007)
18. Harrison, O.: Acceleration of Cryptographic Functions using Graphics Hardware (2010),
https://www.scss.tcd.ie/publications/tech-reports/tr-index.10.php

19. Quisquater, J.-J., Couvreur, C.: Fast Decipherment Algorithm for RSA Public-Key Cryptosystem. Electronics Letters 18(21), 905–907 (2008)
20. Menezes, A., van Oorschot, P., Vanstone, S.: Handbook of Applied Cryptography. CRC Press, Boca Raton (1996)
21. Riffa, M.C., Bonnairea, X., Neveub, B.: A revision of recent approaches for two-dimensional strip-packing problems. Engineering Applications of Artificial Intelligence 22(4-5), 823–827 (2009)

A   OCF Extensions

A.1   New Memory Management Interface

The following lists the interface extension for the Crypto and Cryptodev layers within the OCF to support the new memory management system.

A.1.1   Crypto Layer Interface

crypto_alloc(): pass in the size and optionally the device ID for memory allocation. Returns a kernelspace pointer and the device that performed the allocation. If the allocation fails null is returned.

crypto_free(): pass in a pointer returned by crypto_alloc().

crypto_translate(): pass in a pointer returned by crypto_alloc(). Returns a device space pointer if a mapping is found.

A.1.2   Cryptodev Layer ioctl Interface

CIOCALLOC: this takes in an allocation request structure as a parameter, which specifies the requested buffer size and suggested device ID. The same structure is used to return the userspace pointer and actual device used for the allocation.

A.2   Gpucrypt ioctl Interface

The following ioctls are used to communicate with the Gpucrypt device via /dev/gpucrypt. These are used by the Gpucryptd userspace daemon to co-operate with the OCF to complete cryptographic requests.

GPU_REGISTER_*_REQ_BUF: this is a series of ioctls which Gpucryptd driver executes on startup to register shared request queues for each type of OCF request supported by the GPU driver. * refers to ALLOC, FREE, KPROCESS (asymmetric request processing) and PROCESS (symmetric request processing).

GPU_READY: after the request buffers are created, shared and initialised and all state is ready for operation, the Gpucryptd daemon registers itself as ready for work with the Gpucrypt driver. The driver, on receipt of this ioctl, issues a register command to the OCF to inform it that it is ready to start receiving requests.

GPU_WAIT_FOR_WORK: the Gpucryptd daemon process cycles through the request queues, continually processing any available work. When there are no more requests to process, rather than continually scanning it calls the Gpucrypt driver to wait for work using this ioctl. On receipt of this request the Gpucrypt sleeps the calling process on a kernel wait queue. When work is subsequently received from the OCF, the Gpucryptd is woken by calling wake on this wait queue, thus releasing Gpucryptd to finish the rest of the ioctl and return to userspace to process the new work.

GPU_RETURN_*_REQ: on finishing of a request, the Gpucryptd daemon uses this series of ioctls to deliver the work back to the OCF. This ioctl calls the OCF crypto_done() function, which can either process the registered callback function for the request immediately or allow the OCF return queue kernel thread to do so later. An application that has a long callback function may configure the cryptographic request to not execute an immediate callback as the callback is normally run in interrupt context. However, when the GPU is used, crypto_done is called from within process context, specifically the Gpucryptd context, and thus long callback functions are less problematic and thus the use of the separate return queue kernel thread can be avoided. * refers to ALLOC, FREE, KPROCESS (asymmetric request processing) and PROCESS (symmetric request processing).

GPU_SHUTDOWN: this ioctl is called on shutdown, which in turn unregisters the Gpucrypt driver from the OCF.

From a Generic Framework for Expressing Integrity Properties to a Dynamic MAC Enforcement for Operating Systems

Patrice Clemente, Jonathan Rouzaud-Cornabas, and Christian Toinard

Ensi de Bourges – LIFO
Université d'Orléans
88 bd Lahitolle, 18020 Bourges Cédex, France
{patrice.clemente,jonathan.rouzaud-cornabas,
christian.toinard}@ensi-bourges.fr

Abstract. Protection deals with the enforcement of integrity and confidentiality. Integrity violations often lead to confidentiality vulnerabilities. This paper proposes a novel approach of Mandatory Access Control enforcement for guaranteeing a large range of integrity properties. In the literature, many integrity models are proposed such as the Biba model, data integrity, subject integrity, domain integrity and Trusted Path Execution. There can be numerous integrity models. In practice, an administrator needs to combine various integrity models. The major limitations of existing solutions deal first with the support of indirect activities aiming at violating integrity and second with the impossibility to extend existing models or even define new ones.

This paper proposes a novel framework for expressing integrity requirements associated with direct or indirect activities, mostly in terms of information flows. It presents a formalization for the major integrity properties of the literature. The formalization of the required security is efficient and a straightforward enforcement is proposed. In contrast with our previous work, an information flow graph provides a dynamic analysis of the requested properties.

The paper also provides a MAC implementation that enforces every integrity property supported by our formalization. Thus, a system call fails if it could violate the required security properties.

A large scale experiment on high interaction honeypots shows the relevance, robustness and efficiency of our approach. This experimentation sets up two kinds of hosts. Hosts with our solution in IDS mode detect the violation of the requested properties. That IDS allows us to verify the completeness of our MAC protection. Hosts with our MAC protection guarantee all the required properties.

Keywords: Integrity models, security properties, MAC enforcement, information flows.

1 Introduction

Protection of computer systems is often seen in terms of availability, integrity and confidentiality. Those fields have already been widely investigated by researchers. However, everyday systems lack facilities of configuration. Moreover, they lack expressiveness

M.L. Gavrilova et al. (Eds.): Trans. on Comput. Sci. XI, LNCS 6480, pp. 131–161, 2010.

in the way the security policies are defined. Finally, existing security enforcements are still vulnerable to complex attacks, such as sequences of authorized system calls. This paper focuses on the field of integrity properties. General integrity aims at protecting entities on a system against any kind of modification. We survey the most classical integrity models of the literature. We provide a theoretical model of operating systems, and activities occurring on it, in terms of information flows, transition sequences[1] and direct or indirect executions. We propose a generic framework to express any existing or new integrity property, under our model of the system. We go a step further into the analysis of the classical integrity properties and provide more accurate definitions, especially considering indirect violations of those security properties. The various security properties considered here are mainly related to data integrity, subject integrity and binary integrity. We also consider more specific and complex so-called integrity properties such as Trusted Path Execution and domain integrity (or Virtual Chroot).

After defining our model of the operating system entities and activities, we present how it can cover any integrity property that one (e.g. an administrator) may want to express. We then explain how our MAC mechanism can enforce any of the properties our framework can allow us to model.

In order to evaluate the usefulness of our approach, we provide experiment results with the protection against a complex attack case synthesizing numerous attacks instances gathered on our honeypots. This long term experiment (six months) provides strong assessment of the relevance, robustness, completeness and efficiency of our approach.

The paper is organized as follows. Section 2 surveys research work done in the field of integrity properties. It also positions and motivates the paper regarding that work. Section 3 presents our model of the system entities and operations. It focuses on the combination of those operations into information flows, as such flows enable to define and enforce multiple integrity properties. It also presents other complex activities such as sequences of transitions and indirect executions. Using that formalism, Sect. 4 presents the modeling of major integrity properties of the literature. It also provides more accurate and complete definitions of those security properties. It introduces new and more specific security properties, to which we come back later to show the relevance of our approach for real systems. Section 5 presents the algorithm used to enforce the proposed security properties. It also provides a description of our enforcement architecture, with our PIGA-KERNEL module and our PIGA-DYN decision engine. Section 6 presents results obtained on our honeypot hosts. We compare classical DAC linux protection against our MAC solution. This section also considers correctness and efficiency issues. Section 7 sums up the paper and gives further ways to investigate.

2   Related Work and Motivations

Integrity models have been widely studied. Our approach is based on the analysis of information flows, which have been also deeply studied in the past.

As explained below, classical integrity models often lack, in their implementation and sometimes also in the principle itself, the notion of indirect integrity violations.

---

[1] i.e. role transitions, that we call here context transitions.

Indeed, the integrity of a resource can be directly altered by a process, or indirectly by a user running a process that transits to another process that finally modifies the resource.

Data integrity has been informally defined in [1]. That general definition means that data cannot be modified without authorization. In practice, such a general definition is not sufficient to protect all the resources according to the user needs. That is why various types of integrity properties and models were proposed.

2.1 Theoretical Approaches

Biba Integrity Model introduced in [2] proposed three policies. The Strict Integrity Policy is the mathematical dual of the Bell-La Padula Model [3] (BLP). As BLP, the system consists of a set of subjects, a set of objects and a set of integrity levels (respectively security levels in BLP). Biba Model enforces No-Read-Down/No-Write-Up policy. The model guarantees both direct and indirect integrity. [4] proposed to use Biba Model to build a kind of MLS model for commercial systems. However, the Biba model cannot be extended to support other protection models such as non interference or domain integrity.

In [5], the authors provide a model aiming at preventing integrity violation of subjects through interferences, called subject integrity. In [6], the authors consider the actions that can violate the integrity. Those actions include interferences but also any type of direct and indirect violations of the integrity. However, the theoretical modeling language provided is not adapted to the enforcement of an Operating System, and implementation perspectives are very partial. In [7], the author presents a Trusted Path Execution property making it possible to execute only safe binaries, such as guaranteeing another form of integrity of the subjects involved.

In [8], the authors provide a model in which integrity and confidentiality are both related to the notion privileges separation. The proposed model is based on the formal verification of the code of the application, which is not always possible.

In [9], Fred B. Schneider introduces an automata based approach to enforce security properties. In contrast with common approaches, the author proposed a protection oriented method. However, the author states that not all security properties can be enforced because the future interactions cannot be predicted. Only the safety properties can be enforced. Moreover, the author states that information flows cannot be protected. The author uses Communicating Sequential Processes language [10]. The method works by verifying security properties on a set of execution traces. The automata is potentially unbound. Advanced enforceable security properties, i.e. availability properties, were later introduced in [11,12], but as far as we know, no implementation was provided.

In [13], the authors proposed a security policy model based on state transition to guarantee both integrity and confidentiality. In contrast with [3], they consider an entity as an information repository. The data contained in the entities are dubbed attributes. A state transition takes place between a requestor and a destination entity known as the observer. However, the integrity deals with the fact that the confidentiality controls are correctly enforced. So, the paper deals rather with confidentiality than integrity.

In [14], the authors describe the Domain Integrity property allowing to confine some subject in a virtual chroot.

2.2 Operating System Approaches

Several studies address how to support indirect information flows within an Operating System. The HiStar Operating System [15] associates each object or subject with an information flow level. The problem of HiStar is that it is very close to the Biba integrity model and suffers from the same limitations. The Flume Operating System [16] is very close to HiStar. However, Flume does not control efficiently the information flows. Asbestos [17] and HiStar both consider four different levels of information. The protection rules can only express pairwise relationship patterns. Again, information flows involving multiple interactions and processes cannot be controlled easily.

In [18], a tool, called 'Apol' (analyse policy) has been provided as part of a complete software suite for the administration of SELinux. Apol can search for transitive information flows within the static SELinux Type Enforcement (TE) policy. But they cannot prevent such transitive flows, nor even dynamically detect them during execution. Apol thus comes as an assistant for analyzing and writing policies. Moreover, all transitive flows detected in the policy cannot be removed by modifying the policy. It is due to the TE's principle itself: each interaction is treated individually from others. Transitivity is not covered by the TE under Selinux.

In [19], a method to enforce all the security properties related to integrity and confidentiality was introduced. It is based on a Security Properties Language (SPL) used to describe the security properties. The approach reuses existing MAC policies, such as SELinux, enforcing direct security properties. In contrast to SELinux, the proposed approach adds protections against indirect interactions. The method is able to compute in advance all the forbidden indirect interactions i.e. the sequences of interactions that could break the required security properties. However, the solution requires existing SELinux MAC policies for classical TE.

Since a few years ago, a hybrid approach between DAC and MAC systems has emerged: augmented DAC approach. DAC is the most commonly implemented and easier to use access control model. The main drawback of a DAC system is the impossibility to guarantee any security property [20]. The main drawback of classical MAC systems (SELinux [21], GRSecurity, RSBAC) is the difficulty in writing a safe security policy and the impossibility to guarantee system-wide integrity. Thus, [22,23,24] proposed to use hybrid DAC and MAC models for mandatory integrity. However, [22], like the others, reuses the DAC protection policy to control the flows between the users, so they do not solve the impossibility problem to guarantee a security property starting from a DAC policy.

2.3 Other Approaches

Language-Based Information-Flow Security [25] is one of the main research fields related to information flow security. It is built on static analysis of source or binary code and language semantics. But Language-based Information-Flow security requires that any program on the system has been deeply studied in terms of information flow. More than that, each new program must be analyzed, even without its source code. In addition, information flows between processes and programs are sometimes impossible to prevent. Moreover, programs corrupted after their analysis can lead to new information flows.

Nowadays, with the increasing number of virtualization technologies, another kind of integrity checking is returning to the forefront: fingerprints. Classically, a database of fingerprints of objects [26,27] e.g. critical files, kernel structures and objects, manages the modifications. The objects are checked against their fingerprints. The integrity of the fingerprints themselves is critical and unless TPM (Trusted Platform Module) is used [28], it is possible to tamper with them. Integrity Enforcement through virtualization is a new field. Fingerprints [29] are resistant because they are outside the operating system scope and used vTPM [30]. But such fingerprints require virtualizing the operating system and causes overhead. However, fingerprints do not allow the control of any kind of object, mainly files.

In [31], the authors use virtual machine introspection to isolate the policy decision point from the kernel, which cannot be attacked any more by a malicious user gaining root privileges. This work focuses on protecting the kernel integrity itself specifically against well-known rootkits. The authors argue that their system is highly flexible and handles the evolution of the system over time (e.g. objects attributes and the context of the system). While the authors give an interesting usage of their model to define a generic Chinese Wall policy, they do not consider any real-world usage of it, nor any real enforcement perspectives. The work presented still lacks expressiveness: e.g. dynamic separation of duty is not covered by the proposed solution. Moreover, the authors do not provide any performance evaluation of any of their work, questioning the usability of the solution. Given those weaknesses that we address in this paper, the authors do not propose a solution to replace existing access control enforcement systems but rather a complementary solution to keep those enforcement mechanisms secure.

2.4 Motivations

Our purpose is to propose a general framework for defining complex and mixed security properties. In this paper, we focus on integrity properties. Thus, one needs a formal language that is powerful enough. A straightforward support of that framework is required within an Operating System. We aim at providing a powerful but easy and comprehensive way to define ad-hoc required security properties and allow their automatic enforcement using our PIGA-DYNdynamic protection module. Our approach tracks the information flows onto the target system. Any elementary operation (i.e. system call) on the target operating system is thus monitored. The major advantage of our approach is that it can monitor both direct and indirect flows while providing an immediate implementation that controls the formalized properties. Thus, a security administrator can reuse different integrity canvases. He can also propose new integrity canvases using our language. The new canvases are processed straightforward enough by our MAC module. In contrast with low-level MAC mechanisms, the required integrity policies can be defined easily. Finally, our solution does not require any existing MAC policies. Our solution computes dynamically the relevant system activities in a very efficient and relevant manner. That efficiency must be evaluated. This is done with large scale experimentations using real and unknown attacks.

In a preliminary work [32], we have presented an information flow approach in order to prevent *Race Condition* (*RC*) based attacks. In contrast, this new paper details and extends the model, and addresses integrity. In addition, it provides a strong real world example of attacks gathered on our honeypots, on which our protection system blocked every instance.

3   System Modeling

In order to formalize the integrity properties in terms of activities on the Operating System, let us first define the model of the target system. The first requirement is to be able to associate a unique security label (also called security context) to each system resource. A security context can be a file name or the binary name executed by a process. Our system fits well for DAC OS (GNU Linux, Microsoft Windows) or MAC ones such as SELinux whereas security contexts are special entities controlled by the kernel.

3.1   System Time, Entities and Operations

In essence, an operating system is defined by a set of entities performing operations on other entities. Those entities are referred here as 'security contexts'. Active contexts are called subject contexts while passive ones are called object contexts. Such contexts can carry information.

Formally, an operating system consists of the following elements:

- A set of system timestamps $\mathcal{T}$, representing any possible time given by the system clock, from 0 to ∞.
  Any $t \in \mathcal{T}$ is a timestamp representing a given system time.
- A set of subject security contexts $\mathcal{SSC}$.
  Each $ssc \in \mathcal{SSC}$ characterizes an active entity, i.e. processes, that can perform actions, i.e. system calls.
- A set of object security contexts $\mathcal{OSC}$.
  Each $osc \in \mathcal{OSC}$ characterizes a passive entity (file, socket, . . .) on which system calls can be performed.
- A set of all security contexts $\mathcal{SC} = \mathcal{SSC} \cup \mathcal{OSC}$, with $\mathcal{SSC} \cap \mathcal{OSC} = \emptyset$.
  For example, let us consider the apache webserver reading an HTML file. The apache process is identified as a subject ($/usr/bin/apache \in \mathcal{SSC}$ in a classical Linux system or $apache\_t \in \mathcal{SSC}$ in a SELinux environment) and the file is considered as an object ($/var/www \in \mathcal{OSC}$ in a classical Linux system or $var\_www\_t \in \mathcal{OSC}$ in a SELinux environment).
- A set of elementary operations $\mathcal{EO}$.
  $\mathcal{EO}$ denotes all the elementary operations, i.e. system calls, that can occur on the system (i.e. $read\_like$ and $write\_like$ operations).
- A set of interactions: $\mathcal{IT} : \mathcal{SSC} \times \mathcal{EO} \times \mathcal{SC} \times \mathcal{T} \times \mathcal{T}$.
  Each element of $\mathcal{IT}$ is thus a 5-uple that formally represents an interaction on the system. We will use the following notation for such an interaction: $ssc \xrightarrow{eo} tsc$, where $ssc \in \mathcal{SSC}, tsc \in \mathcal{SC}, eo \in \mathcal{EO}, ssc \neq tsc$. In essence, an interaction $it \in \mathcal{IT}$ represents a subject $ssc \in \mathcal{SSC}$ invoking an operation $eo \in \mathcal{EO}$ on a given context $tsc \in \mathcal{SC}$, starting at a system time t_s and ending at a system time t_e.

- Three functions: $src : \mathcal{IT} \rightarrow \mathcal{SSC}$, $tgt : \mathcal{IT} \rightarrow \mathcal{SC}$ and $op : \mathcal{IT} \rightarrow \mathcal{EO}$, that return respectively the *source* context, the *target* context and the *operation* involved in an interaction.
- A system trace S.
  The execution of an operating system can be seen as a set of invoked interactions. The executed interactions modify the OS state [33]. When we consider prevention, we work with invocation trace. The invocation trace thus contains all tried interactions, even those which are finally not allowed to be performed. Thus, each time an interaction it_i occurs on a given system (before being allowed, in case of a prevention system), the corresponding system trace becomes $S_i \leftarrow S_{i-1} \cup it_i$.

3.2   Information Flows

As the purpose of this paper is to present security properties defined using information flows, we first define precisely what those information flows are. We consider information flows at the operating system level: when an interaction occurs (i.e. an elementary operation is performed between two entities on the system), there is one potential consequence: that interaction can produce an information flow from one security context to another. For example, when a process reads in a file, the memory of the process receives information from the file read. On the other hand, when a process writes some information in a file (or in memory, or in a socket, etc.), it transfers some information to this resource. The resource thus receives new information.

With our modeling of the system, flows can be described as the following. An *information flow* transfers some information from a security context sc_1 to a security context sc_2 using a *write_like* operation or to sc_1 from sc_2 using a *read_like* operation[2].

The formal modeling of the system is then extended with the following sets:

- A subset of $\mathcal{EO}$ of *read_like* operations $\mathcal{REO}$.
- A subset of $\mathcal{EO}$ of *write_like* operations $\mathcal{WEO}$.

Let us specify various cases of information flows in order to define relevant security properties.

3.3   Direct Information Flows

Let us first give various definitions of direct flows between two security contexts.

Definition 1 (Single Direct Information Flow). *Given a system trace S, a single direct information flow from a subject context ssc performing a single write_like operation to a target context tsc, starting at a system time t_s and ending at a system time*

---

[2] To be able to decide if an interaction produces an information flow between two security contexts, we use a mapping table, derived from the one coming with 'Apol', from Tresys [18]. That mapping table says for each $eo \in \mathcal{EO}$ if it can flow information – and in what direction (possibly both) – between the two security contexts, or not. It also says to which criticality level the interaction is set.

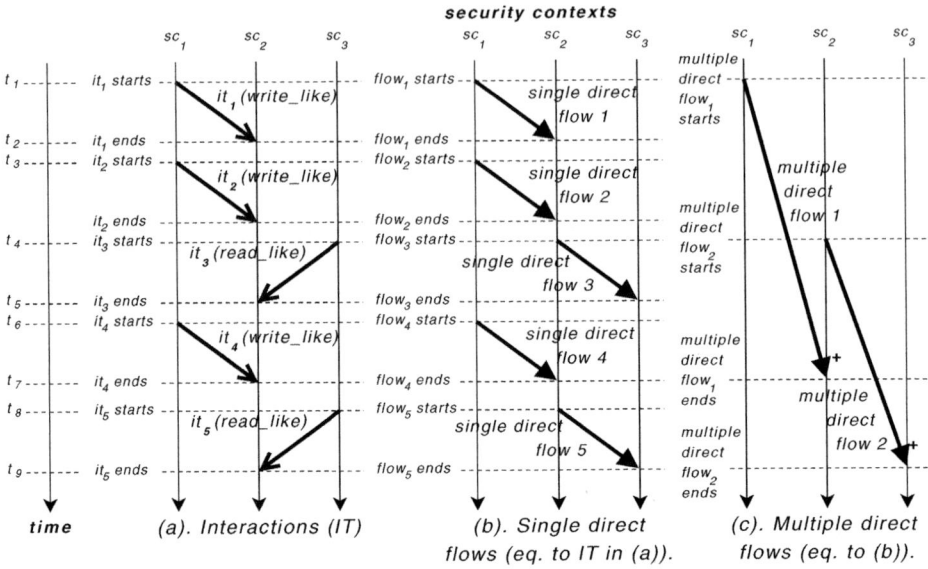

Fig. 1. r/w_like interactions and corresponding single or multiple direct information flows

t_e *(formally an interaction:* $(ssc, weo, tsc, t_s, t_e)$, *where* $ssc \in \mathcal{SSC}, tsc \in \mathcal{SC}, weo \in \mathcal{WEO}, \{t_s, t_e\} \in \mathcal{T}$, *having* $t_s \leq t_e$), *is denoted by:*

$$ssc \overset{S}{\blacktriangleright}_{[t_s, t_e]} tsc.$$

Symmetrically, a single direct information flow from a subject context ssc performing a $read\_like$ operation to a target context tsc starting at t_s, ending at t_e is denoted by:

$$tsc \overset{S}{\blacktriangleright}_{[t_s, t_e]} ssc.$$

Figure 1.(a). shows five $write\_like$ or $read\_like$ operations performed by the security contexts sc_1 and sc_3 on sc_2. The resulting five direct single flows of those interactions are given in Fig. 1.(b). In some specific situations, we may use the following notation $\overset{S}{\blacktriangleright}_{it_i}$ (e.g., in § 4.2) in order to be able to refer to the (unique) interaction related to the single direct flow.

Definition 2 (Multiple Direct Information Flow). *Given a system trace S, a multiple direct information flow from a subject context ssc performing several $write\_like$ operations to a target context tsc, with the first flow starting at t_{s_1}, ending at t_{e_1}, and the last flow starting at t_{s_k}, ending at t_{e_k}, where $ssc \in \mathcal{SSC}, tsc \in \mathcal{SC}, weo \in \mathcal{WEO}, \{t_{s_1}, t_{s_k}, t_{e_1}, t_{e_k} \in \mathcal{T}\}$, having $t_{s_1} \leq t_{e_k}$), denoted by $ssc \overset{S}{\blacktriangleright^+}_{[t_{s_1}, t_{e_k}]} tsc$, is defined by:*

$$ssc \overset{S}{\blacktriangleright^+}_{[t_{s_1}, t_{e_k}]} tsc \overset{def}{\equiv} \left(\begin{array}{l} \exists k, k > 1, \forall i \in [1..k], \\ \bigwedge \left(ssc \overset{S}{\blacktriangleright}_{[t_{s_i}, t_{e_i}]} tsc \right) \end{array} \right)$$

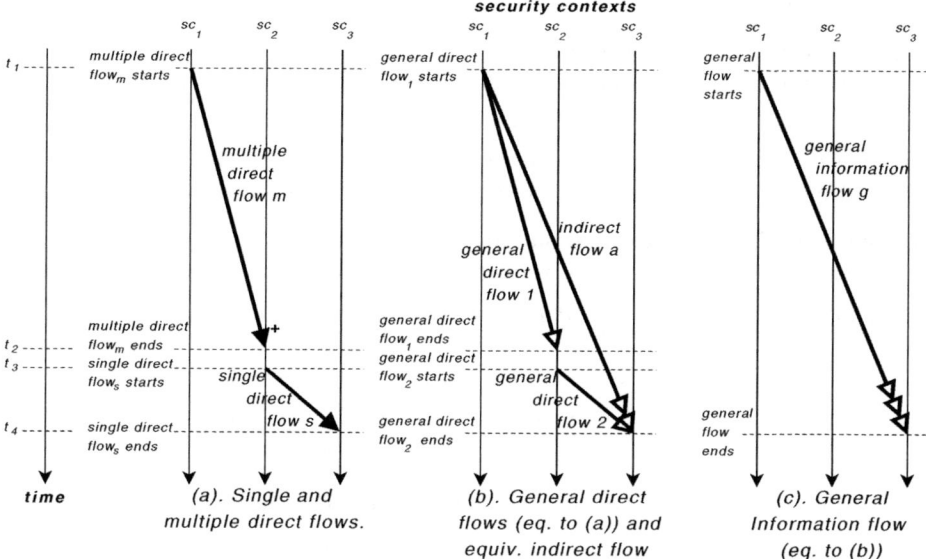

Fig. 2. Single or multiple direct information flows and related indirect information flows

Symmetrically, a multiple direct information flow from a subject context ssc performing several $read\_like$ operations to a target context tsc starting at t_{s_1} and ending at t_{e_k} is denoted by: $tsc \blacktriangleright^{+S}{}_{[t_{s_1}, t_{e_k}]} ssc$. That second definition introduces the direct information flows possibly occurring several times between the same contexts. It aims at abstracting multiple flows between two contexts in one global flow between those contexts. For example, in Fig. 1.(b)., there are three flows from sc_1 to sc_2: $flow_1$, $flow_2$ and $flow_4$. In terms of information transfers, one can consider that given those three information flows, information has started to go from sc_1 to sc_2 with the beginning of the first flow ($flow_1$), in t_1, and has stopped at the end of the third one ($flow_4$ in the figure), in t_7. The resulting multiple direct flows of those three flows is the multiple direct flow 1, $sc_1 \blacktriangleright^{+S}{}_{[t_1,t_7]} sc_2$, shown in Fig. 1.(c).

Definition 3 (General Direct Information Flow). *Given a system trace S, a general direct information flow from a security context ssc to a security context tsc occurring when either a single direct flow or a multiple flow occurs from ssc to tsc, starting at t_{s_i}, ending at t_{e_j}, where $ssc \in \mathcal{SSC}, tsc \in \mathcal{SC}, weo \in \mathcal{WEO}, \{t_{s_i}, t_{e_j} \in \mathcal{T}\}$, having and $t_{s_i} \leq t_{e_j}$, denoted by $ssc \,\triangleright^{S}{}_{[t_{s_i}, t_{e_j}]}\, tsc$, is defined by:*

$$ssc \,\triangleright^{S}{}_{[t_{s_i}, t_{e_j}]}\, tsc \stackrel{def}{\equiv} \left(\begin{array}{c} \exists i, j \in \mathbb{N}, i \leq j, \\ \left((ssc \,\blacktriangleright^{S}{}_{[t_{s_i}, t_{e_i}]}\, tsc) \vee (ssc \,\blacktriangleright^{+S}{}_{[t_{s_i}, t_{e_j}]}\, tsc)\right) \end{array} \right)$$

3.4 Indirect Information Flows

As said previously, an information flow can occur directly between two security contexts. But it can also occur in many indirect ways. For example, there may be a first flow $ssc \ \overset{S}{\triangleright}_{[t_{s_i},t_{e_i}]} \ osc$ and then a second flow $osc \ \overset{S}{\triangleright}_{[t_{s_j},t_{e_j}]} \ tsc$, having $t_{s_i} \leq t_{e_j}$. We consider this as an *indirect information flow* from ssc to tsc. Transitively, there may theoretically be an infinite number of intermediary contexts between ssc and tsc.

Definition 4 (Indirect Information Flow). *Given a system trace S, an indirect information flow from one context sc_1 to another context sc_k, starting at a system time t_{s_1} and ending at a system time t_{e_k}, denoted by $sc_1 \ \overset{S}{\triangleright}_{[t_{s_1},t_{e_k}]} \ sc_k$, is defined by:*

$$sc_1 \ \overset{S}{\triangleright}_{[t_{s_1},t_{e_k}]} \ sc_k \ \overset{def}{\equiv} \ \left(\begin{array}{c} \exists k \in [3..+\infty], \forall i \in [1..k-2], sc_i \in \mathcal{SC} \\ (sc_i \ \overset{S}{\triangleright}_{[t_{s_i},t_{e_i}]} \ sc_{i+1}) \wedge (sc_{i+1} \ \overset{S}{\triangleright}_{[t_{s_{i+1}},t_{e_{i+1}}]} \ sc_{i+2}) \\ \wedge (t_{s_i} \leq t_{e_{i+1}}) \end{array} \right),$$

where k represents the total number of contexts involved in the indirect flow.

Figure 2.(b). shows an example of such an indirect information flow where $k = 3$. There are thus $k - 1 = 2$ general direct flows involved, as visible in Fig. 2.(b). The first general direct flow $(flow_1)$ is equivalent to the multiple direct flow $(flow_m)$ of Fig. 2.(a). The second general direct flow $(flow_2)$ is equivalent to the single direct flow $(flow_s)$ of Fig. 2.(a). Given those two general direct flows $sc_1 \ \overset{S}{\triangleright}_{[t_1,t_2]} \ sc_2 \ (flow_1)$ and $sc_2 \ \overset{S}{\triangleright}_{[t_3,t_4]} \ sc_3 \ (flow_2)$, there is an indirect information flow $sc_1 \ \overset{S}{\triangleright}_{[t_1,t_4]} \ sc_3$ $(flow_a$ in Fig. 2.(b).).

Definition 5 (General Information Flow). *Given a system trace S, an information flow from one context sc_1 to another context sc_k, starting at the system time t_{s_1} and ending at the system time t_{e_k}, denoted by $sc_1 \ \overset{S}{\triangleright\!\triangleright}_{[t_{s_1},t_{e_k}]} \ sc_k$, is formally defined by:*

$$sc_1 \ \overset{S}{\triangleright\!\triangleright}_{[t_{s_1},t_{e_k}]} \ sc_k \ \overset{def}{\equiv} \ \left(sc_1 \ \overset{S}{\triangleright}_{[t_{s_1},t_{e_k}]} \ sc_k \right) \vee \left(sc_1 \ \overset{S}{\triangleright}_{[t_{s_1},t_{e_k}]} \ sc_k \right)$$

Figure 2.(c). gives an example of a general information flow: $sc_1 \ \overset{S}{\triangleright\!\triangleright}_{[t_1,t_4]} \ sc_3 \ (flow_g)$. It is equivalent to the indirect information flow $(flow_a)$ in Fig. 2.(b).

3.5 Other Kinds of Interactions and Flows

Transitions and executions. In some particular security properties we present at the end of the next section, such as Trusted Path Execution ([7]), we need to exploit specific kinds of interactions, such as *transition_like* system calls and *exec_like* ones. The formal modeling of the system is then extended with the two following sets:

- A set $\mathcal{TEO}$ of *transition_like* operations, which is a subset of $\mathcal{WEO}$.
  Indeed, a transition is performed when a subject security context transits to another subject security context, in order to gain its privileges. For example, *user_t* can

transit to *root_t* in order to access root privileges. Following that sense, any transition is assumed to be a *write_like* operation, as the source context comes with its own information to another context.
 – A set $\mathcal{XEO}$ of *exec_like* operations, which is a subset of $\mathcal{REO}$.
   An *exec_like* operation is performed when a subject security context runs an object binary context. To do that, it loads the binary data into its own memory. That is why an *exec_like* operation is seen as a *read_like* operation.

Transition Sequences and Indirect Executions. In terms of information flows, before the current section, the two previous kinds of interactions were implicitly included in the *read_like* and *write_like* ones. We only need their use for the TPE security property. In other security properties, we stay at the above level of detail: only *read_like* and *write_like* operations, that provide all the information needed for their definitions and application. For the sake of conciseness, we only briefly introduce three new operators:

1. a first operator for the transitive sequence of subjects transitions;
2. a second operator for 'direct executions';
3. a third operator for 'general executions', including both notions of direct and indirect executions. So-called indirect execution are transition sequences followed by direct executions.

Definition 6 (General Transition Sequence). *Given a system trace S, a general transition sequence from one subject context ssc_1 to another subject context ssc_k, starting at the system time t_{s_1} and ending at the system time t_{e_k}, is denoted by:*

$$ssc_1 \overset{S}{\underset{tr\ [t_{s_1},t_{e_k}]}{\rrbracket\!\!\triangleright}} ssc_k.$$

Definition 7 (Direct Execution). *Given a system trace S, a direct execution performed by a subject context ssc_1 on an object context ssc_2, starting at the system time t_{s_1} and ending at the system time t_{e_1}, is denoted by:* $ssc_1 \overset{S}{\underset{x\ [t_{s_1},t_{e_1}]}{\blacktriangleright}} ssc_2.$

The definition of the two previous operators follows the same construction pattern as for the 'Direct Single Information Flow' operator.

Definition 8 (Indirect Execution). *Given a system trace S, an indirect execution performed by a subject context ssc_1 on an object context osc, starting at the system time t_{s_1} and ending at the system time t_{e_k}, is denoted by* $ssc_1 \overset{S}{\underset{x\ [t_{s_1},t_{e_k}]}{\rrbracket\!\!\triangleright}} osc.$ *In essence, it is a transitive sequence of transitions from ssc_1 to another subject context sc_k, followed by the execution of osc invoked by sc_k. It is formally defined by:*

$$ssc_1 \overset{S}{\underset{x\ [t_{s_1},t_{e_k}]}{\rrbracket\!\!\triangleright}} osc \overset{def}{\equiv} \left(\begin{array}{c} \exists j, k \in \mathbb{N}, 1 \leq j \leq k, \\ \left(ssc_1 \overset{S}{\underset{tr\ [t_{s_1},t_{e_j}]}{\rrbracket\!\!\triangleright}} sc_k\right) \wedge \left(sc_k \overset{S}{\underset{x\ [t_{s_k},t_{e_k}]}{\blacktriangleright}} osc\right) \end{array} \right).$$

Definition 9 (General Execution). *Given a system trace S, a general execution performed by a subject context ssc_1 on an object context osc, starting at the system time*

t_{s_1} and ending at the system time t_{e_k}, is either a direct or an indirect execution of osc by ssc_1, denoted by $ssc_1 \overset{S}{\underset{x\ [t_{s_1},t_{e_k}]}{\boxed{\triangleright}}} osc$. It is formally defined by:

$$ssc_1 \overset{S}{\underset{x\ [t_{s_1},t_{e_k}]}{\boxed{\triangleright}}} osc \overset{def}{\equiv} \left(\begin{array}{c} \exists k \in \mathbb{N}, k \geq 1, \\ (ssc_1 \overset{S}{\underset{x\ [t_{s_1},t_{e_k}]}{\blacktriangleright}} osc) \vee (ssc_1 \overset{S}{\underset{x\ [t_{s_1},t_{e_k}]}{\boxed{\triangleright}}} osc) \end{array} \right).$$

3.6   Sequential Flows

The general notion of flows includes interleaving of interactions. Sometimes one needs to manage a sequential flow that is a case of the general flow. A sequential flow is a transitive closure where each flow ends before the next one starts. For example, a transitive closure of $(sc_1 \overset{S}{\boxed{\triangleright}}_{[t_{s_1},t_{e_i}]} sc_2)$ ending before $(sc_2 \overset{S}{\underset{tr\ [t_{s_j},t_{e_k}]}{\boxed{\triangleright}}} sc_3)$ ending before $(sc_3 \overset{S}{\underset{x\ [t_{s_l},t_{e_m}]}{\boxed{\triangleright}}} sc_4)$ corresponds to a sequential flow denoted without timestamp information as follows: $sc_1 \overset{S}{\boxed{\triangleright}} sc_2 \overset{S}{\underset{tr}{\boxed{\triangleright}}} sc_3 \overset{S}{\underset{x}{\boxed{\triangleright}}} sc_4$.

4   Integrity Properties Modeling

In this section, we present the modeling of the major integrity properties found in literature, which uses information flows. In addition to giving a canonical expression framework for any existing or non-existing integrity property, the main motivation of this part, based on the formal model given previously is to allow the direct compilation of any modeled security property into a protection algorithm. In the first subsection, we present the data integrity followed by non interference. We end this subsection by introducing a general integrity property. In the second subsection, we present other properties such as Domain Integrity (or Virtual Chroot (VChroot)) and Trusted Path Execution which is a particular property requiring other kinds of flows such as transition flows and $exec\_like$ interactions.

4.1   Data Integrity

The purpose of the Data Integrity Property (informally defined in [1] as 'data integrity') is to guarantee that no modification of a given object will be done on the system. It can be defined as follows:

> Given X, an entity and I, an information or a resource, the Data Integrity Property for X on I is respected if X is not able to modify I.

For example, one should want to define such a property between a user and the $/etc/shadow$ file. Thus the user with the subject context $user\_t$ is not allowed to modify a file with the object context $shadow\_t$.

Under our system modeling, given a trace S, the Data Integrity property definition below expresses that a subject context cannot modify (the data contained in) an object context:

Property 1 (Direct Data Integrity). *Given a system trace S, a subject security context ssc and an object security context osc, the Direct Data Integrity (DDI) property for ssc and osc is respected iff ssc is not able to directly transfer information to osc:*

$$DDI(T, ssc, osc) \stackrel{def}{\equiv} \neg(ssc \stackrel{S}{\triangleright} osc)$$

The previous definition conforms with the common acceptance of the definition of 'non-interference' given by Ko and Redmond [5]:

> A group of users X, using a certain set of commands, is non-interfering with a set of data D if what the group does with those commands has no effect on the value of D.

Under our model, this definition leads to exactly the same security property as DDI, where any member of X is abstracted into a subject context, and any data of D into an object context. However, the previous definition explicitly does not consider indirect flows. The next definition we propose hereafter tackles that limitation by taking into account situations where a subject transits to another subject which may have the authorization to directly or transitively modify osc:

Property 2 (General Data Integrity). *Given a system trace S, a subject security context ssc and an object security context osc, the General Data Integrity (GDI) property for ssc and osc is respected iff ssc is not able to directly or indirectly transfer information to osc:*

$$GDI(T, ssc, osc) \stackrel{def}{\equiv} \left(DDI(T, ssc, osc) \wedge \neg(ssc \stackrel{S}{\triangleright\!\!\!\triangleright} osc)\right)$$
$$\Leftrightarrow \neg(ssc \stackrel{S}{\triangleright\!\!\!\triangleright\!\!\!\triangleright} osc)$$

which says that to enforce the GDI property between ssc and osc, no flow must occur from ssc to osc.

Subjects Integrity. We present here security properties aiming at preventing subject contexts against modifications. The main ones are often referred to as 'non-interference' in the literature. They were proposed by Goguen and Meseguer [6].

Goguen and Meseguer. Goguen and Meseguer defined the following non-interference security property:

> Given X, a set of subjects, Y a second set of users and D a set of data. The 'non-interference' security property between X and Y is respected if no member of X is able to modify D or if no member of Y is able to read data from D.

Under our formalism, the common acceptance of that definition can be modeled as the following:

Property 3 (GM-Integrity). *Given a system trace S, two sets of subject security contexts SSC_1 and SSC_2, the Goguen and Meseguer (GMI) integrity property for SSC_1 against SSC_2 is respected iff no member of SSC_1 is able to directly transfer information to a shared object osc and no member of SSC_2 is able to later receive information from that shared object osc:*

$$GMI(T, SSC_1, SSC_2) \stackrel{def}{\equiv} \begin{pmatrix} \forall ssc_1 \in SSC_1, \forall ssc_2 \in SSC_2, osc \in \mathcal{OSC}, \\ \forall t_{s_i}, t_{e_i}, t_{s_j}, t_{e_j} \in \mathcal{T}, (t_{s_i} \leq t_{e_j}), \\ \neg\big((ssc_1 \stackrel{S}{\triangleright}_{[t_{s_i}, t_{e_i}]} osc) \wedge (osc \stackrel{S}{\triangleright}_{[t_{s_j}, t_{e_j}]} ssc_2)\big) \end{pmatrix}.$$

That common acceptation does not allow to take indirect information flows into account. To go a step further, we propose the more accurate definition that follows:

Property 4 (General Integrity of Subjects). *Given a system trace S, two sets of security subjects SSC_1 and SSC_2, the General Integrity of Subjects (GSI) property for SSC_1 against SSC_2 is respected iff no member of SSC_1 is able to transfer information to any member of SSC_2:*

$$GSI(T, SSC_1, SSC_2) \stackrel{def}{\equiv} \begin{pmatrix} \forall ssc_1 \in SSC_1, \forall ssc_2 \in SSC_2, \\ \neg(ssc_1 \stackrel{S}{\Longrightarrow} ssc_2) \end{pmatrix}.$$

4.2 Domain Integrity

Domain Integrity. The Domain Integrity Property allows the confinement of a set of subjects into a subset of entities. We also call it a virtual chroot, or $VCHROOT$. It is generally seen [14] as a strict confinement of a set of contexts: no interaction from contexts of this set can actively be initiated from those contexts to other contexts outside of this set. In terms of information flows, this security property can be defined as follows:

Property 5. *Given a system trace S and a set SC of security contexts, the Domain Integrity for SC is enforced iff each member of SC is only able to share information with other members of SC.*

$$VCHROOT(T, SC) \stackrel{def}{\equiv} \begin{pmatrix} \forall sc \in SC, \\ ((sc \stackrel{S}{\Longrightarrow} sc_k) \vee (sc_k \stackrel{S}{\Longrightarrow} sc)) \\ \Rightarrow \\ (sc_k \in SC) \end{pmatrix}.$$

In order to allow information flows generated by contexts outside of the $VCHROOT$, we propose a more precise and practically accurate definition of the $VCHROOT$ property:

Property 6. *Given a system trace S and a set SC of security contexts, the General Domain Integrity for SC is enforced iff each member of SC is only able to actively* *share information with other members of its domain SC, or not be able to actively exchange information with entities outside of SC.*

$$GVCHROOT(T, SC) \stackrel{def}{\equiv} \left(\begin{array}{c} \forall sc \in SC, \\ ((sc \stackrel{S}{\blacktriangleright}_{it_i} sc_k) \vee (sc_k \stackrel{S}{\blacktriangleright}_{it_i} sc)) \\ \Rightarrow \\ ((sc_k \in SC) \vee (src(it_i) \neq sc)) \end{array} \right).$$

The property above is optimal: it does not use any indirect flow. It is thus useless to track indirect flows, single direct flows are sufficient. Thus, if any information exchange occurs between a context of SC and another, the property is respected if the other context is in SC or if sc is not the actor of the exchange.

Trusted Path Execution (TPE). The purpose of the Trusted Path Execution is to guarantee that a set of subjects is only able to make system calls using data that comes from a set of objects known and trusted [7]. For example, the subject *user_t* is only able to execute binaries that are coming from the object *bin_t*. Generally, the classically used so-called TPE security property only deals with direct executions. It thus could be defined as follows:

Property 7 (Direct TPE). *Given a system trace S, a subject security context ssc and a set of object contexts OSC, the Direct Trusted Path Execution (DTPE) Security Property for ssc and OSC is respected iff ssc is only able to directly execute (binary) objects of OSC:*

$$DTPE(T, ssc, OSC) \stackrel{def}{\equiv} \left(\begin{array}{c} \forall osc \in \mathcal{SC}, ssc \in \mathcal{SSC}, \\ (ssc \stackrel{S}{\underset{x}{\blacktriangleright}} osc) \Rightarrow (osc \in OSC) \end{array} \right).$$

That definition is sometimes related to integrity because of its potential effects on the subject performing the execution. Indeed, executing a malicious binary can lead to the violation of the integrity for this subject. Unfortunately, this 'direct' definition does not cover indirect executions, such as *user_t* transiting to *root_t* in order to execute privileged programs. We thus introduce a more general definition of TPE.

Property 8 (General TPE). *Given a system trace S, a subject security context ssc and a set of object contexts OSC, the General Trusted Path Execution (GTPE) for ssc and OSC is respected iff ssc is only able to directly or indirectly execute (binary) objects of OSC:*

$$GTPE(T, ssc, OSC) \stackrel{def}{\equiv} \left(\begin{array}{c} \forall osc \in \mathcal{SC}, ssc \in \mathcal{SSC}, \\ (ssc \stackrel{S}{\underset{x}{\Rrightarrow}} osc) \Rightarrow (osc \in OSC) \end{array} \right).$$

The logical contraposition of the previous implication is strictly equivalent to the following formula:

$$\neg(osc \in OSC) \Rightarrow \neg(ssc \stackrel{S}{\underset{x}{\Rrightarrow}} osc)$$

which is another way of expressing (and allowing the enforcement of) the same property: if *osc* is not in the OSC set (i.e. in the list of the trusted paths), *ssc* cannot execute it.

5   Implementation

5.1   Architecture

As depicted in Fig. 3, our implementation for enforcing security properties on GNU/Linux system is divided into two main parts: kernel-space and user-land. When a user-land application requests a system call (edge #1), the system call is hooked by the DAC protection (edge #2). For simplicity for our first prototype, as we need labels for

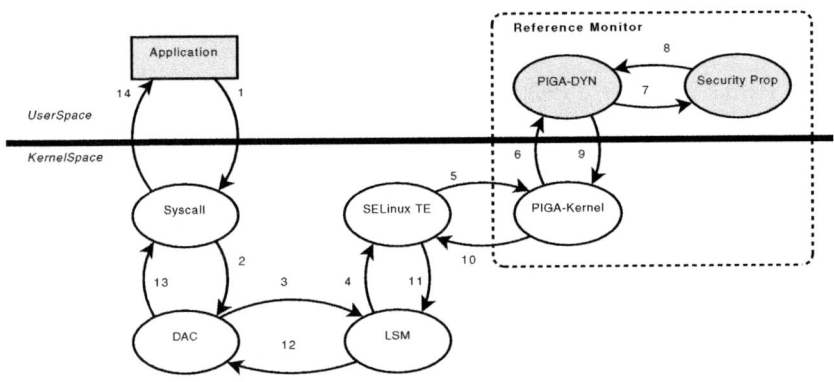

Fig. 3. Architecture of PIGA-DYN Control Model

all entities on the operating system, we used SELinux Type Enforcement that provides such labels. SELinux uses LSM[3] hooks. Thus, LSM is also called when a system call occurs (edge #3). However, LSM is not a security mechanism by itself, it is just a way to hook the system call to implement a security monitor. Accordingly, LSM calls the security monitor, in our case, the SELinux Type Enforcement (edge #4). To plug our solution within the protection mechanisms, we modify the SELinux Type Enforcement to make it call (edge #5) our kernel module: PIGA-kernel. PIGA-kernel is an interface making it possible to plug our MAC monitor PIGA-DYN. PIGA-DYN runs in user-land and listens (edge #6) for new system calls to be approved (or denied). It retrieves the required security properties (edge #7) and computes each system call against them (edge #8). The permission decision is forwarded back to kernel space (edge #9). Then, it is returned to the SELinux Type Enforcement (edge #10) that returns it to LSM (edge #11). LSM sends the decision back to DAC (edge #12)that does a logical "AND" between its decision and the one made by PIGA-DYN. Based on this computation, it allows or denies the system call (edge #13). Finally, the system call is executed (or not) and the corresponding response is sent back to the application that requested it in the user-land (edge #14).

---

[3] Linux Security Modules (LSM) is a Linux kernel framework allowing the support of numerous security models while avoiding favoritism toward any single security implementation.

5.2 Kernel-Space

System Call Hooks. As described in Sect. 3, our model requires two timestamps per system call to manage parallel information flows. However, LSM cannot hook the end of a system call to allow (or deny) the return of the result to the application in user-land. As a consequence, in our testbed prototype, we used the only timestamp provided by LSM: the starting time of the syscalls. Thus, for the moment, the start and end time are assumed to be equal. This will be fixed in the future with a complete implementation. In practice, as ending and starting times of direct information flows are stored in the IFG all flows involving more than one syscall are already properly covered.

PIGA-kernel. The PIGA-kernel module is an interface between the access control in kernel-space and our decision engine PIGA-DYN in user-land. Its sole purpose is to generate a trace representing the corresponding interaction including the entity that makes the system call and its destination but also the type of the system call itself and its timestamp. Once the trace is generated, PIGA-kernel waits for the PIGA-DYN's decision and forwards it back to LSM. The communication between PIGA-DYN and PIGA-kernel is done using a *seq_file*, which is a special *procfs* interface. It is often used to transfer data between kernel and user land, while avoiding many issues like maximum size of the buffer.

That implementation is consistent with the LSM and SELinux approaches. Indeed, our implementation uses LSM. Thus, each system call generates a single trap within the kernel. During that trap execution, DAC permissions are processed first, then SELinux permissions and finally PIGA permissions are processed. Since that current implementation aims at promoting portability and security of the PIGA protections, the decision engine of PIGA runs in user space as a Java application. It is a safe approach since the decision engine is a complex piece of code that cannot be easily integrated within the Linux kernel as a C piece of code. PIGA's decision within the kernel would be very unsafe since it could not take advantage, first, of the safety of the Java language and, secondly, of the SELinux protections for the Java application. The presented performance in Sect. 6.3 show an acceptable overload due to the Java application.

5.3 User-Land

Expressing Security Properties. Each security property described in Sect. 4 is implemented as a function. Each function requires one or more parameters. These parameters are security contexts. Accordingly, the set of required security properties is simply a set of functions with specific parameters. For example, to enforce the data integrity between a user with the *user_t* context and a file with *shadow_t* context, the required function is $GDI(T, user\_t, shadow\_t)$.

As a list of security properties can be enforced on the operating system and they must all be respected at any time, a logical "AND" is done between each item (i.e. security property) on that list. The result of this logical combination is the decision taken by PIGA-DYN for the system call.

For example, the three following security properties, $GTPE(T, user\_t,$ $\{usr\_bin\_t\})$, $GVCHROOT(T, \{user\_t, usr\_bin\_t, user\_home\_t\})$ and

$GDI(T, user\_t, \{usr\_bin\_t\})$, are required. If one of them is broken by a system call, the system call is denied. Using our formal framework, we can create a security property that combines those three security properties into a more general one, e.g. $Restricted\_User(\{user\_t, usr\_bin\_t, user\_home\_t\})$. $Restricted\_User$ property is simply a combination of the three security properties with a logical "AND".

Our framework can also be used to create new custom security properties. Accordingly, an administrator can implement a very large array of security properties just by using our language. For example, an administrator may want to guarantee that the $user\_t$ subject cannot directly or indirectly modify the $shadow\_t$ object at any time except if the modification is done directly through the $passwd\_t$ subject context. Using our formal language, it could be defined as follows:

$$GDI_{custom}\ (T, user\_t, passwd\_t, shadow\_t) \overset{def}{\equiv}$$
$$\left(\begin{array}{c} (user\_t \overset{S}{\rhd}_{[t_{s_i}, t_{e_j}]} shadow\_t) \Rightarrow \\ \left(\begin{array}{c} (user\_t \overset{S}{\rhd}_{[t_{s_i}, t_{e_x}]} passwd\_t) \wedge \\ (passwd\_t \overset{S}{\blacktriangleright}_{[t_{s_y}, t_{e_j}]} shadow\_t) \wedge \\ (t_{s_i} \leq t_{e_j}) \end{array} \right) \end{array} \right) .$$

PIGA-DYN dynamically computes those different operators using the following Information Flow Graph.

Information Flow Graph. PIGA-DYN uses an Information Flow Graph (IFG) to keep the knowledge of previously allowed information flows. Within the IFG, each node represents a single security context and each edge represents a direct (single or multiple) information flow. The temporal relationships between information flows are managed on the IFG through a parameter on each edge. This parameter contains the couple of timestamp that represents the first and the last occurrence of the general direct information flow. Thus, the IFG is able to store the different information flows we have defined in Sect. 3.

Direct Single Information Flow. A direct single information flow, $\blacktriangleright$ (see definition 1), is stored as an edge between the two nodes that represent the related security contexts. The two timestamps contained on the edge are equal and set by the direct single information flow occurrence time.

Direct Multiple Information Flow. As explained in definition 2, the direct multiple information flow, $\blacktriangleright^+$, is composed of several direct single information flows. Thus, as $\blacktriangleright$, $\blacktriangleright^+$ is stored as an edge between two nodes. But, the two timestamps contained on the edge are not equal. The first one contains the starting time of the first occurrence of the first flow. The second one contains the ending time of the last flow.

General Direct Information Flow. As described in definition 3, a general direct information flow , $\rhd$, can be either $\blacktriangleright$ or $\blacktriangleright^+$. Thus, the IFG stores $\rhd$ as an edge between two nodes where the couple of timestamp is defined through the same approach as for $\blacktriangleright$ or $\blacktriangleright^+$.

Indirect Information Flow. $\rhd\!\!\rhd$ are not explicitly stored in the IFG. As introduced in definition 4, it is composed of several general direct information flows. Thus,

$\triangleright\!\!\!\triangleright$ is a combination of several edges representing a path on the IFG. Path computations enable the enumeration of all the indirect flows in the trace providing thus a dynamic computation of the indirect information flows.

General Information Flow. As $\triangleright\!\!\!\triangleright$ is either a $\triangleright$ or a $\triangleright\!\!\!\triangleright$ (see definition 5), the IFG implements it as an edge (direct flow) or a path (indirect flow) between the two related nodes (security contexts).

Extending IFG to store transitions. As the TPE security property requires to manage the transitions, the transition flows are stored as special edges. The computation of indirect transition flows is close to information flows. Moreover, the transition interactions are also added to the IFG as $write\_like$ interactions and thus can be computed as any other type of information flows. Thus, our graph enables us to manage efficiently covert channels associated with the transitions.

Complexity. Obviously, using multiple direct single flows instead of single general direct flows clearly provides an over-approximation of the flows. But this has the great advantage of highly reducing the IFG's size. The number of nodes is theoretically bounded by $n = \mathcal{O}(|\mathcal{SC}|)$, whereas the number of edges is theoretically bounded by $v = \mathcal{O}((|\mathcal{SC}| \times |\mathcal{SSC}| - 1) + (|\mathcal{SSC}| \times |\mathcal{SSC}| - 1))$. This complexity corresponds to the worst case, where each subject exchanges information with any other entity on the system ($|\mathcal{SC}| \times |\mathcal{SSC}| - 1$) and each subject transits to any other subject ($|\mathcal{SSC}| \times |\mathcal{SSC}| - 1$). However impossible in practice, v is really bounded as presented. The graph thus fits easily in memory. For example, with a Gentoo Linux OS, $n < 800$ and $v < 60,000$, occupying less than 128Mo of memory. Thus, the graph can be efficiently processed permitting a real time enforcement of the required properties.

PIGA-DYN. PIGA-DYN is divided into three main steps: construction of the graph, enforcing of the security properties and updating the IFG.

Construction. The purpose is to maintain the IFG that represents all the information flows that previously occurred. It is built by analyzing each trace received from PIGA-kernel. A trace is translated to a direct single information flow ($\blacktriangleright$). Then it modifies the IFG in two different ways:

1. If it is the first occurrence of the information flow, the edge storing $\blacktriangleright$ is added to the graph.
2. If it is not the first occurrence, the last occurrence time of the information flow is updated with the time of the trace. Within our model, a flow such as $\blacktriangleright_{[t_{s_d}, t_{e_d}]}$ leads to having $\triangleright_{[t_{s_g}, t_{e_g}]}$ modified into $\triangleright_{[t_{s_g}, t_{e_d}]}$. The previous ending timestamp is backed up into a temporary variable.

For example, the set of traces in the listing 1.1 is used to build the IFG shown in Fig. 4.a. where:

- timestamp t_{210}: a user's process starts reading a file in the user's home directory;
- t_{212}: the user's process ends reading a file in the user's home;
- t_{214}: the user's process transits to the root context;
- t_{219}: root writes into the /etc/shadow file.

Listing 1.1. "A set of traces representing system interactions"

```
210:  user_t −file : read−> user_home_t
212:  user_t −file : read−> user_home_t
214:  user_t −process : transition −> root
219:  root −file : write−> shadow_t
```

Enforcement. Each security property is abstracted to flows associated to the edges of the IFG. PIGA-DYN is able to enforce a security property by searching for direct and/or indirect paths within the IFG. Accordingly, searching for the occurrence of a flow is similar to searching for a path between two nodes. For example, in Fig. 4.(a), the data integrity between $user\_t$ and $shadow\_t$ can be abstracted to an indirect information flow between $user\_t$ and $shadow\_t$. As a path $user\_t \blacktriangleright^+ root\_t \blacktriangleright^+ shadow\_t$ (i.e. an indirect information flow $user\_t \rhd shadow\_t$) exists in Fig. 4.(a), the property could be broken and PIGA-DYN will refuse the corresponding call (i.e. 219: root −file : write−> shadow_t) as described in the following section.

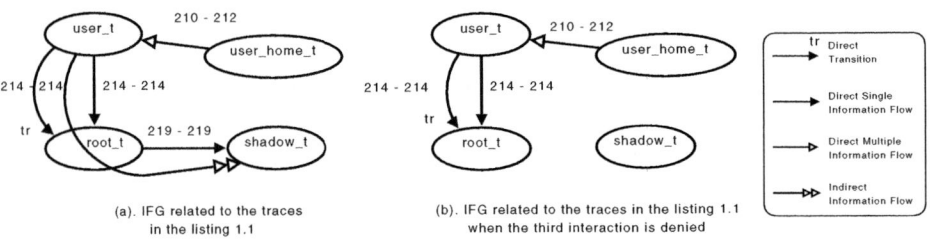

(a). IFG related to the traces
in the listing 1.1

(b). IFG related to the traces in the listing 1.1
when the third interaction is denied

Fig. 4. IFG related to the traces in listing 1.1

Updating the IFG. If a flow breaks a security property, the corresponding system call is denied. Accordingly, the system call does not occur on the system. Thus, the IFG needs to be restored to its previous state. This leads to two update cases. If the edge corresponding to the flow:

1. has the same first and last occurrence timestamps: the flow occurred once. The edge can then be simply deleted from the IFG.
2. has different first and last occurrence timestamps, the flow occurred multiple times. The last occurrence timestamp is then restored to its previous value from a temporary backup variable.

For example, the IFG shown in Fig. 4.(a) is modified into the one in Fig. 4.(b). once the data integrity violation is detected and the corresponding system call is denied (i.e. the interaction with the timestamp 219 is deleted in the IFG). As a consequence, on the target OS, the corresponding syscall is denied.

6   Experiments

For three years, we have been running multiple high-interaction honeypots. One of the purposes of our honeypots is to test different system configurations and see how the attackers react and modify their attacks scenario to fit these different test cases. The following section describes one of these tests.

6.1 Protecting against Information Harvesting through a Small Binary Path

We designed a test situation on our honeypots to see how an attacker reacts when he has only a small set of commands available. To reach this goal, we limited the executable binaries for the attackers to only the ones that are stored in /usr/bin. Basically, these binaries allow the attackers to run classic shell commands like cd but the execution of binaries allowing device interactions like ifconfig or system update like emerge[4] is denied. Moreover, within this test case, the set of executable binaries by the root account is larger as he can execute any binary stored in the /usr/bin, /usr/sbin, /sbin and /bin directories. We advertised this constraint in the SSH server's banner to direct the attackers and we also advertised a privilege escalation vulnerability in the chsh binary[5] stored in /usr/bin.

Objectives of the Experiment. We aimed at preventing the previously explained attack: a malicious user succeeding in collecting information that only privileged users should have access to, by using illegitimately privileged commands (through *root* role). In order to evaluate our protection for a real system use, we wanted to allow at the same time (i.e., on the same machines as the ones used for the protection against the attack) the legitimate root to use those commands. Our test scenario thus consisted of two scenarios: a *legitimate* scenario allowing root to administrate the system using privileged commands, and *another* one called the 'malicious' one where an attacker tries to use root commands. The first scenario L (legitimate) is fairly basic:

Step L.1. An administrator (with the root login) connects to the system through the TTY device.
Step L.2. He executes a python script by using the python interpreter stored in the /usr /bin directory.
Step L.3. He updates the system by launching the emerge command stored in the /sbin directory.

The second scenario M (malicious) contains the attack:

Step M.1. An attacker connects to the system through the SSH server and gets an interactive shell.
Step M.2. He harvests information about the system by launching a python script through the python interpreter stored in the /usr/bin directory.
Step M.3. He tries to harvest information about the system network configuration by launching the /sbin/ ifconfig command.
Generally, when this fails, the attacker tries the next two steps.

---

[4] The emerge command is similar under Gentoo to the apt−get command under Debian.
[5] The chsh command changes the user login shell. It has the setuid flag on and it is owned by root. Thus, the privilege escalation allows the attacker to gain root privileges.

Step M.4. He uses the local privilege escalation contained in /usr/bin/chsh binary to gain root access.

Step M.5. He tries to harvest information about the system network configuration by launching the /sbin/ ifconfig command.

We describe now what happens when those scenarios are run under a classic Linux coming only with DAC protection and then what happens under a Linux coming with our solution PIGA-DYN.

Protection on a Classic GNU/Linux. For an operating system without our solution, the running of the two scenarios above is described in the listing 1.2 for the legitimate scenario L and in the listing 1.3 for the attack scenario M. The binaries are protected against illegal execution through the DAC.

Listing 1.2. "Running of the legitimate steps of the scenario without our solution"

```
1   [root] /usr/bin/python checkUpdate.py
2     Updates are available for the system. [10] Libraries,
        [21] Applications.
3   [root] /sbin/emerge ——update ——deep ——newuse world
4     [100%] System is up–to–date.
```

Legitimate Scenario (cf. listing 1.2):

Step L.1. The first step is allowed as nothing denied the administrator connection to the TTY device and, after entering his login and password, access to an interactive shell.

Step L.2. The second step is allowed too (line #1-2) as root has the right to execute commands in the /usr/bin directory.

Step L.3. The last step is allowed too (line #3-4) as root has the right to execute commands in the /sbin directory.

So, this legitimate scenario works as it is intended to on a classic GNU/Linux system.

Listing 1.3. "Running of the malicious steps of the scenario without our solution"

```
1   [user] /usr/bin/python harvestInfo.py
2     Information harvested has been saved into harv.txt
3   [user] /sbin/ifconfig
4   /sbin/ifconfig: Permission denied.
5   [user] /usr/sbin/chsh "user%24%80;%42/bin/bash"
6   ********* LOCALE LiNUX EXPLOIT ********* (uid=0(root) gid=0(
        root)): You are know root !
7   [root] /sbin/ifconfig
8   eth0      Link encap:Ethernet  HWaddr 00:1a:0c:d6:c4:d9
9             inet addr:172.29.1.35  Bcast:172.29.1.255  Mask
                :255.255.255.0
```

Malicious Scenario (cf. listing 1.3):

Step M.1. The first step is allowed as nothing denied a user connection to the SSH server and access to an interactive shell.

Step M.2. The second step is allowed too (lines #1-2) as a user has the right to execute commands in the /usr/bin directory.

Step M.3. The third step (lines #3-4) is denied as a user does not have the right to execute commands that are stored in a directory other than /usr/bin.

Step M.4. Thus, to try to get network configuration, the attacker launches a local privilege escalation exploit (lines #5-6) on the /usr/bin/chsh command. As the invoked command is stored in the /usr/bin directory, the attacker is allowed to execute it. Thus, the fourth step is allowed and the privilege escalation succeeds. Accordingly, the attacker has now the root privileges.

Step M.5. The last step (lines #7-8) is allowed as root has the right to execute a command in the /sbin directory. Accordingly, the attack succeeds as the attacker has harvested the network configuration.

Protection with PIGA-DYN.

TPE Security properties. In order to prevent against such attacks, but keeping in mind that legitimate administrators should have the right to execute privileged commands, we set up specific security properties above classical DAC protection protecting binaries against illegal executions. Those were two Trusted Path Execution properties.

The first one is defined to allow the admin scenario whereas the second one aims at preventing the malicious one:

1. **For the Legitimate Scenario**: $GTPE(T, root, \{usr\_bin\_t, usr\_sbin\_t, bin\_t, sbin\_t\})$: staff's GTPE allowing the execution of binaries that come from the /usr/bin but also from /usr/sbin, /sbin and /bin. In terms of flows, the security property expresses that *exec_like* interactions coming from $root\_t$ or from an intermediary context (e.g. *ssc*) that is the end of a transition flow between $root\_t$ and *ssc*, only end with $\{usr\_bin\_t, usr\_sbin\_t, bin\_t, sbin\_t\}$ (i.e. $root\_t \blacktriangleright_x$

   $\{usr\_bin\_t, usr\_sbin\_t, bin\_t, sbin\_t\}$, or $root\_t \overset{S}{\underset{tr}{\rhd\!\!\!\rhd}} ssc \blacktriangleright_x \{usr\_bin\_t, usr\_sbin\_t, bin\_t, sbin\_t\}$).

2. **For the Malicious Scenario**: $GTPE(T, user\_t, \{usr\_bin\_t\})$: user's GTPE that allows the attackers to run binaries that only come from the /usr/bin/ directory. In terms of flows, the security property expresses that (indirect) executions coming from $user\_t$ only end with $\{usr\_bin\_t\}$ i.e.

   $user\_t \blacktriangleright_x \{usr\_bin\_t\}$ or $user\_t \overset{S}{\underset{tr}{\rhd\!\!\!\rhd}} ssc \blacktriangleright_x \{usr\_bin\_t\}$.

Enforcement. For an operating system with PIGA-DYN, the running of the scenario is described in the listing 1.4 for the legitimate scenario L and in the listing 1.5 for the attack scenario M. The corresponding information flow graph is displayed in Fig. 5.

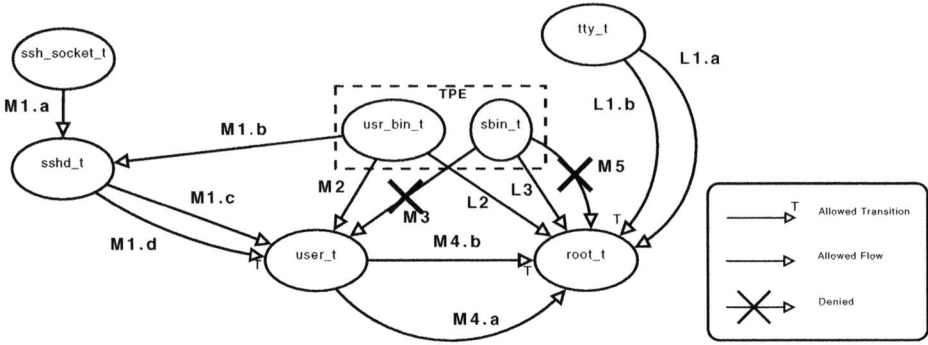

Fig. 5. Information Flow Graph related to the Generic TPE Attack Scenario

Legitimate Scenario (cf. listing 1.4):

   Listing 1.4. "Running of the legitimate steps of the scenario with our solution"

```
1    [root] /usr/bin/python checkUpdate.py
2      Updates are available for the system. [10] Libraries,
         [21] Applications.
3    [root] /sbin/emerge —update —deep —newuse world
4      [100%] System is up-to-date.
```

Step L.1. The first step is allowed as nothing denied the administrator connection to
         the TTY device (edges #L1.a and #L1.b) i.e. tty_t $\blacktriangleright^+$ $root\_t$ and
         tty_t $\underset{tr}{\blacktriangleright}$ $root\_t$, after having entered login and password, access to an in-
         teractive shell (edge #L2) i.e. $root\_t$ $\underset{x}{\blacktriangleright}$ usr_bin_t.
Step L.2. The second step is allowed too (lines #1-2 and edge #B) i.e. $root\_t$ $\underset{x}{\blacktriangleright}$
         usr_bin_t as root has the right to execute commands in the /usr/bin direc-
         tory. Indeed, the second $GTPE$ property allows this step as usr_bin_t is part
         of the set of objects that the subject $root\_t$ is allowed to execute. Moreover,
         there is no transition flow between $user\_t$ and $root\_t$ at this moment.
Step L.3. The last step is allowed too (lines #3-4 and edge #L3) i.e. $root\_t$ $\underset{x}{\blacktriangleright}$ $sbin\_t$
         as root has the right to execute commands in the /sbin directory. Indeed, the
         second $GTPE$ property allows this step as the context is part of the set of
         executable contexts allowed for the $root\_t$ subject. Moreover, there is still no
         transition flow between $user\_t$ and $root\_t$.

Thus, this legitimate scenario works as it is intended to with our solution.
   Malicious Scenario (cf. listing 1.5):

Listing 1.5. "Running of the malicious steps of the scenario with our solution"

```
1   [user] /usr/bin/python harvestInfo.py
2     Information harvested has been saved into harv.txt
3   [user] /sbin/ifconfig
4   /sbin/ifconfig: Permission denied.
5   [user] /usr/sbin/chsh "user%24%80;%42/bin/bash"
6   ********* LOCALE LiNUX EXPLOIT ********* (uid=0(root) gid=0(
        root)): You are know root !
7   [root] /sbin/ifconfig
8   /sbin/ifconfig: Permission denied.
```

Step M.1. The first step is allowed as nothing denied an attacker connection to the SSH server $sshd\_t$ (edge #M1.a) i.e. $ssh\_socket\_t \blacktriangleright^+ sshd\_t$ and to get an interactive shell $user\_t$ (edges #M1.b, #M1.c and #M1.d),

i.e. $sshd\_t \underset{x}{\blacktriangleright} usr\_bin\_t, sshd\_t \blacktriangleright^+ user\_t$ and $sshd\_t \underset{tr}{\blacktriangleright} user\_t$.

Step M.2. The second step is allowed too (lines #1-2 and edge #M2) i.e. $user\_t \underset{x}{\blacktriangleright}$ usr_bin_t. as a user has the right to execute commands in the /usr/bin directory at the DAC level. Moreover, our solution defines a $GTPE$ for the $user\_t$ context that must be applied. The property expresses that $user\_t$ has the right to execute binaries with the usr_bin_t context. Thus, the second step is allowed by our solution too.

Step M.3. The third step (lines #3-4 and edge #M3) i.e. $user\_t \underset{x}{\blacktriangleright} sbin\_t$ is denied as a user does not have the right to execute commands that are stored in a directory other than /usr/bin at the DAC level. Moreover, our solution defines a $GTPE$ for $user\_t$. The property expresses that $user\_t$ does not have the right to execute binaries that do not have the usr_bin_t context. Thus, the third step is denied by our solution too.

Step M.4. To try to get network configuration, the attacker launches a local privilege escalation exploit on the /usr/bin/chsh command (lines #5-6 and edges #M2, #M4.a and #M4.b) i.e.

$user\_t \underset{x}{\blacktriangleright} usr\_bin\_t, user\_t \blacktriangleright^+ root\_t$ and $user\_t \underset{tr}{\blacktriangleright} root\_t$. As the invoked command is stored in /usr/bin directory, the attacker is allowed to execute it at a DAC level. On the other hand, the $GTPE$ enforced for users allows $user\_t$ to execute an object with the usr_bin_t context. Thus, the fourth step is allowed and the privilege escalation succeeds. Accordingly, the attacker has now the root privileges.

Step M.5. The last step (lines #7-8 and edge #M5) is allowed at the DAC level as a user with root privileges has the right to execute a command in the /sbin directory. But, the $GTPE$ property related to $user\_t$ denied such execution. Indeed, the $GTPE$ for $root\_t$ allows it but a transition flow exists between $user\_t$ and $root\_t$ by combining the edge #M4.b and the edge #M5. Thus, the $GTPE$ property for the user also applies as the attacker has passed root through the $user\_t$ context[6]. Accordingly, the last step is denied as the

---

[6] Also, a user gaining root privileges via "su" or "sudo" would face the same denial.

sbin_t context is not part of the user's GTPE and the attack scenario failed as the attacker is not able to harvest network configurations.

We show that our solution blocks the last step of the attack scenario and thus the final objective of the attacker. That cannot be denied by a classic GNU/Linux system. Accordingly, we also show that classic security components like DAC permissions are easily bypassed and are not sufficient to enforce system wide security. Finally, it is worth noting that for the last step, our solution detects a path between the *user_t* context and *root_t* context thus showing that we effectively detect *indirect* attacks.

Our solution does not block the exploit but *the effect* of it on the system i.e. a privilege escalation that allows a user to execute binaries outside of its $GTPE$. Whatever the privilege escalation used, even a legitimate one like entering the valid root password, the $GTPE$ property denies the execution of a binary inside the root's TPE by a subject that has passed through the *user_t* context. Thus, by considering the effect and not the exploit itself, our solution can block previously unknown attacks like 0-Day exploits.

6.2   Completeness

To evaluate the correctness of our approach, we configured our honeypot hosts with PIGA-DYN in both detection and prevention mode. That way, we could verify if every detected attack was prevented or not. As a result, during the six months of experiment, we detected 224 real attacks. Under detection mode, each one generated several security property alarms as shown in Table 1. All attacks were blocked by our protection mechanism. As shown in Table 1, there are more alarms than attacks. Indeed, in detection

Table 1. Number of attacks occurrences per security property

Security Property	Nb of occurrences
$GTPE(T, user\_t, ...)$	1684 (direct: 1427 and indirect: 257)
$GVCHROOT(T, user\_t, ...)$	2735
$GDI(T, user\_t, ...)$	346
$GTPE(T, root\_t, ...)$	12

mode, each attack can generate several alarms since it may break several security properties (even several times for each property). Some of them even generated thousands of alarms for a single attack. There was a large number of $GTPE(user, ...)$ alarms as almost all the attackers download and execute binaries in their home (or in the temporary directory). As we explain in Sect. 6.1, the execution and/or the download of the file is canceled by our solution in protection mode. There is also a large number of $GVCHROOT(user, ...)$ alarms as the attackers tried to harvest information about the system but also tried to interact with local services that was both outside the *user_t* domain. As the $GTPE$ and $GVCHROOT$ properties blocked most of the early steps of the attacks, most of the attackers stopped their attack here. Less than 10% of the attackers tried to modify some binaries, generating thus data integrity alarms. Moreover, in detection mode, only 4 attackers (generating $12\,TPE(root)$ alarms) tried to gain root privileges and used them to execute a binary they downloaded. They were all detected and stopped by our solution in protection mode.

6.3 Performance

In order to evaluate the efficiency of our solution, we used numerous benchmarks to compare three different configurations: 1) a classical Linux system with DAC and SELinux TE; 2) a Linux system with DAC and SELinux TE with our solution PIGA-DYN in detection (IDS) mode for detecting the violation of the required security properties; 3) a Linux system with DAC and SELinux TE with our solution PIGA-DYN in protection (IPS) mode for enforcing the required security properties. Other things (CPU, memory, etc.) were equal: Pentium-4 3Ghz with 1Gb. The tests were performed during the experiments described in Sect. 6.1 in order to obtain relevant results. We use the lmbench [34] suite on the three machines to measure bandwidth and latency. Lmbench attempts to measure performance bottlenecks in a wide range of system applications. These bottlenecks were identified, isolated and reproduced in a set of micro-benchmarks which measure system latency and bandwidth of data movement.

Memory Accesses. First, we focused on the memory subsystem and measured bandwidth with various memory operations. The results are listed in Table 2.

Table 2. Overhead of PIGA-DYN IDS/IPS for memory operations

Operation	Description	IDS	IPS
libc bcopy unaligned	Measuring how fast data blocks are copied when data segments are not aligned with pages using bcopy().	<1%	<1%
libc bcopy aligned	Measuring how fast data blocks are copied when data segments are aligned with pages using bcopy().	<1%	<1%
memory bzero	Measuring how fast memory blocks can be reset using bzero().	<1%	<1%
unroled bcopy unaligned	Measuring how fast data blocks are copied when data segments are not aligned with pages without using bcopy().	<1%	<1%
memory read (> 512Kb)	Measuring time to read x byte word from memory.	<1%	<2%
mem. read (< 512Kb)	Measuring time to read x byte word from memory.	<2%	<3%
mem. write (> 512Kb)	Measuring time to write x byte word to memory.	<1%	<2%
mem. write (< 512Kb)	Measuring time to write x byte word to memory.	2%	3%
mem. r/w (> 512Kb)	Measuring time to read an write x byte word to memory.	2%	3%
mem. r/w (< 512Kb)	Measuring time to read an write x byte word to memory.	4%	5%

The differences between the three different configurations were very small. With data blocks larger than 512Kb, the three configurations have almost the same performance. With data blocks smaller than 512Kb, in most of the cases, the overhead due to our security component is unnoticeable. In the worst case, like memory read-/write, the maximum overhead is about 5%. Consequently, we can state that our security component has little to no influence on data copy to and from the memory.

System Latency. Secondly, we used lmbench to measure latency in five different aspects of the operating system:

1. System call: it measures the time to write one byte to $/dev/null$.
2. Process: it creates four different forms of process and evaluates the time it takes to
   a) invoke a procedure, *b)* fork a process and invoke *execve* system call and *c)* fork
   a process and invoke an interactive shell.
3. Network: it measures the time taken to make a HTTP request ($GET/$) on a LAN
   and a WAN HTTP server.
4. Context switching: it measures context switching time for a small number of processes of a given size (in Kb). The processes are all connected through a ring of
   Unix *Pipes* where each process reads in one, does some work on the read data and
   writes it in the next process's pipe.
5. Filesystem: it measures the time to create and delete files with sizes varying from
   0Kb to 10Kb.
6. We also add an ad-hoc bench for evaluating the reading and writing in files.

Here are the summarized results for those multiple benches that ran lmbench. The

Table 3. Latency average overheads for PIGA-DYN in IDS/IPS modes

Operation	Details	IDS	IPS
1. System call	Writing 1b into $/dev/null$	0%	2%
2. Process creation and deletion	cases *a*, *b* and *c* (see above)	20%	25%
3. Network	WAN/LAN	0.5%	1.5%
4. Context switching	With 2,4,8 processes	1%	1%
	With 16 processes	4%	6%
5. Filesystem: creation/deletion per second	With 0Kb, 1Kb, 4Kb and 10Kb files	10%	11%
6. File read/write	N/A	18%	19%

results show that our approach can sometimes have no impact on performance. But for process forking, we observed some noticeable overheads. However, process creation and deletion is not the most frequently occurring syscall on a system, while context switching, which is one of the main occupations of OS, is weakly impacted by our so-lution. File reading/writing results are more preoccupying as they are massively used on modern OS. That should be addressed in further work, first by optimizing the IFG.

Globally, in prevention mode (i.e. blocking attacks) we had an overhead between 1% and 25%. We think that it is a very reasonable cost compared to the great security improvements it provides. We have to work now on specific situations related to some particularly complex security properties to handle, such as non-interference or general integrity, in order to reduce the impact of our solution.

7   Conclusion

This paper shows that a general canvas is really missing for modeling a wide range of integrity properties. It provides a generic framework that makes it possible to define advanced integrity properties to control direct and indirect flows.

Our framework enables the formalization of the major integrity properties. But, newer security properties can also be easily defined for covering specific protection needs. Those security properties are enforced using an algorithm that dynamically computes an information flow graph related to every system call. The complexity of the graph remains low, thus permitting a real time protection to guarantee the requested properties. Our MAC protection denies a system call if it could break the requested properties.

A large scale experiment manages two concurrent systems. The first one protects against the violations of the requested security properties. The second one uses our approach in Intrusion Detection Mode for evaluating the efficiency of the protection system. During several months of experimentation, our protection system has always prevented all attempts to violate our integrity properties. Further work will then deal with the enforcement of confidentiality. Several novel protection models will be proposed and evaluated in order to provide a large set of predefined canvases that will ease the life of security administrators and the security of end-users.

References

1. Committee on National Security Systems. National Information Assurance Glossary, CNSS Instruction No. 4009, 23 (April 2010)
2. Biba, K.J.: Integrity considerations for secure computer systems, tech. rep., MITRE Corp., 04 (1977)
3. Bell, D., LaPadula, L.: Secure computer systems: Mathematical foundations, tech. rep., Technical Report MTR-2547 (1973)
4. Lee, T.: Using mandatory integrity to enforce 'commercial' security. In: Proceedings of IEEE Symposium on Security and Privacy, pp. 140–146 (April 1988)
5. Ko, C., Redmond, T.: Noninterference and intrusion detection. In: Proceedings of IEEE Symposium on Security and Privacy, pp. 177–187 (2002)
6. Goguen, J., Meseguer, J.: Security policies and security models. In: Proc. 1982 IEEE Symp. Security and Privacy, Oakland, CA, pp. 11–20. IEEE, Los Alamitos (1982)
7. Rahimi, N.A.: Trusted path execution for the linux 2.6 kernel as a linux security module. In: ATEC 2004: Proceedings of the Annual Conference on USENIX Annual Technical Conference, Berkeley, CA, USA, pp. 34–34. USENIX Association (2004)
8. Clark, D.D., Wilson, D.R.: A Comparison of Commercial and Military Computer Security Policies. In: IEEE Symposium on Security and Privacy, pp. 184–194. IEEE Computer Society Press, Los Alamitos (1987)
9. Schneider, F.B.: Enforceable security policies. ACM Trans. Inf. Syst. Secur. 3(1), 30–50 (2000)
10. Roscoe, A.W., Hoare, C.A.R., Bird, R.: The Theory and Practice of Concurrency. Prentice Hall PTR, Upper Saddle River (1997)
11. Clarkson, M.R., Schneider, F.B.: Hyperproperties. In: IEEE 21st Computer Security Foundations Symposium, CSF 2008, pp. 51–65 (June 2008)

12. Bauer, L., Ligatti, J., Walker, D.: More Enforceable Security Policies. Foundations of Computer Security, 95 (2002)

13. Terry, P., Wiseman, S.: A 'new' security policy model. In: Proceedings of IEEE Symposium on Security and Privacy, pp. 215–228 (May 1989)

14. Briffaut, J., Lalande, J.-F., Toinard, C.: Formalization of security properties: enforcement for mac operating systems and verification of dynamic mac policies. International Journal on Advances in Security 2, 325–343 (2009)

15. Zeldovich, N., Boyd-Wickizer, S., Kohler, E., Mazières, D.: Making information flow explicit in histar. In: OSDI 2006: Proceedings of the 7th USENIX Symposium on Operating Systems Design and Implementation, Berkeley, CA, USA, pp. 19–19. USENIX Association (2006)

16. Krohn, M., Yip, A., Brodsky, M., Cliffer, N., Kaashoek, M.F., Kohler, E., Morris, R.: Information flow control for standard os abstractions. SIGOPS Oper. Syst. Rev. 41(6), 321–334 (2007)

17. Efstathopoulos, P., Kohler, E.: Manageable fine-grained information flow. SIGOPS Oper. Syst. Rev. 42(4), 301–313 (2008)

18. TRESYS., Setools–policy analysis tools for selinux (2010)

19. Briffaut, J., Rouzaud-Cornabas, J., Toinard, C., Zemali, Y.: A new approach to enforce the security properties of a clustered high-interaction honeypot. In: Guha, R.K., Spalazzi, L. (eds.) Workshop on Security and High Performance Computing Systems, Leipzig, Germany, June 2009, pp. 184–192. IEEE Computer Society, Los Alamitos (2009)

20. Harrison, M.A., Ruzzo, W.L., Ullman, J.D.: Protection in operating systems. Commun. ACM 19(8), 461–471 (1976)

21. Spencer, R., Smalley, S., Loscocco, P., Hibler, M., Andersen, D., Lepreau, J.: The flask security architecture: system support for diverse security policies. In: SSYM 1999: Proceedings of the 8th Conference on USENIX Security Symposium, Berkeley, CA, USA, pp. 11–11. USENIX Association (1999)

22. Mao, Z., Li, N., Chen, H., Jiang, X.: Trojan horse resistant discretionary access control. In: SACMAT 2009: Proceedings of the 14th ACM Symposium on Access Control Models and Technologies, New York, NY, USA, pp. 237–246. ACM, New York (2009)

23. Liang, H., Sun, Y.: Enforcing mandatory integrity protection in operating system. In: ICCNMC 2001: Proceedings of the 2001 International Conference on Computer Networks and Mobile Computing (ICCNMC 2001), Washington, DC, USA, p. 435. IEEE Computer Society, Los Alamitos (2001)

24. Li, N., Mao, Z., Chen, H.: Usable mandatory integrity protection for operating systems. In: IEEE Symposium on Security and Privacy, SP 2007, pp. 164–178 (May 2007)

25. Sabelfeld, A., Myers, A.: Language-based information-flow security. IEEE Journal on Selected Areas in Communications 21, 5–19 (2003)

26. Mohay, G., Zellers, J.: Kernel and shell based applications integrity assurance. In: Proceedings of 13th Annual Computer Security Applications Conference, pp. 34–43 (December 1997)

27. Iglio, P.: Trustedbox: a kernel-level integrity checker. In: 15th Annual Proceedings of Computer Security Applications Conference (ACSAC 1999), pp. 189–198 (1999)

28. Sailer, R., Zhang, X., Jaeger, T., van Doorn, L.: Design and implementation of a tcg-based integrity measurement architecture. In: SSYM 2004: Proceedings of the 13th Conference on USENIX Security Symposium, Berkeley, CA, USA, pp. 16–16. USENIX Association (2004)

29. Quynh, N.A., Takefuji, Y.: A real-time integrity monitor for xen virtual machine. In: ICNS 2006: Proceedings of the International Conference on Networking and Services, Washington, DC, USA, p. 90. IEEE Computer Society, Los Alamitos (2006)

30. Berger, S., Cáceres, R., Goldman, K.A., Perez, R., Sailer, R., van Doorn, L.: virtualizing the trusted platform module. In: USENIX-SS 2006: Proceedings of the 15th Conference on USENIX Security Symposium, Berkeley, CA, USA, USENIX Association (2006)
31. Xu, M., Jiang, X., Sandhu, R., Zhang, X.: Towards a VMM-based usage control framework for OS kernel integrity protection. In: Proceedings of the 12th ACM Symposium on Access Control Models and Technologies, p. 80. ACM, New York (2007)
32. Rouzaud Cornabas, J., Clemente, P., Toinard, C.: An Information Flow Approach for Preventing Race Conditions: Dynamic Protection of the Linux OS (best paper award). In: Fourth International Conference on Emerging Security Information, Systems and Technologies SECURWARE 2010, Venise Italy (July 2010)
33. Uppuluri, P., Joshi, U., Ray, A.: Preventing race condition attacks on file-systems. In: SAC 2005: Proceedings of the 2005 ACM Symposium on Applied Computing, pp. 346–353. ACM, New York (2005)
34. McVoy, L., Staelin, C.: lmbench: portable tools for performance analysis. In: ATEC 1996: Proceedings of the 1996 Annual Conference on USENIX Annual Technical Conference, Berkeley, CA, USA, pp. 23–23. USENIX Association (1996)

Performance Issues on Integration of Security Services

Fábio Dacêncio Pereira[1] and Edward David Moreno[1,2]

[1] University of São Paulo (USP), Polytechnic School, Sao Paulo - SP, Brazil
[2] Federal University of Sergipe (UFS), DCOMP, Aracaju - SE, Brazil
prof.fabiopereira@gmail.com, edwdavid@gmail.com

Abstract. The integration of security services is an important solution to combat anomalies and attacks on computer systems, assuming that possible difficulties of a security service may be compensated by others. The current works that aim to integrate two or more security services are usually focused on a particular implementation strategy, because the systematic approach to integrated security systems requires the analysis of relations between security data. In our work was proposed and developed a Security Services Integrated Layer (SSIL), consisting of an organization pattern of information security, as well as behavioral models to analyze the occurrence of abnormality identified. The Hidden Markov Model and the proposed solutions as subHMM and Sequential Model allowed the integration of security services based on behavior. In this article we highlight the rates of detection of anomalies and a critical analysis of results.

Keywords: Hidden Markov Model, anomalies detection, behavior models.

1 Introduction

The anonymity, the weakness and other factors often encourage individuals to create malicious tools and attacks techniques on information and computer systems. This can generate from minor inconveniences up to moral and financial damage. Intrusion detection combined with other security tools can protect and prevent malicious attacks and anomalies in computer systems. However, considering the complexity and robustness of such systems, the security services are often not able to examine and audit all information flow, causing defective points of security that can be discovered and exploited [1].

It is inevitable that malicious individuals organize themselves to create attacks more efficient and intelligent. Likewise, should be created integrated security systems, especially knowing the importance of integration of security services to improve the prevention, detection and performance in anomalous situations [4]. Generally, research on security systems have typically focused on creating new services or improving the performance and reliability of a single technique, algorithm or mechanism.

The approaches to integrate two or more security services usually focused on a particular implementation strategy, assuming that the systematic approach to integrated security systems requires the analysis of relations between data.

M.L. Gavrilova et al. (Eds.): Trans. on Comput. Sci. XI, LNCS 6480, pp. 162–178, 2010.
© Springer-Verlag Berlin Heidelberg 2010

The current models for integration of security services determine what the relationship between a specific set of security services, so these can integrate information to prevent or treat deficiencies of the system. However, current models propose solutions on limited set of services, ignoring or disregarding the existence of other [5] [7] [4].

Despite the difficulty in defining a strategy for creating a security services integrated model, there are some works that have satisfactory solutions. The works highlighted in section 2 using techniques for analyzing the behavior of computing systems to distinguish those situations considered anomalous and normal, using data models to format and organize this information.

The work described in this paper shows how solution a Security Services Integrated Layer (SSIL), which may include security services proprietary, open source, applications that want to provide security as necessary or differential.

The SSIL has a common structure capable of containing the security services belonging to a particular computer system. The information stored in the SSIL can be analyzed using behavioral models (section 3) and generate graphs that represent the behavior of different anomalies. These behavioral models can be used effectively to detect attacks early, reduce false positives, classifying the attack power to define the techniques used for defense, among other advantages when has possession of such information.

The objectives of this work are project and develop the SSIL for allowing the integration of security services and for investigating the efficiency and impact of behavioral models used in SSIL specialized for detecting anomalies. Finally, we propose and develop improvements using the special techniques as subHMM and Sequential Model. The paper is organized into 6 sections. In the section 2 are presented the behavioral models used. The characteristics of the SSIL are in section 3. In the section 4 the test environment and validation, while the results are in section 5. Finally, section 6 presents the conclusions.

2   Related Works

The work was developed based on models proposed by RASHEED and CHOW [4], which was adopted IDMEF standard for formatting the anomalous activities reports, and model YASAMI et al. [6] as a behavioral analyzer using Hidden Markov Model (HMM).

2.1 Model RASHEED

The information model for integration of security services proposed by Rasheed and Chow [4] explores three security services: access control, intrusion detection and response to intrusion. In the model, the authors characterized the tasks and responsibilities for each security tool addressed. They concluded that often security tools may be different or mutually exclusive, but it is possible to create integration and cooperation at different levels of integration. Two models of classification and organization of data were highlighted by the authors:

- XACML (eXtensible Access Control Markup Language) is an OASIS standard, and was published in version 2.0 in February 2005. XACML specifies a language for defining policies for access control, as well as requests and responses to access, in XML [13].
- IDMEF (Intrusion Detection Message Exchange Format): The Internet Engineering Task Force's Intrusion Detection Work Group has an experimental RFC to formatting of intrusion detection messages, the IDMEF. The messages containing the specification of alert classes that are sent to a decision-making system, creating an event that combines the response criteria to specific intrusion. These are sent asynchronously and a single message can be more than one anomalous event [8].

The IDMEF model was used in our work as a mechanism for reporting anomalies detected by the security services present on a computer system, assuming that the diversity of applications and security services requires a uniform standard for recording such information, as presented and discussed in the thesis of Pereira [9].

2.2 Model Yasami

Yasami et al. [6] presents an algorithm for anomaly detection based on analysis of ARP requests, using the Hidden Markov Model (HMM). This model can analyze ARP traffic on a computers network to create graphs representing the normal behavior of a computer system. For this, require a training period. According to the authors, the longer the training period, more specific and representative will be the graph of the model.

In this case, there wasn't integration of any security service and the detection system works on low level of the OSI model, between the physical layer and data link. However, the work clearly presents a real application using the HMM model.

The HMM is being used in other research and work, such as [2] and [11] to model the behavior of computing systems, classified operation mode as normal or abnormal. In this paper, the HMM is used to assist in intrusion detection, helping in the classification of attacks levels, among other features.

According Yasami et al [6], the training period is a decisive stage for the creation of the behavioral model. It's generated as a representation of events of a computer system. The normal model built during the training period includes states and transitions that are defined follows:

- *States*: States are to identify the node corresponding to destination IP address of ARP request.
- *Transitions*: Any ARP request produces a transition from one state, and the transition can occur for the same state.

According to the author, these graphs become more complex and robust as the training environment and transitional rules.

In the training period, parameters are calculated for each state as: probability of be in state S; Steady State Duration Average in the state S; Steady State Duration Variance in State S. These parameters, as well as, the graph of transitions comprising the behavioral model of the computer system will be analyzed.

After the training period, this behavioral model is compared with the model generated in real time in order to detect and distinguish the different actions accumulated during the training period. Based on a factor that indicates the percentage of deviant behavior, the system indicates an abnormality or not.

In our work, the HMM was adopted to analyze the behavior of the anomalies reported by security services present on computer system. The equations for comparison and behavior analysis are highlighted in section 3.1.

3 The Security Service Integrated Layer (SSIL)

The main structure of the SSIL has (i) a model for formatting anomalies reported by security services (IDMEF) and (ii) a behavioral model to detect anomalies (HMM). The SSIL is based on the fact that a determined attack on a computer system passes through various stages and tests before this succeed. This attack trajectory can be mapped, i.e., failures can be transformed into models, and eventually can be prevented. The failure in this case refers to the attacks attempts detected by security services and notified to SSIL.

The services of the SSIL are organizing information from security services present on computer system, creating a behavioral model or import custom models of behavioral abnormalities. In a second step, using algorithms to detect anomalies described in this section, examines the behavior of the system, classify and detect abnormalities.

After reviewing the initial results achieved by HMM was proposed and developed a simplified behavioral model, called the Sequential Model, as objective of improving the results obtained by HMM, when it has difficulty in detecting an anomalous sequence. The description of the behavioral models in SSIL is following.

3.1 Hidden Markov Model (HMM)

The HMM can be represented by a graph composed of states (vertices) and transitions (edges). The states of HMM are characterized under the SSIL as unique elements that have three attributes: (i) IP source of the anomaly, (ii) security service and (iii) classification of the security attribute.

The classification of security attributes is predefined as: access control, confidentiality, nonrepudiation, authentication, integrity and availability. Other classifications can be created and customized in the SSIL, as explained in more detail in [9].

The creation of HMM occurs as the detection of anomalies identified by security services in the system. Figure 1 shows the stages of the lifecycle of HMM applied to SSIL.

Note that in Figure 1a is only the start of construction of the behavioral model as the detection three occurrences. Over time, events will be happening and the annotations in the HMM are cumulative (Fig. 1b and 1c). At the end of a training period, the graph of occurrences will be robust enough to identify attacks characterized.

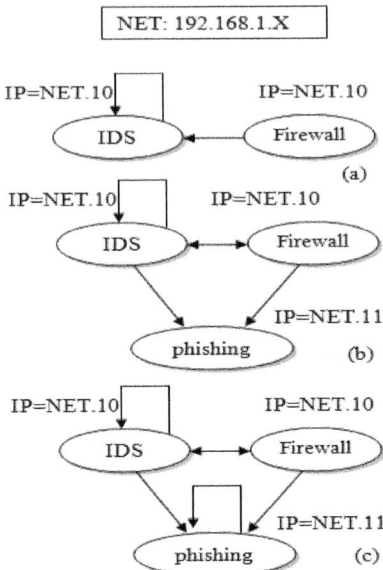

Fig. 1. Example of Creation of the Behavioral Model

The construction of the behavioral model occurs early in system configuration, called the training period. Thus, the behavior graph is created during a training period where the SSIL is stimulated by arrival of attacks that can be artificial or real to creation of the behavioral model.

The SSIL allows continuous training, therefore, the real-time system can feed the HMM with new states and transitions. Thus, the model can adapt at runtime to new attacks or new applications installed in computer system.

According Yasami et al [6], the approach of continuous training in the first instance, seems to be interesting since the system would learn in real time after the training period. However, there are some problems that discouraged this practice, and the most obvious is the distortion of the model based on behavioral models and the inclusion of abnormal or normal in error. The direct effect is the degradation capacity of characterization and detection of anomalous situations.

In this context, the behavioral model does not express the functional characteristics of normality or abnormality, this means that the period of training was insufficient or all possible anomalous events have not occurred during this period. However, the SSIL advantage anomalous events not occurs in the training period to create the subHMM, idealized in this work for treat attacks not known during training period (see section 3.2).

3.1.1 Basic Measures

The basic measures are crucial to characterize the HMM built. Hence, any change in the behavioral model has an impact on all the basic measures, which are in the training period for each state.

The formula 1 determines the probability of being in a particular state E (PE), composed of three terms: IP source, security service and security attribute which recorded the occurrence of an anomaly, after a security event. The calculation of PE is the number of occurrences in the state, divided by the sum of all occurrences of states present in the graph [6].

E = represents a state (vertex) in the HMM graph behavior.

$$P_E = \frac{N_E}{\sum_{i=1}^{k} N_i} \tag{1}$$

Where, N_E = number of occurrences E; N_i = number of occurrences of state i ; K = total number of states in graph behavior

The formula 2 determines the average permanence in specific state E. This formula, although simple, has an important factor: the time. Through this we can determine the behavior of events on Elapsed time. Within the scope of implementation is required counting time between last occurrence and the current instance. In attacks by malicious software, this factor is critical to identify the behavior of attack in the elapsed time.

$$\varepsilon_E = \frac{\sum_{i=1}^{N_E} T_{Ei}}{N_E} \tag{2}$$

The formula 3 determines the population variance after the elapsed time in a specific state E. The variance is significant because it represents the statistical dispersion of an occurrence set, indicating the distances between values typically and expected values.

$$\sigma^2 = \frac{1}{N} \sum_{i=1}^{N} (y_i - \mu)^2 , \tag{3}$$

Where: N = number of terms of population; μ = Population average; y_i = Terms of population.

Finally, it is necessary to calculate the conditional probabilities of transition from a state i to state j. An array of transitions assists in the calculation, and is represented in the formula 4.

The occurrence of a new event has direct impact on the calculation of probability PE, average ε_E, variance and conditional probability of all nodes in graph. Thus, the calculations these four basic measures should be performed for every occurrence and nodes in graph.

$$P_{ij} = \begin{array}{c} \\ S1 \\ S2 \\ S3 \\ \cdots \\ S_{m+1} \\ S_M \end{array} \begin{array}{cccccc} S1 & S2 & S3 & S_{n+1} & \cdots & S_N \\ \begin{bmatrix} P_{11} & P_{12} & P_{13} & P_{1n+1} & \cdots & P_{1N} \\ P_{21} & P_{22} & P_{23} & P_{2n+1} & \cdots & P_{2N} \\ P_{31} & P_{32} & P_{33} & P_{3n+1} & \cdots & P_{3N} \\ \cdots & \cdots & \cdots & \cdots & & \cdots \\ P_{m+11} & P_{m+12} & P_{m+13} & P_{m+1n+1} & \cdots & P_{m+1N} \\ P_{M1} & P_{M2} & P_{M3} & P_{Mn+1} & \cdots & P_{MN} \end{bmatrix} \end{array} \qquad (4)$$

In Formula 4 has the transition matrix that represents the behavior of HMM graph. From matrix can calculate the conditional probabilities of transitions from state i to state j [9].

The basic measures presented are calculated on training period and used at runtime normal. In the next section shows the Anomaly Detection Algorithm (ADA) that uses these basic measures to identify an anomalous situation.

3.1.2 Anomaly Detection Algorithm (ADA)

The anomaly detection algorithms are used to interpret the behavior graph and produce an index that determines the level of anomaly detected. For this the basic measures calculated during the training period are compared with the parameters generated in runtime.

Basically, the algorithm compares the behavioral model generated in the training period with the model created in runtime normal. Therefore, scores are created, and when their values are included in the same interval, characterize a particular abnormality. Initially the equations are presented for calculating the scores of training period (ST) (formula 5) and the scores of the runtime normal period (SE) (formula 6).

$$ST = \frac{PST_j}{P_{E_j}} + \sum_{k=2}^{N} \frac{PST_j}{P_{E_j} P_{ij}} \qquad (5)$$

$$SE = \frac{PSE_j^k}{P_{E_j}} + \sum_{k=2}^{N} \frac{PSE_j^k}{P_{E_j} P_{ij}} \qquad (6)$$

In the formula 6, the variable SE accumulates the PSE (Partial SE) that are partial results of ADA. Likewise, are calculated PST (Partial ST), which are explained later. In these equations has j as state current of behavioral model generated in the period of training after k anomalous events. N is the number of security events reported in SSIL during the process of anomaly detection.

PST_j/PEj and PSE_i^k/P_{Ei} are calculated for the first anomalous event for both the SE and TS. If the first event of anomalous $PSE_i^k = PSE_i^1$. In the formula 5 e 6 has Pij, this variable describes the conditional probability of a transaction of state i to state j, described by transition matrix.

Finally, there is the PST (Partial ST) and PSE (Partial SE), which are calculated for each anomalous event recorded in SSIL using the formula 7 and 8 respectively.

$$PST_j = \frac{(\mathcal{E}_j)^2}{\sigma_j} \qquad (7)$$

$$PSE_j^{\ k} = \frac{(t_j^{\ k})^2}{\sigma_j} \qquad (8)$$

The PSE and PST use the same basic measures of training period. Where t_{jk} found in formula 8 is the time interval between k e $(k+1)$ abnormality occurrence.

After calculating ST and SE, you can compare them, and how much more near the values, greater the indication of anomaly (formula 9). However, must establish a threshold to distinguish the abnormality or normality in system. This parameter is the Threshold (Th). The higher Th, more sensitive is the ADA, and may increase the number of false positives [6].

For initial testing and following the suggestion of [9], was set a value of 15% for this parameter. This means that all events that have 85% of compatibility as the model built in training period, will notify the identification of an anomaly.

$$SE(1-Th) \leq ST \leq SE(1+Th) \qquad (9)$$

Thus, a decision-making system will identify with a percentage of accuracy the occurrence of an anomaly. Setting a value for the threshold can be done by simulation or real testing, calculating the number of false positives generated. In this case, the threshold of 15% was set based on the development and testing in [9].

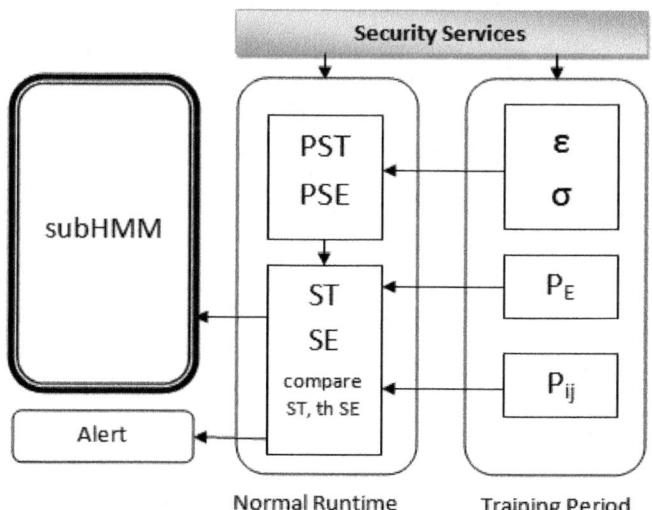

Fig. 2. ADA - Anomaly Detection Algorithm of SSIL

Figure 2 shows the ADA using the equations cited in this section. The basic metrics used in the process of detecting anomalies. At the end of the process are generated alerts in IDMEF format that are treated in SSIL using the subHMM technique, described in next section.

3.2 subHMM Technique

A situation that may occur during the process of anomaly detection is the occurrence of the State X. This problem occurs during normal runtime, when a state or transition unscheduled occurs. In other words, an event is not provided during the training period, and occurs in normal runtime, making this not represented on the initial behavioral model.

In practice, the occurrence of a new event is possible, but it should not be frequent, because this situation means that the behavioral model is not robust enough and, consequently, the period of training is insufficient.

Some solutions can be found: how to assign low levels of probability for the State X, for example, replicating the lowest probability of states in the model behavior. This solution was proposed by YASAMI [6]. However, this does not solve the problem; only mask a weakness of the method.

Another way is to simply ignore in analysis, if an event occurs to state X, which is again a mask for the problem. Figure 3 illustrates the occurrence of State X, where there is a transition from state S2 known, to a state not defined in the behavioral model built during the training period.

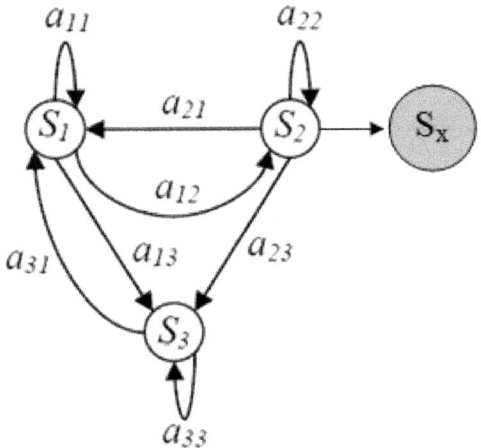

Fig. 3. State X Problem

The SSIL uses strategically the occurrence of the state X. This means that the model can be modified or not in normal runtime, if rules for occurrences of the state X have been created and contemplated in normal runtime. This technique was called of subHMM [9].

In SSIL, when a new sequence is found, alerts are amortized using the method of assigning the state X the lowest probability found in behavioral model. But it is also made the registration and classification of these unexpected events into subHMM.

If a particular occurrence relapse, a new weight is assigned, reclassifying the new sequence detected. Rules can be created to incorporate the new anomalous sequence (subHMM) into model created in training period, for example, after three relapses, the new sequence is inserted into original behavioral HMM model. So the weights assigned to basic measures and the ADA when occurs the State X are:

If Pij, j does not exist (arrival at the State X)

$$Pij = MIN\{P_{ij}\}$$
$$P_{Ej} = MIN\{ P_{Ej} \}$$

If Pij, i do not exist (starting state X)

$$Pij = MIN\{P_{ii}\}$$
$$P_{Ej} = P_{Ej} \text{ (maintains)}$$

Other effect is possible, where the states i and j exist, but the transition between them does not exist.

$$Pij = MIN\{P_{ii}\}$$
$$P_{Ej} = P_{Ej} \text{ (maintains)}$$

For all cases of State X or no transition, the calculation of the PST and PSE is defined by:

$$MIN\{t_j^k\} \text{ and } MAX \{\sigma^2\}$$

3.3 Sequential Model

The use HMM in the SSIL was an effective step to integrate important safety information in order to detect, prevent and act in anomalous situations. This model allows warning on possible anomalous situations, reducing the occurrence of false positives [9].

However, despite the good results generated using the HMM, presented in the next section, the behavior model HMM based has strong dependence on time factor. Therefore, the sequence of security events in a time classifies an anomaly. Hence, the same sequence of events, but distributed differently during the time in the HMM is treated as a separate anomaly, making the training period be exhaustive or endless to predict the possible variations of anomalies.

The proposal to improve this situation was the creation of the Sequential Model. This model neglects the time factor and considers only the anomalous sequence events. Its structure is simple and allows you to generate an additional parameter to better categorize the anomalies alert.

The Sequential Model can be structured through of a graph, where states are the occurrence characteristics and the transitions are the anomalous events detected. The model has two stages of implementation, as well as the HMM. The training period of the sequential model should be combined with the HMM training period.

The entire anomalous sequence events are recorded in graph ASG (Anomalous Sequences Graph). During the training period many ASGs are generated and can be incorporated into new format signature sequences of attacks. Figure 4 illustrates an example of ASG, where the proprietary application notified 10(ten) attempts to access invalid (S1) then the IDS (S2) was able to identify abnormalities and finally Firewall (S3) has been acting. The integers associated with the transitions are number of detected attacks.

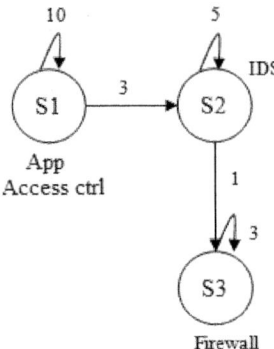

Fig. 4. ASG: Anomalous Sequence Graphs

Signatures are ASGs that cover common attacks and specify the security tools, security attributes and the source IP of occurrences. After the training period, has not only the behavioral model HMM, but also the anomalous sequences graphs.

3.3.1 ADA for Sequential Model
During the normal runtime, all anomalous occurrence recorded is set as new state and transition, creating the so-called Total Sequence Graph (TSG).

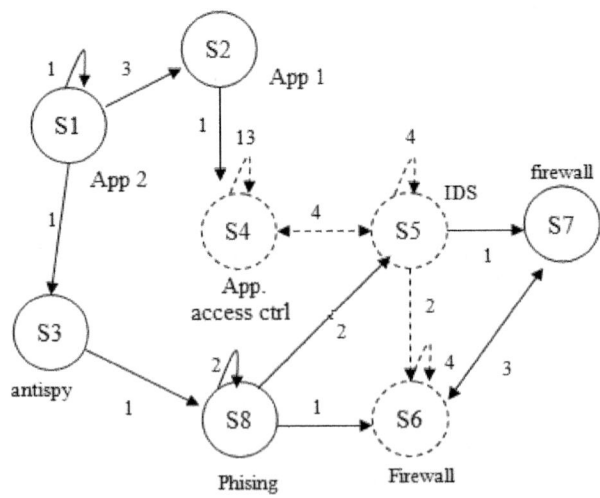

Fig. 5. TSG: Identification of a subgraph in a graph

The ADA aims is finds an ASG in a TSG. If this occurs, the fault is notified. This can be identified search of subgraph in a graph. The complexity of search depth is proportional to number of vertices added, the number of edges of graphs, making it impracticable to search subgraphs with number n of vertices very large.

The creation of the best algorithm is not the focus this work. Thus, we used a depth search to find the subgraphs. In practice, the subgraphs of attacks are not great and with an acceptable computational effort, it is possible identify a subgraph.

The figure 5 shows the identification of a subgraph G (X) (dashed) in a graph G (Y). The anomaly was identified the sample stored in the ASG (portscan) shown in Figure 4.

4   Testing and Validation Environment

This section presents as was designed the anomalies vectors for test and validation in SSIL. Initially we present the types of tests and subsequent construction methods of test vectors.

4.1   Simulator

To assess the efficiency of the SSIL in detecting anomalies created a simulator that includes the routine storage of anomalous occurrences in the IDMEF standard, incorporates behavioral models as HMM, Sequential Model and subHMM, besides generating artificial attacks for the training period and enforcement normal runtime.

4.2   Test Methodology

Both HMM and Model Sequential generate graphs of abnormal behavior of safety occurrences recorded in the SSIL. After the training period, many anomalous routines were recorded, building a representative model of the abnormalities common to computational system in focus.

The routine tests to validate the robustness and efficiency of the generated models are based on variations of anomalous routines implemented in the training period. These variations are classified into four types:

- *Full*: In this case, the same routine tests performed on the training period were performed in normal runtime.
- *Sequence partial*: In this case variations among 10% to 50% of the original sequences were generated for validating the detection of anomalies partially known.
- *Time partial*: The time factor affects the variation among 10% to 50%. The time factor is important to HMM. In the results presented in this paper can be observed that the greatest difficulty lies in detecting routines modified.
- *Time and Sequence partial*: in this case are routines modified by time and sequence, with a rate of 10% to 50%. In this work, these routines are not assumed as unknown by the behavioral models. In section 5, the rates of identification are smaller than those noted with other types of variations.

The parameter described as *threshold*, has direct impact on the sensitivity of detection. The higher the threshold, the algorithm is less sensitive for detecting abnormalities (ADA), generating higher rates of identification. However, it may be more vulnerable to misidentifications.

The determination of the threshold was based on the tests, especially with the implementation of routines of type, *full*, *time partial* and *sequence partial*. In this case, we adopted a threshold of 15%, following the recommendations found in [9].

4.3 Methods for Construction of Test Vectors

Each routine attacks created during the training period consists in two or more anomalous occurrences (events). The anomalous occurrence of a routine are organized in fixed order and separated by time interval between occurrences. In Figure 6, we have the representation this concept.

Fig. 6. Example of organization of routine attack

To generate the tests described above, we should keep the same routines for testing complete; vary time for the routines partly modified by time, vary the sequence for the modified routines partially sequence; vary time and sequence, to create a new routine.

5 Results Analysis

During the training period ten security applications were simulated, 170 anomaly occurrences and 320 variations of routine occurrences. In table 1 have the data of the models built.

Table 1. Results after the Period of training

Number of Sequence Attacks	**320**
Number of States (HMM)	541
Number of Transaction (HMM)	758
Number of Sequences (Sequential Model)	290

The Figure 7 shows the evolution of the HMM model as the performance of routine occurrences. For the data considered, it was observed that the number of states and transitions stabilized after 280 events performed. This means that the model can generate good results in detecting anomalies, because the reported events after this period already are contained in the behavioral model.

But not necessarily all possibilities are contained in the behavioral model created, so the occurrence of unexpected events should be treated case by case as presented in section 3.2.

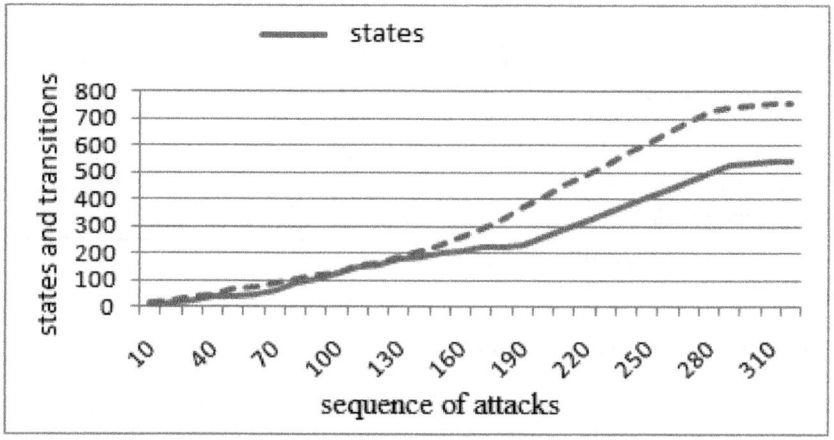

Fig. 7. Evolution of the creation of states and transitions

5.1 Anomaly Detection

The occurrences routines used in the training period and its variations as described in section 4.2 were performed in normal runtime, in order to evaluate the efficiency of SSIL.

Initially, we have the discussion of results for full events (i.e, the same used in training period). For all tests were selected 100 routine occurrences. In this case, the HMM is able to identify an anomalous sequence of more than 95% certainty in 96% of cases.

The time factor is difference between the computation performed during the training and in normal runtime. As the HMM algorithm is strongly coupled to the time factor, this may contribute to the error of 4% found in the detection of full routine occurrences.

However, such occurrences are not identified were marked by the ADA-HMM. By observing the sequential model, it was found that 98% of anomalous routines were detected with 97% average rate of certainty. Thus, two routines have not been identified, but are not the same behavioral model of HMM. In this context, when the HMM did not demonstrate the anomaly, the sequential model was able to identify it.

In Table 2, presents the results after the sequences of events partially and fully modified.

Routines differentiated by time and sequence were created to assess the behavior of the SSIL for attacks not known. However, were kept the same security services and not dealt with the state X.

It is important to stress that this percentage is directly linked to the factor of sensitivity (threshold) of the SSIL, which is set to 15%. The rate of identification varied, depending on the routines of facts not known. And the average was made of 56% detection using HMM and 36% for the sequential model.

Table 2. Results after routine occurrences partially modified

Type of Attacks	Number of attacks	HMM	Sequential Model
Sequence partial	100	PIA=96% CAR=89,8%	PIA=80% CAR=90,1%
Time partial	100	PIA=84% CAR=85,5%	PIA=92% CAR=95,7%
Time and Sequence partial	100	PIA=56% CAR=87,2%	PIA=36% CAR=88,8%

* PIA, percentage of identified attacks.
* CAR, certainty average rating.

5.2   Anomalies Detection with subHMM

In this test in normal runtime, the subHMM module was activated and the results achieved were satisfactory. After ten repetitions of the unknown routine attacks the rate of identifying attacks was from 56% to 91% in average. Table 3 shows this evolution. The rules for incorporated a subHMM in an original HMM were:

- Threshold = 40% higher
- Number of repetition: largest equal 7.

These rules will inhibit the incorporation of anomalous sequences without relationship to model created in training period to avoid a distortion of the original behavioral model.

Table 3. subHMM Results

Attempts	Number of attacks	HMM
1º a 6º	100	PIA=56% CAR= 87,2%
7º	100	PIA=62% CAR=89,1%
8º	100	PIA=74% CAR=89,5%
9º	100	PIA=86% CAR =91,2%
10º	100	PIA=91% CAR= 93,7%

6   Conclusions

This works concludes that there are advantages in become a set of security services in a single integrated system, since the possible fragility of a service can be compensated by others. Despite this necessity, most of the approaches found in literature aim the integration of two or more security tools, but they emphasize generally in a single strategy, because the inherent complexity of integration make to necessary there are a relationship between the data of different security services analyzed.

Thus, this study developed a Security Services Integrated Layer, called SSIL. The SSIL can be defined as an application responsible for storing, organizing and relating anomalous events, reported by security services present on computer system.

For this reason, these events should be reported in the IDMEF format and analyzed by behavioral models as HMM and Sequential Model, in order to detect anomalies as efficient, early and preventive.

Through a simulator of artificial anomalies the behavioral model HMM can be tested and evaluated and the results show satisfactory rates of recognition of anomalous sequences known (98%), partially known (84%) and unknown (56%).

The design of new solutions combined, cited in this work as subHMM and Sequential Model improved the detection of anomalies, as shown in the results achieved, in which attacks partially modified by time the HMM achieved a rate of 84%, but with use of the Sequential Model the rate up to 92%.

When the attack is unknown the HMM hit rates up to 56%, including the solution subHMM, the behavioral model can be incremented during the performance, incorporating new anomalous situation to original model. The unknown attacks were detected at rates of up to 91%

In this paper, it can be seen that the relationship between the security services was built through the behavioral model. Thus, it is possible to relate the different security services even if they have not shared scopes. Finally it is important to note that the IDMEF standard should be promoted, as this will facilitate the desired integration of security services.

References

1. Androulidakis, G., Papavassiliou, S.: Improving network anomaly detection via selective flow-based sampling. Institution of Engineering and Technology (IET) 2(3), 399–409 (2008)
2. Joshi, S.S., Phoha, V.V.: Investigating hidden Markov models capabilities in anomaly detection. In: ACM Southeast Regional Conference Proceedings of the 43rd Annual Southeast Regional Conference (2005)
3. Rabiner, L.R.: A tutorial on Hidden Markov Models and Selected Applications in Speech Recognition. Proc. IEEE 77(2) (1989)
4. Rasheed, H., Chow, Y.C.R.: An Information Model for Security Integration. In: 11th IEEE International Workshop on Future Trends of Distributed Computing Systems (FTDCS 2007), pp. 41–47 (2007)
5. Zilys, M., Valinevicius, A., Eidukas, D.: Optimizing strategic control of integrated security systems. In: 26th International Conference on Information Technology Interfaces (2004)

6. Yasami, Y., Farahmand, M., Zargari, V.: An ARP-based Anomaly Detection Algorithm Using Hidden Markov Model in Enterprise Networks. In: IEEE Second International Conference on Systems and Networks Communications (ICSNC 2007) (2007)
7. Jonsson, E.: Towards an integrated conceptual model of security and dependability, Availability, Reliability and Security, ARES (2006)
8. Debar, H., Curry, D., Feinstein, B.: The intrusion detection message exchange format (2007), http://www.rfc-editor.org/rfc/rfc4765.txt
9. Pereira, F.D.: Approach and Design of SSIL – Security Services Integration Level in SoC and Software (in Portuguese), PhD Thesis, University of Sao Paulo (USP) (2009)
10. Pereira, F.D., Ordonez, E.D.M.: A Hardware Architecture for Integrated-Security Services. In: Gavrilova, M.L., Tan, C.J.K., Moreno, E.D. (eds.) Transactions on Computational Science IV. LNCS, vol. 5430, pp. 100–114. Springer, Heidelberg (2009)
11. Yang, C., Deng, F., Haidong, Y.: An Unsupervised Anomaly Detection Approach using Subtractive Clustering and Hidden Markov Model. In: IEEE International Conference on Communications and Networking in China, CHINACOM (2007)
12. Nissanke, N.: An integrated security model for component-based systems. In: IEEE Conference on Emerging Technologies and Factory Automation, ETFA 2007, pp. 638–645 (2007)
13. Moses, T.: eXtensible Access Control Markup Language(XACML) Version 2.0. OASIS (February 2005)
14. Cappé, O., Moulines, E.: Inference in Hidden Markov Models, Ed. Springer, Heidelberg (2005)
15. Bunke, H., Caelli, T.: Hidden Markov Models: Applications in Computer Vision. World Scientific Publishing, Singapore (2001)
16. Olzoni, D.: Revisiting Anomaly-based Network Intrusion Detection Systems. PhD thesis, University of Twente. CTIT Ph.D.-thesis series No. 09-147 (2009) ISBN 978-90-365-2853-5

Statistical Model Applied to NetFlow for Network Intrusion Detection

André Proto, Leandro A. Alexandre, Maira L. Batista,
Isabela L. Oliveira, and Adriano M. Cansian

UNESP – Universidade Estadual Paulista "Júlio de Mesquita Filho" Cristóvão Colombo Street,
2265, Jd. Nazareth, S. J. do Rio Preto, S. Paulo, Brazil. Departamento de Ciências de
Computação e Estatística, ACME! Computer Security Research Lab.
{andreproto,leandro,maira,isabela,adriano}@acmesecurity.org

Abstract. The computers and network services became presence guaranteed in
several places. These characteristics resulted in the growth of illicit events and
therefore the computers and networks security has become an essential point in
any computing environment. Many methodologies were created to identify
these events; however, with increasing of users and services on the Internet,
many difficulties are found in trying to monitor a large network environment.
This paper proposes a methodology for events detection in large-scale
networks. The proposal approaches the anomaly detection using the NetFlow
protocol, statistical methods and monitoring the environment in a best time for
the application.

Keywords: Security, network, statistical, NetFlow, intrusion detection,
anomaly.

1 Introduction

1.1 Motivation and Objectives

It is possible found many communication software for different purposes on the
Internet: instant messengers, voice and video applications, distributed applications
among others. These applications and the growing Internet users number contribute to
the amount traffic increased in computer networks and the security incidents number
in such environments.

The biggest challenge for network administrators is how to monitor the perimeter
of a large network in a scalable way. Some methodologies and tools were created to
protect a computer connected in a network, such as antivirus, personal firewalls,
antispyware and Intrusion Detection System (IDS) based on signatures. These tools
protect the users of certain attacks types, such as worm propagation, vulnerabilities
exploitation, among others. However, another attacks types, such as Distributed
Denial of Service (DDoS) and brute force attacks on user passwords [1], may involve
not only a computer but also several of them (groups of computers on the Internet can
attack others computers or network infra-structures). These attacks are not detectable

M.L. Gavrilova et al. (Eds.): Trans. on Comput. Sci. XI, LNCS 6480, pp. 179–191, 2010.

by these kinds of tools. Moreover, they use the payload packet analysis to detect events. This approach for a large-scale network perimeter requires large computational resources and may disturb the network traffic.

This paper proposes a new methodology for detection attacks on large-scale networks using detection by traffic anomaly, the NetFlow protocol (IPFIX standard) [2][3] and statistical techniques. The aim is to detect anomalies in certain traffic of network services (such as web, FTP, SSH, telnet, among others), using few computational resources and without disturb the network traffic.

1.2   Related Works

There are some methodologies for perimeter defense in computer networks. The Snort [4], a tool with large acceptance in community, uses the signature detection technique to identify network events. This kind of detection uses as a basic concept the exact description of an attack behavior (it is called attack signature). This tool can be installed in devices such as firewalls and gateways, identifying events in network environments with some dozens of connected devices. However, in larger environments, this tool may have performance problems, because its methodology analyzes each data packet passed by such devices. The large number of devices in these environments results in a large amount of data packets.

The work of [5] discusses a new methodology for event detection in computer networks using the NetFlow protocol and storage information on a relational database. SQL queries are applied to the storage data to identify some events in networks including attacks. The work only proposes a new architecture for storage information, leaving to further works to research for more robust intrusion detection methods combined with this architecture. The storage architecture is used in this paper to support the process of event detection network.

The work of [6] proposes an IDS that uses NetFlow protocol to detect attacks such as DDoS and worm propagation. The paper compares the flows content searching for similarities between the flows that represents the environment at specific moment and flows classified as attacks. It also proposes countermeasure techniques to these attacks (access control rules in routes or firewalls for example), but the methodology used in [6] has problems with false positives in its detection.

The authors of [7] propose an anomaly detection methodology to identify worm traffics in computer networks. The methodology consists to characterize the normal and worm traffic. It analyzes the behavior of network, summarizing the number of connections in its subnets and calculating the standard deviation of traffic behavior. The work uses NetFlow data storage in a relational database and shows that the anomaly detection using NetFlow information is promised and has good results. However, the work is limited to detect worm behavior and has problems with performance and the time of attack detection.

Finally, the work in [8] proposes a robust monitoring system based on NetFlow data analysis. It storages NetFlow information on a relational database (Oracle) and uses statistical methods to detect anomalous events. Two algorithms are applied: one of them is based on variance similarity and the other is based on Euclidian distance. The system has a web interface to support the events monitoring. The results are promising, but the algorithms applied are complexes. Developers require advanced knowledge to implement it.

2 General Concepts

2.1 Attacks and Anomaly Detection

According to [9], an attack is any action in order to subvert at least one of the pillars of information security: confidentiality, authenticity, integrity or availability of a computer system. Several techniques to subvert systems are currently widespread on the Internet and they are accessible by anyone.

Statistically, any network perimeter has a determinate traffic pattern based on user behavior belonging in the network. Some kinds of attacks result in a considerable traffic quantity in a network, showing an anomaly when compared to a normal pattern behavior. This difference in traffic can be detected with intrusion detection methodologies by anomaly. Their fundamental concepts are based in a normal traffic pattern and identification of variances on network traffic, showing an anomaly.

Some related anomaly events used in this paper are cited follow:

- **Scan:** This is usually the first phase of an attack. The scan or network mapping is used to recognize available services in one or more networks, identifying vulnerable services and hosts;
- **DDoS (Distributed Denial of Service):** When a computer group attempt to make a host or network resource unavailable to its users;
- **Worms propagation:** Worm is a malicious program able to spread automatically through networks, sending copies of itself to another hosts [1]. Usually the worm explore vulnerabilities in programs used by user;
- **Dictionary attack:** This attach uses the technique of "attempt and error" to guess user passwords in services like SSH (Secure Shell Client) or Web. Usually this attack has success, because many users use weak and easy deduction passwords;
- **Spamming:** Spams are unwanted e-mails sending by peoples whose intention is spread malicious code, products advertisement or inadequate content [1].

The attacks cited previously have as common features the communication with several computers or servers in a short time, resulting in large network traffic. The approach in this paper uses this specific feature to detect the kinds of attacks cited previously.

2.2 Network Flows

The *Cisco Systems* defines a network flow as a unidirectional sequence of packets between source and destination hosts. The NetFlow provides a summarization of information about the router or switch traffic. Network flows are highly granulated; they are identified by source and destination IP address, as well as by ports number of transport layer. To identify uniquely a flow, the NetFlow also uses the fields "*Protocol type*" and "*Type of Service*" (*ToS*) from IP header and the input logical interface of router or switch. The flows kept in the router/switch cache are exported to a collector in the following situations: remains idle more than 15 seconds; its duration exceeds 30 minutes; a TCP connection is finished with the flag FIN or RST; the flows table is full or the administrator resets the flow configurations. It is important note the maximum time that one flow is kept in the cache device before be exported is 30 minutes.

The Fig. 1 shows the NetFlow v5 protocol fields, as well as your header. The fields that really important to this paper are described in "*Flow Record Format*". They are responsible to represent the information summarized in a connection/session between two hosts, describing source and destination address, source and destination port, input and output interface in the router or switch, number of packets and octets involved in connection, flow creation timestamp and last update timestamp (fields first and last), TCP flags, and others.

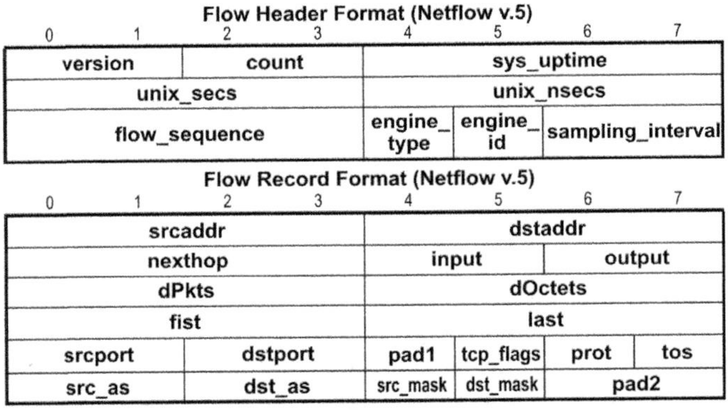

Flow Header Format (Netflow v.5)

0	1	2	3	4	5	6	7
version		count		sys_uptime			
unix_secs				unix_nsecs			
flow_sequence				engine_type	engine_id	sampling_interval	

Flow Record Format (Netflow v.5)

0	1	2	3	4	5	6	7
srcaddr				dstaddr			
nexthop				input		output	
dPkts				dOctets			
fist				last			
srcport		dstport		pad1	tcp_flags	prot	tos
src_as		dst_as		src_mask	dst_mask	pad2	

Fig. 1. NetFlow datagram format

2.3 Statistical Concepts

For this paper some descriptive statistics concepts [10] are used to identify an anomaly in the computers network traffic. The descriptive statistic is used to describe and summarize one data set. The main elements used in this paper are:

- **Median:** The element that separates the higher half of an ordered sample from the lower half is called median. In some cases where the sample has even numbers (don't have a unique central element), two central elements are added and divided by 2, resulting in the median.
- **Quartile:** It is one of three values that divide the ordered sample in four equal parts. Each one represents exactly one quarter of data sample. The second quartile, for example, represents the median. This paper uses the first and third quartile.
- **Outliers:** They are elements of sample that are distant from the rest of sample. The outliers represent anomalies in a network data sample.

These concepts are fundamental to understand the proposal of this paper. The next section describes the methodology in this project and its main features.

3   Methodology

As described previously, the aim in this project is identify security events in a computers network using the NetFlow protocol and detection methodologies by anomaly based on statistical techniques. Thus, the following items should be defined:

- Collection and storage of data provided by NetFlow;
- Model to define the network traffic pattern;
- Outliers detection model.

3.1   Data Architecture Storage

The collection and storage of NetFlow data are very important to provide information about the network that will be defended. We used the storage architecture proposed in [5], enabling robustness and versatility in data storage and SQL queries on data provided by NetFlow.   The architecture enables the flows storage in a relational database. A special table stores a window with the last thirty minutes of flows generated by environment. This table is essential to monitoring of the network traffic in the shortest time possible. More details about the architecture are described in [5].

3.2   Defining the Pattern Traffic in a Network

All detection by anomaly needs the traffic pattern definition of a computers network. This definition should be based on the behavior of each environment at a specific time. For example, the network traffic at daytime has not the same behavior than the same network at night. So, the network traffic behavior is different in certain times of the day. The Fig. 2 shows the network traffic behavior in one day.

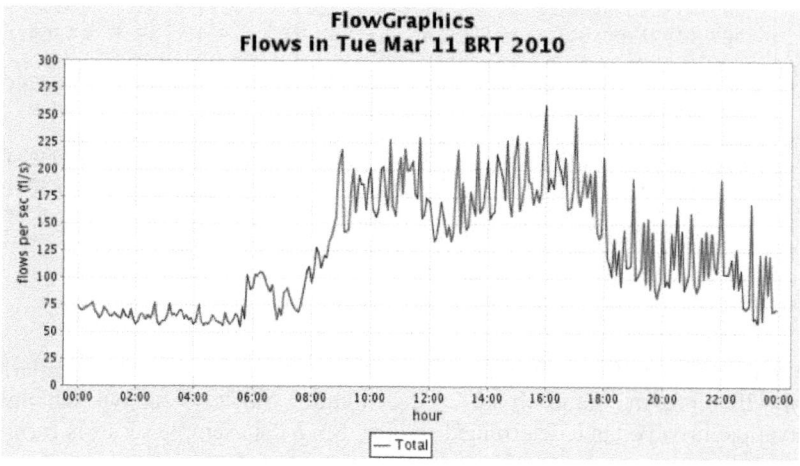

Fig. 2. Example of network traffic behavior

Especially in this project it is necessary to define all network traffic as well as each services used in it. For perform this task, data traffic were collected for three months. For each day in this interval, we selected a time windows of five minutes of flows. This means for each time window were computed the flows average per second related for each service in the network environment. For this, we used a query in SQL language [11] executed in database. The query is described below:

```
SELECT date_sub(subtime(first,second(first)), interval
mod(minute(first),5) minute) as Time, dstport as
Service, input, count(*) as Flows, sum(dPkts) as Pkts,
sum(dOctets) as Bytes FROM TableDay WHERE dstport <
1024 GROUP BY month(first), day(first), hour(first),
(minute(first) div 5), dstport, input ORDER BY Time,
dstport, input;
```

The query result can be seen in Fig. 3. It shows flows averages of services running on TCP/UDP ports less than 1024. This ports interval was selected because includes the most network services used in a computational environment. This data were stored in a new table, in order to be used again in the second step.

Time	Service	input	Flows	Pkts	Bytes
2009-01-01 00:00:00	0	28	24	484	188311
2009-01-01 00:00:00	21	28	2	3	144
2009-01-01 00:00:00	22	28	2	6	304
2009-01-01 00:00:00	25	28	35	285	103863
2009-01-01 00:00:00	53	28	633	665	47678
2009-01-01 00:00:00	80	28	48	688	50636
2009-01-01 00:00:00	123	28	65	170	12920
2009-01-01 00:00:00	137	28	75	75	16575
2009-01-01 00:00:00	143	28	4	22	1364
...	...	...	...	...	...
2009-01-01 23:55:00	445	28	10	14	560
2009-01-01 23:55:00	496	28	95	100	4800
2009-01-01 23:55:00	500	28	2	2	1352
2009-01-01 23:55:00	769	28	117	128	9928
2009-01-01 23:55:00	771	28	72	141	12288

5282 rows in set (3.62 sec)

Fig. 3. SQL query result

With the data collected is possible to define the network traffic pattern. This pattern will represent, for each set of five minutes traffic, a sample containing the flow average involved in a determined service. So, a data sample series is formed, and each sample represents a traffic service in determined time (considering the five minutes set). The Fig. 4 shows an example of data sample collected concerning a network service.

Average flow per second of service X.

Date	00:00	00:05	00:10	00:15	00:35		···	23:45	23:50	23:55
05/31	10	3	3	3	8			21	10	10
05/30	40	33	63	13	10			28	17	8
05/29	1	13	6	7	9			25	18	13
···										
03/03	9	6	3	3	7			20	20	15
03/02	5	8	3	10	2			35	15	10
03/01	10	4	7	6	7			20	15	12

Time

Fig. 4. Example of data sample of a network service

A question that should be considered is how to remove possible anomalous traffics in the three months of data collected to pattern creation. This is resolved using the same outlier identification formula discussed in the next subsection.

3.3 Outlier Identification Model

According to [10], a data sample can be divided in five summarization points: the minimum, the maximum, the median sample and the 25th and the 75th data empirical percentage. The 25th empirical percentage is called 1st quartile (Q_1); the 75th empirical percentage is called 3rd quartile (Q_3). The distance between Q_3 and Q_1 is called interquartile range (IQE) and can be viewed in (1). The author in [10] define the minimum point in a sample is 1.5*IQR faraway of Q_1(2) and the maximum point is 1.5*IQR faraway of Q_3 (3). Any element in the sample that is outside of the minimums and maximums limits is considered an outlier [10].

$$IQR = Q_3 - Q_1 \tag{1}$$

$$Min = Q_1 - 1.5\ IQR \tag{2}$$

$$Max = Q_3 + 1.5\ IQR \tag{3}$$

For this paper only the maximum point of a sample is used. Considering the data samples described in the previously subsection, the elements of a determinate sample whose the flows quantity exceed the maximum value are considered anomalous.

The methodology cited in the previously paragraph is used as base to: identify anomalies in the traffic and; remove the outliers in the collected sample during three months of flows. Thus, the following algorithm to remove outliers was used:

```
Do {
  exist_outlier = 0;
  Calculate Q1 and Q3 of the sample X;
  Do MAX = Q3 + 1.5*(Q3-Q1);
  Scroll all element of the sample X {
    If element > MAX
    Remove element;
    Do exist_outlier = 1;
  }
} While exist_outlier = 1;
```

The algorithm cited calculates the maximum point and remove the outliers until this points no longer exists. The algorithm is executed for each sample collected of each network service. For each outlier removed should be calculate Q1 and Q3 again, because the sample will have a smaller quantity of points in relation of previously iteration and, therefore, a new MAX value should be calculate. When the iteration doesn't find any outliers, the algorithm is finished.

Finished the algorithm, the MAX value will be used as threshold to define the anomaly of a traffic. So, a system monitors the flow average of last five minutes of the current date and each collected values is compared with the MAX value corresponding to its service. For example, assuming the SSH service data was collected in the interval between 10h00min and 10h05min. The flow average per second in this interval will be compares with the sample MAX value corresponding in the same interval. If higher than MAX value, it will be possible conclude that there is an anomaly in the SSH traffic in this time interval.

4   Results

This section describes tests and results obtained with the new anomalies detection model proposed in this paper. The subsection 4.1 describes the environment used for the tests. The subsection 4.2 describes which services were monitored, their threshold obtained after the calculation of pattern traffic and the outliers removal. The subsection 4.3 describes the obtained results. Finally, the subsection 4.4 describes the system performance analyze.

4.1   Environment of Tests

The tests environment is located at a university with more than thousand network devices, including computers, routers, and mobile devices. The environment has a CISCO 7200 VXR router that exports NetFlow flows version 5. The collector computer is a PC x86 Pentium D 3.4 GHz, 2 GB of RAM and 200GB SATA HD, dedicated to collect, store and analyze the flows in database. To define the traffic pattern on environment, flow samples were collected in a period of three months (March to May of 2009). The tests environment is represented in Fig. 5.

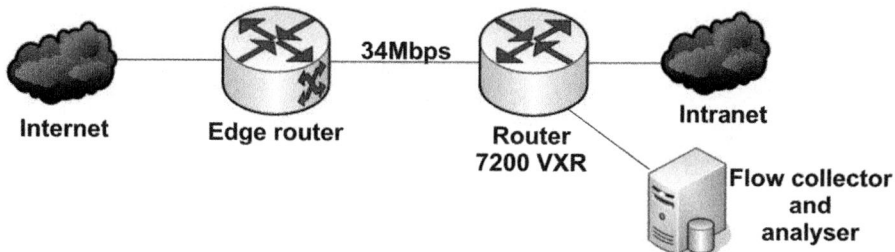

Fig. 5. Institute's network structure which the proposed system was executed

4.2 Monitored Services

For this paper, four services in particular were monitored by the proposed system:

- FTP (File Transfer Protocol): protocol to transfer files;
- SSH (Secure Shell): protocol for remote access;
- SMTP (Simple Mail Transfer Protocol): protocol for mail service;
- HTTP (Hypertext Transfer Protocol): protocol for web services;

The maximum points calculated on the traffic pattern definition phase are showed in Fig. 6 (FTP and HTTP) and Fig. 7 (SMTP and SSH). These graphics already show the services traffic threshold representation, excluding the outliers of the sample.

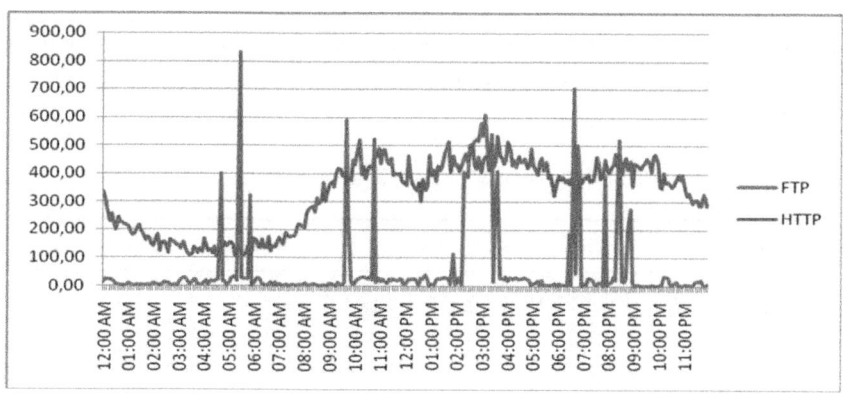

Fig. 6. Top limits of HTTP and FTP services by time

In Fig. 6 and Fig. 7 we note on the HTTP and SMTP service a large traffic amount during the day (7:00AM to 6:00PM), different from that presented during the night (6:00PM to 7:00AM). This difference is explained by the user's number in activity on the network during the cited periods. However, what draws more attention is the FTP traffic representation, which has points in time whose its values are too high if compared to other values. This occurs because, based on the sample collected, the attacks number to the FTP service on these points has been so frequent that they were

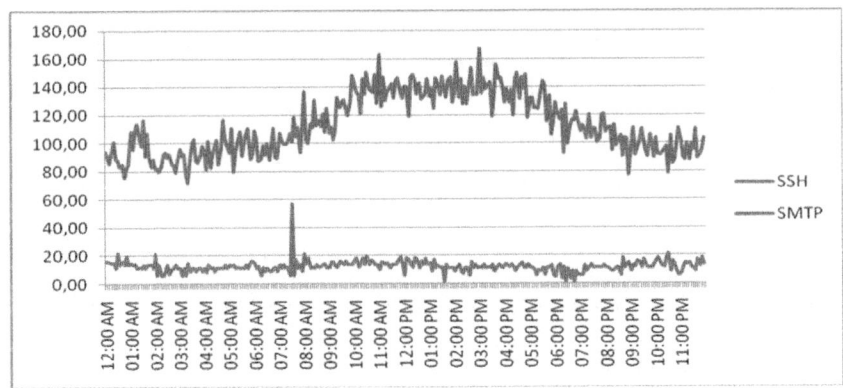

Fig. 7. Top limits of SSH and SMTP services by time

not considered outliers and therefore they were not removed. Finally, the SSH service has just one threshold value excessive if compared to the others by the same reasons presented before in FTP.

4.3   Detection Results

The detection system monitored the proposed environment for a period of six days, which is equivalent to analyze 1728 periods of five minutes each one. The table 1 shows the number of detected events in the proposed services and the number of false positives for them. The calculation of false positives was based in log system analysis on the servers under attack or using the techniques described in [5] for the proof or not of each detected event.

Table 1. Number of anomalous events detected

Service	Events detected	False-positives	Percentage
FTP	10	0	100%
SSH	615	2	99,67%
SMTP	153	4	97,38%
HTTP	33	14	57,57%

For the FTP service, ten events were detected and no false positive was found. All these attacks are characterized as a service scan, whereupon the attacker wants to know if in the network there is an FTP server running.

The events detected in SSH service are quite numerous and only two false-positive were found. These attacks, in general, are characterized by "dictionary attacks" or scans on the service. The large number of events can be explained by the large number of attackers using this intrusion technique, beyond the fact that "dictionary attack" is a kind of attack that usually are performed for several hours, even days. Curiously, one of this false-positive found is explained because researchers from the institute used the service several times in an unconventional time (between 1:00AM and 3:00AM) for servers maintenance.

About the SMTP service, 153 events were detected, with only four false-positives. The attack mode in this service can be divided between scans and spamming. To identify spams, we were used techniques to verify if the computer that sent a lot of mails is mentioned by DNS domain [12] to which it belongs and whether it has a SMTP service running all the time (characteristics of a legitimate mail provider). Among the 149 events detected and verified, 5 are scans, 139 are spams and 5 are both.

Finally, the events detection in the HTTP service showed the highest number of false-positives. Of the 33 events detected, 14 were false-positive, a total of almost 57% of correct detections. The scan was the only kind of attack detected in this service. Two factors explain this result: the HTTP behavior service varies with high frequency between days or in a year; the use of HTTP version 1.0 [13] implies the generation of one connection for each object requested, resulting in a large flows number (because each connection is represented by a flow). In fact, the traffic amount on the HTTP service measured few months ago is very different from the current month, because the popularity of such service and the growing number of users who have been using this service.

4.4 System Performance

In order to obtain the system performance, the training period (sampling to define the traffic pattern) and the environment monitoring (events detection) were computed. The time for sample collection and traffic pattern calculation is about 74 minutes (considering the 92 days of data). It means, on average, about 48 seconds to calculate a day. It is important say this collection occurs only once and the results are stored in the database. The query responsible for monitoring the flows amount is performed in about 2.57 seconds to analyze almost 1024 services (transport layer ports). With the sum of time required to do a query with the processing time to identify anomalies, the average is approximately 3.72 seconds. At certain moments which there was a large number of requests in database (others applications that are using the database at same time), the processing average time for anomalies identification was approximately 25.13 seconds.

5 Conclusion and Future Works

This paper presented a proposal for event detection in computer networks using statistical methods and analysis of NetFlow data flows. The aim is to use this proposal to monitor a computer network perimeter, detecting attacks in the shortest time possible through anomalies identification in traffic and alerting the administrator when necessary. The project proposes to model the traffic pattern of one or more network services through traffic samples based on time intervals. An algorithm removes the outliers of a sample. Using the same algorithm, the maximum values to separate normal and anomalous traffic are defined. The interval used in the tests to summarize data flows was five minutes, but it can be implemented with shorter intervals.

Tests were performed with the monitoring system to four services widely used by users on the Internet: FTP, SSH, SMTP and HTTP. Among them, the FTP, SSH and SMTP had positive results related to events detection and the false-positives number. However, the HTTP service had the highest false-positives number, explained by the service characteristics and scalability on environment. Despite the presented tests are related with only four services types, this model can be applied to any other service that the administrator wants.

System performance was satisfactory, especially because it run in a feasible time to monitory the environment. This measure reflects the low computational cost for the traffic analysis of a large-scale network. It may also be noted that most execution time are performed with queries in the database, involving in and out operations data on hard disk. Therefore, a possible solution to further increase system performance is to migrate this database to another database manager, considered more effective.

The counting of false-negatives number was the most difficult task on this study. This happens because to measure this quantization is necessary to perform other methods or using tools in the environment and compare the results, but the administrative restrictions on environment does not allow this tests.

Finally as future work, the traffic sampling should be improved, so it can be update over time, solving problems cited such as HTTP service. Other works may compare new outliers detection techniques in order to improve the detection performance.

Acknowledgements. The authors acknowledge Jorge Luiz Corrêa by important discussions about the theme in this paper and the support granted by CNPq and FAPESP to the INCT-SEC (National Institute of Science and Technology – Critical Embedded Systems – Brazil), processes 573963/2008-8 and 08/57870-9.

References

1. Cole, E., Krutz, R., Conley, J.R.: Network Security Bible, 2nd edn. Wiley Publishing Inc., Indianapolis (2009)
2. Quittek, J., Zseby, T., Claise, B., Zender, S.: RFC 3917: Requirements for IP Flow Information Export: IPFIX, Hawthorn Victoria (2004), http://www.ietf.org/rfc/rfc3917.txt
3. Claise, B.: RFC 3954: Cisco Systems NetFlow Services Export Version 9 (2004), http://www.ietf.org/rfc/rfc3954.txt
4. Sourcefire, Snort.org, http://www.snort.org
5. Corrêa, J. L., Proto, A., Cansian, A. M.: Modelo de armazenamento de fluxos de rede para análises de tráfego e de segurança. In: VIII Simpósio Brasileiro em Segurança da Informação e de Sistemas Computacionais (SBSeg), Gramado (2008)
6. Zhenqi, W., Xinyu, W.: NetFlow Based Intrusion Detection System. In: International Conference on Multimedia and Information Technology, Phuket (2008)
7. Roche, V.P., Arronategui, U.: Behavioural Characterization for Network Anomaly Detection. In: Gavrilova, M.L., Tan, C.J.K., Moreno, E.D. (eds.) Transactions on Computational Science IV. LNCS, vol. 5430, pp. 23–40. Springer, Heidelberg (2009)

8. Bin, L., Chuang, L., Jian, Q., Jianping, H., Ungsunan, P.: A NetFlow based flow analysis and monitoring system in enterprise networks. Computer Networks 5, 1074–1092 (2008)
9. Gollmann, D.: Computer Security, 1st edn. John Wiley & Sons, New York (1999)
10. Dekking, F.M., Kraaikamp, C., Lopuhaä, H.P., Meester, L.E.: A modern introduction to probability and statistics, 1st edn., pp. 234–244. Springer, Heidelberg (2005)
11. Elmasri, R.E., Navathe, S.: Fundamentals of Database Systems, 4th edn. Addison-Wesley, Reading (2005)
12. Mockapetris, P.: RFC 1034: Domain Names – Concepts and Facilities (1987), http://www.ietf.org/rfc/rfc1034.txt
13. Kurose, J.F., Ross, K.W.: Computer Networking: A Top-Down Approach, 5th edn., pp. 65–80. Addison-Wesley, Reading (2003)

J-PAKE: Authenticated Key Exchange without PKI

Feng Hao[1] and Peter Ryan[2]

[1] Thales E-Security, Cambridge, UK
[2] Faculty Of Science, University of Luxembourg

Abstract. Password Authenticated Key Exchange (PAKE) is one of the important topics in cryptography. It aims to address a practical security problem: how to establish secure communication between two parties solely based on a shared password without requiring a Public Key Infrastructure (PKI). After more than a decade of extensive research in this field, there have been several PAKE protocols available. The EKE and SPEKE schemes are perhaps the two most notable examples. Both techniques are however patented. In this paper, we review these techniques in detail and summarize various theoretical and practical weaknesses. In addition, we present a new PAKE solution called J-PAKE. Our strategy is to depend on well-established primitives such as the Zero-Knowledge Proof (ZKP). So far, almost all of the past solutions have avoided using ZKP for the concern on efficiency. We demonstrate how to effectively integrate the ZKP into the protocol design and meanwhile achieve good efficiency. Our protocol has comparable computational efficiency to the EKE and SPEKE schemes with clear advantages on security.

Keywords: Password-Authenticated Key Exchange, EKE, SPEKE, key agreement.

1 Introduction

Nowadays, the use of passwords is ubiquitous. From on-line banking to accessing personal emails, the username/password paradigm is by far the most commonly used authentication mechanism. Alternative authentication factors, including tokens and biometrics, require additional hardware, which is often considered too expensive for an application.

However, the security of a password is limited by its low-entropy. Typically, even a carefully chosen password only has about 20-30 bits entropy [3]. This makes passwords subject to dictionary attacks or simple exhaustive search. Some systems willfully force users to remember cryptographically strong passwords, but that often creates more problems than it solves [3].

Since passwords are weak secrets, they must be protected during transmission. Currently, the widely deployed method is to send passwords through SSL/TLS [29]. But, this requires a Public Key Infrastructure (PKI) in place; maintaining a PKI is expensive. In addition, using SSL/TLS is subject to man-in-the-middle

M.L. Gavrilova et al. (Eds.): Trans. on Comput. Sci. XI, LNCS 6480, pp. 192–206, 2010.

attacks [3]. If a user authenticates himself to a phishing website by disclosing his password, the password will be stolen even though the session is fully encrypted.

The PAKE research explores an alternative approach to protect passwords without relying on a Public Key Infrastructure (PKI) at all [10, 16]. It aims to achieve two goals. First, it allows zero-knowledge proof of the password. One can prove the knowledge of the password without revealing it to the other party. Second, it performs authenticated key exchange. If the password is correct, both parties will be able to establish a common session key that no one else can compute.

The first milestone in PAKE research came in 1992 when Bellovin and Merrit introduced the Encrypted Key Exchange (EKE) protocol [10]. Despite some reported weaknesses [16, 20, 23, 25], the EKE protocol first demonstrated that the PAKE problem was at least solvable. Since then, a number of protocols have been proposed. Many of them are simply variants of EKE, instantiating the "symmetric cipher" in various ways [7].

The few techniques that claim to resist known attacks have almost all been patented. Most notably, EKE was patented by Lucent Technologies [12], SPEKE by Phoenix Technologies [18] and SRP by Stanford University [28]. The patent issue is arguably one of the biggest brakes in deploying a PAKE solution in practice [13].

2   Past Work

2.1   Security Requirements

Before reviewing past solutions in detail, we summarize the security requirements that a PAKE protocol shall fulfill (also see [10, 11, 16, 28]).

1. **Off-line dictionary attack resistance** – It does not leak any information that allows a passive/active attacker to perform off-line exhaustive search of the password.
2. **Forward secrecy** – It produces session keys that remain secure even when the password is later disclosed.
3. **Known-session security** – It prevents a disclosed session from affecting the security of other established session keys.
4. **On-line dictionary attack resistance** – It limits an active attacker to test only one password per protocol execution.

First, a PAKE protocol must resist off-line dictionary attacks. An attacker may be passive (only eavesdropping) or active (directly engaging in the key exchange). In either case, the communication must not reveal any data – say a hash of the password – that allows an attacker to learn the password through off-line exhaustive search.

Second, the protocol must be forward-secure. The key exchange is authenticated based on a shared password. However, there is no guarantee on the long-term secrecy of the password. A well-designed PAKE scheme should protect past session keys even when the password is later disclosed. This property also implies that if an attacker knows the password but only passively observes the key exchange, he cannot learn the session key.

Third, the protocol must provide known session security. If an attacker is able to compromise a session, we assume he can learn all session-specific secrets. However, the impact should be minimized such that a compromised session must not affect the security of other established sessions.

Finally, the protocol must resist on-line dictionary attacks. If the attacker is directly engaging in the key exchange, there is no way to prevent such an attacker trying a random guess of the password. However, a secure PAKE scheme should mitigate the effect of the on-line attack to the minimum – in the best case, the attacker can only guess exactly one password per impersonation attempt. Consecutively failed attempts can be easily detected and thwarted accordingly.

Some papers add an extra "server compromise resistance" requirement: an attacker should not be able to impersonate users to a server after he has stolen the password verification files stored on that server, but has not performed dictionary attacks to recover the passwords [7, 17, 28]. Protocols designed with this additional requirement are known as the augmented PAKE, as opposed to the balanced PAKE that does not have this requirement.

However, the so-called "server compromise resistance" is disputable [24]. First, one may ask whether the threat of impersonating users to a *compromised* server is significantly realistic. After all, the server had been compromised and the stored password files had been stolen. Second, none of the augmented schemes can provide any real assurance once the server is indeed compromised. If the password verification files are stolen, off-line exhaustive search attacks are inevitable. All passwords will need to be revoked and updated anyway.

Another argument in favor of the augmented PAKE is that the server does not store a plaintext password so it is more secure than the balanced PAKE [28]. This is a misunderstanding. The EKE and SPEKE protocols are two examples of the balanced PAKE. Though the original EKE and SPEKE papers only mention the use the plaintext password as the shared secret between the client and server [10, 16], it is trivial to use a hash of the password (possibly with some salt) as the shared secret if needed. So, the augmented PAKE has no advantage in this aspect.

Overall, the claimed advantages of an augmented PAKE over a balanced one are doubtful. On the other hand, the disadvantages are notable. With the added "server compromise resistance" requirement that none of the augmented PAKE schemes truly satisfy [7, 17, 28], an augmented PAKE protocol is significantly more complex and more computationally expensive. The extra complexity opens more opportunities to the attacker, as many of the attacks are applicable on the augmented PAKE [7].

2.2   Review on EKE and SPEKE

In this section, we review the two perhaps most well-known balanced PAKE protocols: EKE [10] and SPEKE [16]. Both techniques are patented and have been deployed in commercial applications.

There are many other PAKE protocols in the past literature [7]. Due to the space constraint, we can only briefly highlight some of them. Goldreich and Lindell first provided a formal analysis of PAKE, and they also presented a PAKE protocol that satisfies the formal definitions [33]. However, the Goldreich-Lindell protocol is based on generic multi-party secure computation; it is commonly seen as too inefficient for practical use [34, 35]. Later, there are Abdalla-Pointcheval [1], Katz-Ostrovsky-Yung [34], Jiang-Gong [35] and Gennaro-Lindell [39] protocols, which are proven secure in a common reference model (Abdalla-Pointcheval additionally assumes a random oracle model [1]). All these protocols require a "trusted third party" to define the public parameters: more specifically, the security of the protocol relies on the "independence" of two group generators selected honestly by a trusted third party [1, 34, 35][1]. Thus, as with any "trusted third party", the party becomes the one who can break the protocol security [3]. (Recall that the very goal of PAKE is to establish key exchange between two parties without depending on any external trusted party.) Another well-known provably secure PAKE is a variant of the EKE protocol with formal security proofs due to Bellare, Pointcheval and Rogaway [5] (though the proofs are disputed in [7, 32], as we will explain later). In general, all the above protocols [33, 34, 35, 1, 39, 5] are significantly more complex and less efficient than the EKE and SPEKE protocols. In this paper, we will focus on comparing our technique to the EKE and SPEKE protocols.

First, let us look at the EKE. Bellovin and Merrit introduced two EKE constructs: based on RSA (which was later shown insecure [23]) and Diffie-Hellman (DH). Here, we only describe the latter, which modifies a basic DH protocol by symmetrically encrypting the exchanged items. Let α be a primitive root modulo p. In the protocol, Alice sends to Bob $[\alpha^{x_a}]_s$, where x_a is taken randomly from $[1, p-1]$ and $[\ldots]_s$ denotes a symmetric cipher using the password s as the key. Similarly, Bob sends to Alice $[\alpha^{x_b}]_s$, where $x_b \in_R [1, p-1]$. Finally, Alice and Bob compute a common key $K = \alpha^{x_a \cdot x_b}$. More details can be found in [10].

It has been shown that a straightforward implementation of the above protocol is insecure [20]. Since the password is too weak to be used as a normal encryption key, the content within the symmetric cipher must be strictly random. But, for a 1024-bit number modulo p, not every bit is random. Hence, a passive attacker can rule out candidate passwords by applying them to decipher $[\alpha^{x_a}]_s$, and then checking whether the results fall within $[p, 2^{1024} - 1]$.

There are suggested countermeasures. In [10], Bellovin and Merrit recommended to transmit $[\alpha^{x_a} + r \cdot p]_s$ instead of $[\alpha^{x_a}]_s$ in the actual implementation, where $r \cdot p$ is added using a non-modular operation. The details on defining r can

---

[1] The Jiang-Gong paper proposes to use a trusted third party or a threshold scheme to define the public parameters [35], while the KOY paper suggests to use a trusted third party or a source of randomness [34]. However, neither paper provides concrete descriptions of the "threshold scheme" and "source of randomness". The Gennaro-Lindell paper suggests to choose a large organization as the trusted party for all its employees [39]. However, such a setup also severely limits the general deployment of PAKE among the public.

be found in [10]. However, this solution was explained in an ad-hoc way, and it involves changing the existing protocol specification. Due to lack of a complete description of the final protocol, it is difficult to assess its security. Alternatively, Jaspan suggests addressing this issue by choosing p as close to a power of 2 as possible [20]. This might alleviate the issue, but does not resolve it.

The above reported weakness in EKE suggests that formal security proofs are unlikely without introducing new assumptions. Bellare, Pointcheval and Rogaway introduced a formal model based on an "ideal cipher" [5]. They applied this model to formally prove that EKE is "provably secure". However, this result is disputed in [7,32]. The so-called "ideal cipher" was not concretely defined in [5]; it was only later clarified by Boyd et al. in [7]: the assumed cipher works like a random function in encryption, but must map fixed-size strings to elements of G in decryption (also see [32]). Clearly, no such ciphers are readily available yet. Several proposed instantiations of such an "ideal cipher" were easily broken [32].

Another limitation with the EKE protocol is that it does not securely accommodate short exponents. The protocol definition requires α^{x_a} and α^{x_b} be uniformly distributed over the whole group $\mathbb{Z}_p^*$ [10]. Therefore, the secret keys x_a and x_b must be randomly chosen from $[1, p-1]$, and consequently, an EKE must use 1024-bit exponents if the modulus p is chosen 1024-bit. An EKE cannot operate in groups with distinct features, such as a subgroup with prime order – a passive attacker would then be able to trivially uncover the password by checking the order of the decrypted item.

Jablon proposed a different protocol, called Simple Password Exponential Key Exchange (SPEKE), by replacing a fixed generator in the basic Diffie-Hellman protocol with a password-derived variable [16]. In the description of a fully constrained SPEKE, the protocol defines a safe prime $p = 2q + 1$, where q is also a prime. Alice sends to Bob $(s^2)^{x_a}$ where s is the shared password and $x_a \in_R [1, q-1]$; similarly, Bob sends to Alice $(s^2)^{x_b}$ where $x_b \in_R [1, q-1]$. Finally, Alice and Bob compute $K = s^{2 \cdot x_a \cdot x_b}$. The squaring operation on s is to make the protocol work within a subgroup of prime order q.

There are however risks of using a password-derived variable as the base, as pointed out by Zhang [31]. Since some passwords are exponentially equivalent, an active attacker may exploit that equivalence to test multiple passwords in one go. This problem is particularly serious if a password is a Personal Identification Numbers (PIN). One countermeasure might be to hash the password before squaring, but that does not resolve the problem. Hashed passwords are still confined to a pre-defined small range. There is no guarantee that an attacker is unable to formulate exponential relationships among hashed passwords; existing hash functions were not designed for that purpose. Hence, at least in theory, this reported weakness disapproves the original claim in [16] that a SPEKE only permits one guess of password in one attempt.

Similar to the case with an EKE, a fully constrained SPEKE uses long exponents. For a 1024-bit modulus p, the key space is within $[1, q-1]$, where q is 1023-bit. In [16], Jablon suggested to use 160-bit short exponents in a SPEKE, by choosing x_a and x_b within a dramatically smaller range $[1, 2^{160} - 1]$. But, this would give a passive attacker side information that the $1023 - 160 = 863$ most significant bits in a full-length key are all '0's. The security is not reassuring, as the author later acknowledged in [19].

To sum up, an EKE has the drawback of leaking partial information about the password to a passive attacker. As for a SPEKE, it has the problem that an active attacker may test multiple passwords in one protocol execution. Furthermore, neither protocol accommodates short exponents securely. Finally, neither protocol has security proofs; to prove the security would require introducing new security assumptions [5] or relaxing security requirements [26].

3   J-PAKE Protocol

In this section, we present a new balanced PAKE protocol called Password Authenticated Key Exchange by Juggling (J-PAKE). The key exchange is carried out over an unsecured network. In such a network, there is no secrecy in communication, so transmitting a message is essentially no different from broadcasting it to all. Worse, the broadcast is unauthenticated. An attacker can intercept a message, change it at will, and then relay the modified message to the intended recipient.

It is perhaps surprising that we are still able to establish a private and authenticated channel in such a hostile environment solely based on a shared password – in other words, bootstrapping a *high-entropy* cryptographic key from a *low-entropy* secret. The protocol works as follows.

Let G denote a subgroup of $\mathbb{Z}_p^*$ with prime order q in which the Decision Diffie-Hellman problem (DDH) is intractable [6]. Let g be a generator in G. The two communicating parties, Alice and Bob, both agree on (G, g). Let s be their shared password[2], and $s \neq 0$ for any non-empty password. We assume the value of s falls within $[1, q - 1]$.

Alice selects two secret values x_1 and x_2 at random: $x_1 \in_R [0, q - 1]$ and $x_2 \in_R [1, q - 1]$. Similarly, Bob selects $x_3 \in_R [0, q - 1]$ and $x_4 \in_R [1, q - 1]$. Note that $x_2, x_4 \neq 0$; the reason will be evident in security analysis.

Round 1. *Alice sends out g^{x_1}, g^{x_2} and knowledge proofs for x_1 and x_2. Similarly, Bob sends out g^{x_3}, g^{x_4} and knowledge proofs for x_3 and x_4.*

The above communication can be completed in one round as neither party depends on the other. When this round finishes, Alice and Bob verify the received knowledge proofs, and also check $g^{x_2}, g^{x_4} \neq 1$.

Round 2. *Alice sends out $\mathcal{A} = g^{(x_1+x_3+x_4) \cdot x_2 \cdot s}$ and a knowledge proof for $x_2 \cdot s$. Similarly, Bob sends out $\mathcal{B} = g^{(x_1+x_2+x_3) \cdot x_4 \cdot s}$ and a knowledge proof for $x_4 \cdot s$.*

When this round finishes, Alice computes $K = (\mathcal{B}/g^{x_2 \cdot x_4 \cdot s})^{x_2} = g^{(x_1+x_3) \cdot x_2 \cdot x_4 \cdot s}$, and Bob computes $K = (\mathcal{A}/g^{x_2 \cdot x_4 \cdot s})^{x_4} = g^{(x_1+x_3) \cdot x_2 \cdot x_4 \cdot s}$. With the same keying material K, a session key can be derived $\kappa = H(K)$, where H is a hash function.

---

[2] Depending on the application, s could also be a hash of the shared password together with some salt.

The two-round J-PAKE protocol can serve as a drop-in replacement for face-to-face key exchange. It is like Alice and Bob meet in person and secretly agree a common key. So far, the authentication is implicit: Alice believes only Bob can derive the same key and vice versa. In some applications, Alice and Bob may want to perform an explicit key confirmation just to make sure the other party actually holds the same key.

There are several ways to achieve explicit key confirmation. In general, it is desirable to use a different key from the session key for key confirmation[3], say use $\kappa' = H(K, 1)$. We summarize a few methods, which are generically applicable to all key exchange schemes. A simple method is to use a hash function similar to the proposal in SPEKE: Alice sends $H(H(\kappa'))$ to Bob and Bob replies with $H(\kappa')$. Another straightforward way is to use κ' to encrypt a known value (or random challenge) as presented in EKE. Other approaches make use of MAC functions as suggested in [36]. Given that the underlying functions are secure, these methods do not differ significantly in security.

In the protocol, senders need to produce valid knowledge proofs. The necessity of the knowledge proofs is motivated by Anderson-Needham's sixth principle in designing secure protocols [2]: "*Do not assume that a message you receive has a particular form (such as g^r for known r) unless you can check this.*" Fortunately, Zero-Knowledge Proof (ZKP) is a well-established primitive in cryptography; it allows one to prove his knowledge of a discrete logarithm without revealing it [29].

As one example, we could use Schnorr's signature [30], which is non-interactive, and reveals nothing except the one bit information: "whether the signer knows the discrete logarithm". Let H be a secure hash function[4]. To prove the knowledge of the exponent for $X = g^x$, one sends $\{\text{SignerID}, V = g^v, r = v - xh\}$ where SignerID is the unique user identifier, $v \in_R \mathbb{Z}_q$ and $h = H(g, V, X, \text{SignerID})$. The receiver verifies that X lies in the prime-order subgroup G and that g^v equals $g^r X^h$. Adding the unique SignerID into the hash function is to prevent Alice replaying Bob's signature back to Bob and vice versa. Note that for Schnorr's signature, it takes one exponentiation to generate it and two to verify it (computing $g^r \cdot X^h$ requires roughly one exponentiation using the simultaneous computation technique [37]).

4   Security Analysis

In this section, we show the protocol fulfills all the security requirements listed in Section 2.1.

---

[3] Using a different key has a (subtle) theoretical advantage that the session key will remain indistinguishable from random even after the key confirmation. However, this does not make much difference in practical security and is not adopted in [10, 16].

[4] Schnorr's signature is provably secure in the random oracle model, which requires a secure hash function.

4.1 Off-Line Dictionary Attack Resistance

First, we discuss the protocol's resistance against the off-line dictionary attack. Without loss of generality, assume Alice is honest. Her ciphertext $\mathcal{A}$ contains the term $(x_1 + x_3 + x_4)$ on the exponent. Let $x_a = x_1 + x_3 + x_4$. The following lemma shows the security property of x_a.

Lemma 1. *The x_a is a secret of random value over $\mathbb{Z}_q$ to Bob.*

Proof. The value x_1 is uniformly distributed over $\mathbb{Z}_q$ and unknown to Bob. The knowledge proofs required in the protocol show that Bob knows x_3 and x_4. By definition x_a is computed from x_3 and x_4 (known to Bob) plus a random number x_1. Therefore x_a must be randomly distributed over $\mathbb{Z}_q$.

In the second round of the protocol, Alice sends $\mathcal{A} = g_a^{x_2 \cdot s}$ to Bob, where $g_a = g^{x_1+x_3+x_4}$. Here, g_a serves as a generator. As the group G has prime order, any non-identity element is a generator [29]. So Alice can explicitly check $g_a \neq 1$ to ensure it is a generator. In fact, Lemma 1 shows that $x_1 + x_3 + x_4$ is random over $\mathbb{Z}_q$ even in the face of active attacks. Hence, $g_a \neq 1$ is implicitly guaranteed by the probability. The chance of $g_a = 1$ is extremely minuscule – on the order of 2^{-160} for 160-bit q. Symmetrically, the same argument applies to the Bob's case. For the same reason, it is implicitly guaranteed by probability that $x_1 + x_3 \neq 0$, hence $K = g^{(x_1+x_3) \cdot x_2 \cdot x_4 \cdot s} \neq 1$ holds with an exceedingly overwhelming probability.

Theorem 2 (Off-line dictionary attack resistance against active attacks). *Under the Decision Diffie-Hellman (DDH) assumption, provided that $g^{x_1+x_3+x_4}$ is a generator, Bob cannot distinguish Alice's ciphertext $\mathcal{A} = g^{(x_1+x_3+x_4) \cdot x_2 \cdot s}$ from a random non-identity element in the group G.*

Proof. Suppose Alice is communicating to an attacker (Bob) who does not know the password. The data available to the attacker include g^{x_1}, g^{x_2}, $\mathcal{A} = g_a^{(x_1+x_3+x_4) \cdot x_2 \cdot s}$ and Zero Knowledge Proofs (ZKP) for the respective exponents. The ZKP only reveals one bit: whether the sender knows the discrete logarithm[5]. Given that $g^{x_1+x_3+x_4}$ is a generator, we have $x_1 + x_3 + x_4 \neq 0$. From Lemma 1, $x_1 + x_3 + x_4$ is a random value over $\mathbb{Z}_q$. So, $x_1 + x_3 + x_4 \in_R [1, q-1]$, unknown to Bob. By protocol definition, $x_2 \in_R [1, q-1]$ and $s \in [1, q-1]$, hence $x_2 \cdot s \in_R [1, q-1]$, unknown to Bob. Based on the Decision Diffie-Hellman assumption [29], Bob cannot distinguish $\mathcal{A}$ from a random non-identity element in the group. □

The above theorem indicates that if Alice is talking directly to an attacker, she does not reveal any useful information about the password. Based on the protocol symmetry, the above results can be easily adapted from Alice's perspective – Alice cannot compute $(x_1 + x_2 + x_3)$, nor distinguish $\mathcal{B}$ from a random element

---

[5] It should be noted that if we choose Schnorr's signature to realize ZKPs, we implicitly assume a random oracle (i.e., a secure hash function), since Schnorr's signature is provably secure under the random oracle model [30].

in the group. However, the off-line dictionary attack resistance against an active attacker does not necessarily imply resistance against a passive attacker (in the former case, the two passwords are different, while in the latter, they are the same). Therefore, we need the following theorem to show if Alice is talking to authentic Bob, there is no information leakage on the password too.

Theorem 3 (Off-line dictionary attack resistance against passive attacks). *Under the DDH assumption, given that $g^{x_1+x_3+x_4}$ and $g^{x_1+x_2+x_3}$ are generators, the ciphertexts $\mathcal{A} = g^{(x_1+x_3+x_4)\cdot x_2 \cdot s}$ and $\mathcal{B} = g^{(x_1+x_2+x_3)\cdot x_4 \cdot s}$ do not leak any information for password verification.*

Proof. Suppose Alice is talking to authentic Bob who knows the password. We need to show a passive attacker cannot learn any password information by correlating the two users' ciphertexts. Theorem 2 states that Bob cannot distinguish $\mathcal{A}$ from a random value in G. This implies that even Bob cannot computationally correlate $\mathcal{A}$ to $\mathcal{B}$ (which he can compute). Of course, a passive attacker cannot correlate $\mathcal{A}$ to $\mathcal{B}$. Therefore, to a passive attacker, $\mathcal{A}$ and $\mathcal{B}$ are two random and independent values in G; they do not leak any useful information for password verification.   □

4.2   Forward Secrecy

Next, we discuss the forward secrecy. In the following theorem, we consider a passive attacker who knows the password secret s. As we explained earlier, the ZKPs in the protocol require Alice and Bob know the values of x_1 and x_3 respectively, hence $x_1 + x_3 \neq 0$ (thus $K \neq 1$) holds with an exceedingly overwhelming probability even in the face of active attacks.

Theorem 4 (Forward secrecy). *Under the Square Computational Diffie-Hellman (SCDH) assumption[6], given that $K \neq 1$, the past session keys derived from the protocol remain incomputable even when the secret s is later disclosed.*

Proof. After knowing s, the passive attacker wants to compute $\kappa = H(K)$ given inputs: $\{g^{x_1}, g^{x_2}, g^{x_3}, g^{x_4}, g^{(x_1+x_3+x_4)\cdot x_2}, g^{(x_1+x_2+x_3)\cdot x_4}\}$.

Assume the attacker is able to compute $K = g^{(x_1+x_3)\cdot x_2 \cdot x_4}$ from those inputs. For simplicity, let $x_5 = x_1 + x_3 \mod q$. Since $K \neq 1$, we have $x_5 \neq 0$. The attacker behaves like an oracle – given the ordered inputs $\{g^{x_2}, g^{x_4}, g^{x_5}, g^{(x_5+x_4)\cdot x_2}, g^{(x_5+x_2)\cdot x_4}\}$, it returns $g^{x_5\cdot x_2\cdot x_4}$. This oracle can be used to solve the SCDH problem as follows. For g^x where $x \in_R [1, q-1]$, we query the oracle by supplying $\{g^{-x+a}, g^{-x+b}, g^x, g^{b\cdot(-x+a)}, g^{a\cdot(-x+b)}\}$, where a, b are arbitrary values chosen from $\mathbb{Z}_q$, and obtain $f(g^x) = g^{(-x+a)\cdot(-x+b)\cdot x} = g^{x^3 - (a+b)\cdot x^2 + ab\cdot x}$. In this way, we can also obtain:

$$f(g^{x+1}) = g^{(x+1)^3 - (a+b)\cdot(x+1)^2 + ab\cdot(x+1)}$$
$$= g^{x^3 + (3-a-b)\cdot x^2 + (3-2a-2b+ab)\cdot x + 1 - a - b + ab}$$

---

[6] The SCDH assumption is provably equivalent to the Computational Diffie-Hellman (CDH) assumption – solving SCDH implies solving CDH, and vice versa [4].

Now we are able to compute $g^{x^2} = \left(f(g^{x+1}) \cdot f(g^x)^{-1} \cdot g^{(-3+2a+2b) \cdot x - 1 + a + b - ab} \right)^{1/3}$. This, however, contradicts the SCDH assumption [4], which states that one cannot compute g^{x^2} from g, g^x where $x \in_R [1, q-1]$. $\qquad\square$

4.3   Known Session Security

We now consider the impact of a compromised session. If an attacker is powerful enough to compromise a session, we assume he can learn all session-specific secrets, including the raw session key K and ephemeral private keys. In this case, the password will inevitably be disclosed (say by exhaustive search). This is an inherent threat and applies to all the existing PAKE protocols [1, 33, 34, 35, 5, 10, 16, 7, 17, 28].

Still, we shall minimize the impact of a compromised session: in particular, a corrupted session must not harm the security of other established sessions. In the J-PAKE protocol, the raw session key $K = g^{(x_1+x_3) \cdot x_2 \cdot x_4 \cdot s}$ is determined by the ephemeral random inputs x_1, x_2, x_3, x_4 from both parties in the session. As we mentioned earlier, the probability has implicitly guaranteed that $K \neq 1$ even in the face of active attacks. The following theorem shows that the obtained session key K is random too – in other words, the session keys are all independent. Therefore, compromising a session (hence learning all session-specific secrets) has no effect on other established session keys.

Theorem 5 (Random session key). *Under the Decision Diffie-Hellman (DDH) assumption, given that $K \neq 1$, the past session key derived from the protocol is indistinguishable from a random non-identity element in G.*

Proof. By protocol definition, $x_2, x_4 \in_R [1, q-1]$, and $s \in [1, q-1]$. Since $K = g^{(x_1+x_3) \cdot x_2 \cdot x_4 \cdot s} \neq 1$, we have $x_1 + x_3 \neq 0$. Let $a = x_1 + x_3$ and $b = x_2 \cdot x_4 \cdot s$. Obviously, $a \in_R [1, q-1]$ and $b \in_R [1, q-1]$. Based on the Decision Diffie-Hellman assumption [29], the value $g^{a \cdot b}$ is indistinguishable from a random non-identity element in the group. $\qquad\square$

4.4   On-Line Dictionary Attack Resistance

Finally, we study an active attacker, who directly engages in the protocol execution. Without loss of generality, we assume Alice is honest, and Bob is compromised (i.e., an attacker).

In the protocol, Bob demonstrates that he knows x_4 and the exponent of g_b, where $g_b = g^{x_1+x_2+x_3}$. Therefore, the format of the ciphertext sent by Bob can be described as $\mathcal{B}' = g_b{}^{x_4 \cdot s'}$, where s' is a value that Bob (the attacker) can choose freely.

Theorem 6 (On-line dictionary attack resistance). *Under the SCDH assumption, an active attacker cannot compute the session key if he chose a value $s' \neq s$.*

Table 1. Summary of J-PAKE security properties

Modules	Security property	Attacker type	Assumptions
Schnorr signature	leak 1-bit: whether sender knows discrete logarithm	passive/active	DL and random oracle
Password encryption	indistinguishable from random	passive/active	DDH
Session key	incomputable	passive	CDH
	incomputable	passive (know s)	CDH
	incomputable	passive (know other session keys)	CDH
	incomputable	active (if $s' \neq s$)	CDH
Key confirmation	leak nothing	passive	–
	leak 1-bit: whether $s' = s$	active	CDH

Proof. After receiving $\mathcal{B}'$, Alice computes

$$K' = (\mathcal{B}'/g^{x_2 \cdot x_4 \cdot s})^{x_2} \tag{1}$$

$$= g^{x_1 \cdot x_2 \cdot x_4 \cdot s'} \cdot g^{x_2 \cdot x_3 \cdot x_4 \cdot s'} \cdot g^{x_2^2 \cdot x_4 \cdot (s'-s)} \tag{2}$$

To obtain a contradiction, we reveal x_1 and s, and assume that the attacker is now able to compute K'. The attacker behaves as an oracle: given inputs $\{g^{x_2}, x_1, x_3, x_4, s, s'\}$, it returns K'. Note that the oracle does not need to know x_2, and it is still able to compute $\mathcal{A} = g^{(x_1+x_3+x_4) \cdot x_2 \cdot s}$ and $\mathcal{B}' = g^{(x_1+x_2+x_3) \cdot x_4 \cdot s'}$ internally. Thus, the oracle can be used to solve the Square Computational Diffie-Hellman problem by computing $g^{x_2^2} = (K'/(g^{x_1 \cdot x_2 \cdot x_4 \cdot s'} \cdot g^{x_2 \cdot x_3 \cdot x_4 \cdot s'}))^{x_4^{-1}(s'-s)^{-1}}$. Here[7], $x_4 \neq 0$ and $s' - s \neq 0$. This, however, contradicts the SCDH assumption [4], which states that one cannot compute $g^{x_2^2}$ from g, g^{x_2} where $x_2 \in_R [1, q-1]$. So, even with x_1 and s revealed, the attacker is still unable to compute K' (and hence cannot perform key confirmation later). □

The above theorem shows that what an on-line attacker can learn from the protocol is only minimal. Because of the knowledge proofs, the attacker is left with the only freedom to choose an arbitrary s'. If $s' \neq s$, he is unable to derive the same session key as Alice. During the later key confirmation process, the attacker will learn one-bit information: whether s' and s are equal. This is the best that any PAKE protocol can possibly achieve, because by nature we cannot stop an imposter from trying a random guess of password. However, consecutively failed guesses can be easily detected, and thwarted accordingly. The security properties of our protocol are summarized in Table 1.

5   Comparison

In this section, we compare our protocol with two other balanced PAKE schemes: EKE and SPEKE. These two techniques have several variants, which follow very

---

[7] This explains why in the protocol definition we need $x_4 \neq 0$, and symmetrically, $x_2 \neq 0$.

Table 2. Computational cost for Alice in J-PAKE

Item	Description	No of Exp
1	Compute $\{g^{x_1}, g^{x_2}\}$ and KPs for $\{x_1, x_2\}$	4
2	Verify KPs for $\{x_3, x_4\}$	4
3	Compute $\mathcal{A}$ and KP for $\{x_2 \cdot s\}$	2
4	Verify KP for $\{x_4 \cdot s\}$	2
5	Compute κ	2
	Total	**14**

similar constructs [7]. However, it is beyond the scope of this paper to evaluate them all. Also, we will not compare with augmented schemes (e.g., A-EKE, B-SPEKE, SRP, AMP and OPAKE [27]) due to different design goals.

The EKE and SPEKE are among the simplest and most efficient PAKE schemes. Both protocols can be executed in one round, while J-PAKE requires two rounds. On the computational aspect, both protocol require each user to perform only two exponentiations, compared with 14 exponentiations in J-PAKE (see Table 2).

At first glance, the J-PAKE scheme looks too computationally expensive. However, note that both the EKE and SPEKE protocols must use long exponents (see Section 2.2). Since the cost of exponentiation is linear with the bit-length of the exponent [29], for a typical 1024-bit p and 160-bit q setting, one exponentiation in an EKE or SPEKE is equivalent in cost to 6-7 exponentiations in a J-PAKE. Hence, the overall computational costs for EKE, SPEKE and J-PAKE are actually about the same.

There are several ways to improve the J-PAKE performance. First, the protocol enumerates 14 exponentiations for each user, but actually many of the operations are merely repetitions. To explain this, let the bit length of the exponent be $L = \log_2 q$. Computing g^{x_1} alone requires roughly $1.5L$ multiplications which include L square operations and $0.5L$ multiplications of the square terms. However, the same square operations need not be repeated for other items with the same base g (i.e., g^{x_2} etc). This provides plenty room for efficiency optimization in a practical implementation. In contrast, the same optimization is not applicable to the EKE and SPEKE. Second, it would be more efficient, particularly on mobile devices, to implement J-PAKE using Elliptic Curve Cryptography (ECC). Using ECC essentially replaces the multiplicative cyclic group with an additive cyclic group defined over some elliptic curve. The basic protocol construction remains unchanged.

6   Design Considerations

One notable feature of the J-PAKE design is the use of the Zero Knowledge Proof (ZKP), specifically: Schnorr Signature [30]. The ZKP is a well-established cryptographic primitive [9]. For over twenty years, this primitive has been playing a pivotal role in general two/multi-party secure computations [38].

Authenticated key exchange is essentially a two-party secure computation problem. However, the use of ZKP in this area is rare. The main concern is on efficiency: the ZKP is perceived as computationally expensive. So far, almost all of the past PAKE protocols have avoided using ZKP for exactly the reason.

However, the use of ZKP does not necessarily mean the protocol must be inefficient. This largely depends on how to effectively integrate this primitive into the overall design. In our construction, we introduced a novel juggling technique: arranging the random public keys in such a structured way that the randomization factors vanish when both sides supplied the same password. (A similar use of this juggling technique can be traced back to [15] and [8]). As we have shown, this leads to computational efficiency that is comparable to the EKE and SPEKE protocols. To our best knowledge, this design is significantly different from all past PAKE protocols. In the area of PAKE research – which has been troubled by many patent arguments surrounding existing schemes [13] – a new construct may be helpful.

With the same juggling idea, the current construction of the J-PAKE protocol seems close to the optimum. Note in the protocol, we used four x terms – x_1, x_2, x_3, x_4. As if one cannot juggle with only two balls, we find it difficult to juggle with two x terms. This is not an issue in the multi-party setting where there are at least three participants (each participant generates one "ball") [15]. For the two-party case, our solution was to let each user create two ephemeral public keys, and thus preserve the protocol symmetry. It seems unlikely that one could improve the protocol efficiency by using a total of only 3 (or even 2) x terms. However, we do not have a proof of minimality on this, so we leave the question open.

7   Conclusion

In this paper, we proposed a protocol, called J-PAKE, which authenticates a password with zero-knowledge and then subsequently creates a strong session key if the password is correct. We showed that the protocol fulfills the following properties: it prevents off-line dictionary attacks; provides forward secrecy; insulates a compromised session from affecting other sessions; and strictly limits an active attacker to guess only one password per protocol execution. As compared to the de facto internet standard SSL/TLS, J-PAKE is more lightweight in password authentication with two notable advantages: 1). It requires no PKI deployments; 2). It protects users from leaking passwords (say to a fake bank website).

Acknowledgments

We thank Ross Anderson and Piotr Zieliński for very helpful comments and discussions.

References

1. Abdalla, M., Pointcheval, D.: Simple password-based encrypted key exchange protocols. In: Menezes, A. (ed.) CT-RSA 2005. LNCS, vol. 3376, pp. 191–208. Springer, Heidelberg (2005)
2. Anderson, R.J., Needham, R.: Robustness principles for public key protocols. In: Coppersmith, D. (ed.) CRYPTO 1995. LNCS, vol. 963, pp. 236–247. Springer, Heidelberg (1995)
3. Anderson, R.J.: Security Engineering: A Guide to Building Dependable Distributed Systems. Wiley, New York (2001)
4. Bao, F., Deng, R.H., Zhu, H.: Variations of Diffie-Hellman problem. In: Qing, S., Gollmann, D., Zhou, J. (eds.) ICICS 2003. LNCS, vol. 2836, pp. 301–312. Springer, Heidelberg (2003)
5. Bellare, M., Pointcheval, D., Rogaway, P.: Authenticated key exchange secure against dictionary attacks. In: Preneel, B. (ed.) EUROCRYPT 2000. LNCS, vol. 1807, pp. 139–155. Springer, Heidelberg (2000)
6. Boneh, D.: The decision Diffie-Hellman problem. In: Buhler, J.P. (ed.) ANTS 1998. LNCS, vol. 1423, pp. 48–63. Springer, Heidelberg (1998)
7. Boyd, C., Mathuria, A.: Protocols for authentication and key establishment. Springer, Heidelberg (2003)
8. Chaum, D.: The dining cryptographers problem: unconditional sender and recipient untraceability. Journal of Cryptology 1(1), 65–67 (1988)
9. Camenisch, J., Stadler, M.: Proof systems for general statements about discrete logarithms, Technical report TR 260, Department of Computer Science, ETH Zürich (March 1997)
10. Bellovin, S., Merritt, M.: Encrypted Key Exchange: password-based protocols secure against dictionary attacks. In: Proceedings of the IEEE Symposium on Research in Security and Privacy (May 1992)
11. Bellovin, S., Merritt, M.: Augmented Encrypted Key Exchange: a password-based protocol secure against dictionary attacks and password file compromise. In: Proceedings of the 1st ACM Conference on Computer and Communications Security, pp. 244–250 (November 1993)
12. Bellovin, S., Merritt, M.: Cryptographic protocol for secure communications, U.S. Patent 5,241,599
13. Ehulund, E.: Secure on-line configuration for SIP UAs, Master thesis, The Royal Institute of Technology (August 2006)
14. Ford, W., Kaliski, B.S.: Server-assisted generation of a strong secret from a password. In: Proceedings of the 9th International Workshops on Enabling Technologies, pp. 176–180. IEEE Press, Los Alamitos (2000)
15. Hao, F., Zieliński, P.: A 2-round anonymous veto protocol. In: Proceedings of the 14th International Workshop on Security Protocols, SPW 2006, Cambridge, UK (May 2006)
16. Jablon, D.: Strong password-only authenticated key exchange. ACM Computer Communications Review 26(5), 5–26 (1996)
17. Jablon, D.: Extended password protocols immune to dictionary attack. In: Proceedings of the WETICE 1997 Enterprise Security Workshop, pp. 248–255 (June 1997)
18. Jablon, D.: Cryptographic methods for remote authentication, U.S. Patent 6,226,383 (March 1997)
19. Jablon, D.: Password authentication using multiple servers. In: Naccache, D. (ed.) CT-RSA 2001. LNCS, vol. 2020, pp. 344–360. Springer, Heidelberg (2001)

20. Jaspan, B.: Dual-workfactor Encrypted Key Exchange: efficiently preventing password chaining and dictionary attacks. In: Proceedings of the Sixth Annual USENIX Security Conference, pp. 43–50 (July 1996)
21. Kobara, K., Imai, H.: Pretty-simple password-authenticated key-exchange under standard assumptions. IEICE Transactions E85-A(10), 2229–2237 (2002)
22. Van Oorschot, P.C., Wiener, M.J.: On Diffie-Hellman key agreement with short exponents. In: Maurer, U.M. (ed.) EUROCRYPT 1996. LNCS, vol. 1070, pp. 332–343. Springer, Heidelberg (1996)
23. Patel, S.: Number theoretic attacks on secure password schemes. In: Proceedings of the IEEE Symposium on Security and Privacy (May 1997)
24. Perlman, R., Kaufman, C.: Secure password-based protocol for downloading a private key. In: Proceedings of the Network and Distributed System Security (February 1999)
25. MacKenzie, P.: The PAK suite: protocols for password-authenticated key exchange, Technical Report 2002-46, DIMACS (2002)
26. MacKenzie, P.: On the Security of the SPEKE Password-Authenticated Key Exchange Protocol. Cryptology ePrint Archive: Report 057 (2001)
27. IEEE P1363 Working Group, P1363.2: Standard Specifications for Password-Based Public-Key Cryptographic Techniques. Draft available at, http://grouper.ieee.org/groups/1363/
28. Wu, T.: The Secure Remote Password protocol. In: Proceedings of the Internet Society Network and Distributed System Security Symposium, pp. 97–111 (March 1998)
29. Stinson, D.: Cryptography: theory and practice, 3rd edn. Chapman & Hall/CRC (2006)
30. Schnorr, C.P.: Efficient signature generation by smart cards. Journal of Cryptology 4(3), 161–174 (1991)
31. Zhang, M.: Analysis of the SPEKE password-authenticated key exchange protocol. IEEE Communications Letters 8(1), 63–65 (2004)
32. Zhao, Z., Dong, Z., Wang, Y.: Security analysis of a password-based authentication protocol proposed to IEEE 1363. Theoretical Computer Science 352(1), 280–287 (2006)
33. Goldreich, O., Lindell, Y.: Session-key generation using human passwords only. In: Kilian, J. (ed.) CRYPTO 2001. LNCS, vol. 2139, pp. 408–432. Springer, Heidelberg (2001)
34. Katz, J., Ostrovsky, R., Yung, M.: Efficient password-authenticated key exchange using human-memorable passwords. In: Pfitzmann, B. (ed.) EUROCRYPT 2001. LNCS, vol. 2045, pp. 475–494. Springer, Heidelberg (2001)
35. Jiang, S.Q., Gong, G.: Password based key exchange with mutual authentication. In: Handschuh, H., Hasan, M.A. (eds.) SAC 2004. LNCS, vol. 3357, pp. 267–279. Springer, Heidelberg (2004)
36. Krawczyk, H.: HMQV: a high-performance secure Diffie-Hellman protocol. In: Shoup, V. (ed.) CRYPTO 2005. LNCS, vol. 3621, pp. 546–566. Springer, Heidelberg (2005)
37. Menezes, A.J., van Oorschot, P.C., Vanstone, S.A.: Handbook of applied cryptography. CRC Press, Boca Raton (1996)
38. Goldreich, O., Micali, S., Wigderson, A.: How to play any mental game or a completeness theorem for protocols with honest majority. In: Proceedings of the Nineteenth Annual ACM Conference on Theory of Computing, pp. 218–229 (1987)
39. Gennaro, R., Lindell, Y.: A framework for password-based authenticated key exchange. In: Biham, E. (ed.) EUROCRYPT 2003. LNCS, vol. 2656, pp. 524–543. Springer, Heidelberg (2003)

Distance Based Transmission Power Control Scheme for Indoor Wireless Sensor Network

P.T.V. Bhuvaneswari, V. Vaidehi, and M. Agnes Saranya

Department of Electronics, Madras Institute of Technology, Anna University,
Chennai-44, Tamilnadu, India
ptvbmit@annauniv.edu, vaidehi@annauniv.edu,
saran_1131@yahoo.co.in

Abstract. This paper proposes a Distance Based Transmission Power Control (DBTPC) scheme for selecting Optimal Transmission Power for Indoor Wireless Sensor Network. The proposed work consists of two phases namely Localization phase and data transfer phase. In Localization phase, the relative coordinate of the unknown sensor node with respect to anchor sensor node is estimated by the proposed Received Signal Strength (RSS) based localization algorithm. By performing neighbor discovery process, each node obtains the distance information of its neighboring nodes. Based on this information, in data transfer phase it dynamically controls its transmission power level to reach their neighboring node with acceptable RSS value. This is achieved by the proposed distributed Distance Based Transmission Power Control (DBTPC) scheme. The Optimal Transmission Power (OTP) can be adaptively selected by the proposed DBTPC scheme. This ensures energy efficiency in sensor node and thereby increases the lifetime of the network.

Keywords: Received Signal Strength based localization algorithm, Distance Based Transmission Power Control scheme and Optimal Transmission Power.

1 Introduction

In Wireless Sensor Network, location awareness is the key factor for many potential real time applications such as monitoring, target tracking, person tracking, and context-aware application [1]. The sensor nodes in the network are usually powered by limited batteries which in turn influences the network lifetime. In order to increase the lifetime of the network, localization has to be made power efficient and accurate. The ability of sensors to locate themselves using limited energy and computational resources pose new challenges that require novel solution.

The traditional Global Positioning System (GPS) method of localization is not suitable for Indoor Wireless Sensor Network (IWSN), as it is not accurate and cost-effective [2]. Hence an alternate method like Time of Arrival (ToA), Time Difference of Arrival (TDoA) and Received Signal Strength (RSS) may be used [2]. From the literatures [2], it is found that RSS-based localization method is cost-effective, but not very accurate. However, using appropriated error minimization techniques, this problem can be resolved.

M.L. Gavrilova et al. (Eds.): Trans. on Comput. Sci. XI, LNCS 6480, pp. 207–222, 2010.
© Springer-Verlag Berlin Heidelberg 2010

In this paper, an energy efficient RSS based distributed localization algorithm and Distance Based Transmission Power Control (DBTPC) scheme are proposed. The principal objective of the proposed algorithm is to improve the accuracy in relative coordinate estimation and minimize the energy cost incurred for transmitting information between nodes. The proposed localization algorithm consists of two stages namely, distance estimation and coordinates estimation. Estimation of accurate distance is done by one-dimensional Kalman filter estimator. The number of iterations of the Kalman filter estimator is limited by Cramer Rao Bound (CRB). Min-max bounding box algorithm estimates the coordinates of the unknown nodes more accurately as it considers the overlapping issues prevailing in tri-lateration techniques. The proposed DBTPC scheme, aims to minimize the energy consumption in the network by selecting the Optimal Transmission Power (OTP) for each node to reach their neighboring nodes.

The rest of the paper is organized as follows. Section 2 discusses the related work in preceding power aware localization methods for Wireless Sensor Network. The proposed RSS –based localization algorithm and DBTPC scheme is presented in section 3. Results and discussion of the proposed work are presented in section 4 and Section V concludes the paper with future work.

2   Related Work

The goal of localization algorithms in Wireless Sensor Network (WSN) is to determine the node's position. Number of approaches [3, 4, 5, 6, 7, and 8] has been proposed that formulate the localization problem as joint estimation problem. The estimators determine the unknown node's locations with reference to the anchor node's positions. [9] proposes RSS based localization algorithm with weighted centroid method for indoor Wireless Sensor Network which offers low communication overhead and low computational complexity. But the reduction in RSS measurement errors is achieved by antenna diversity technique. This requires two antennas thus results in increase of hardware complexity.

[5] Proposes a hop-distance algorithm for self-localization in WSN which is based on RSS and uses maximum likelihood estimator to achieve accuracy. [10] Proposed a localization scheme based on RSS and distributed weighted multidimensional scaling algorithm which allows sensors to calculate their own location by means of iterative optimization thus decreases the cost and improves the global coordinate estimate. This method is robust to large errors.

As sensor networks are mainly operated by batteries, transmission of data between nodes needs to consume less power. Dynamic transmission power control mechanisms are required. In [11], a collaborative energy efficient target-tracking algorithm is developed, where transmission power is adjusted based on the amount of mutual information a node wants to share with their neighbors. In this method, the power adjustment scheme depends on the network querying technique and it performs well only if the most informative node is queried. A decomposition algorithm which optimizes the sensor power levels to achieve higher utility factor for scheduling transmissions in Wireless Sensor Networks is proposed in [12]. In [13], an optimal common transmit power for Wireless Sensor Networks is investigated. The optimal transmit power derived in this paper can be applied only to some specific scenarios and also it does not consider the multipath effects. In [14], an optimum selective forwarding policy for data transmission is introduced that selects the messages with higher priority and discards that of low-priority. This method considers only the energy requirements and not the neighbor node's information.

In this paper, an efficient RSS based Localization Algorithm and Distance Based Transmission Power Control (DBTPC) scheme are proposed. The proposed RSS based localization algorithm, aims to estimate an accurate relative coordinate of a sensor node. The proposed DBTPC scheme concentrates on energy optimization in the node, by adaptively selecting the Optimal Transmission Power of a node to reach its neighboring nodes based on their distance.

3 Proposed RSS-Based Localization Algorithm and DBTPC Scheme

Consider a sensor network randomly deployed in an indoor environment with $(M+N)$ nodes, where M denotes the number of anchor nodes and N denotes the number of unknown nodes. All the nodes are assumed to be static and possess the capability of transmitting and receiving information by means of uni-cast communication.

The objective of the proposed RSS-based Localization Algorithm is to determine the relative co-ordinates of N unknown nodes using the distance and location information of M anchor nodes. The accuracy of the proposed algorithm is enhanced by one-dimensional Kalman filter estimator in distance estimation stage and min-max bounding box algorithm in relative coordinates estimation stage. The block diagram of the proposed RSS based localization algorithm is presented in Figure 1.

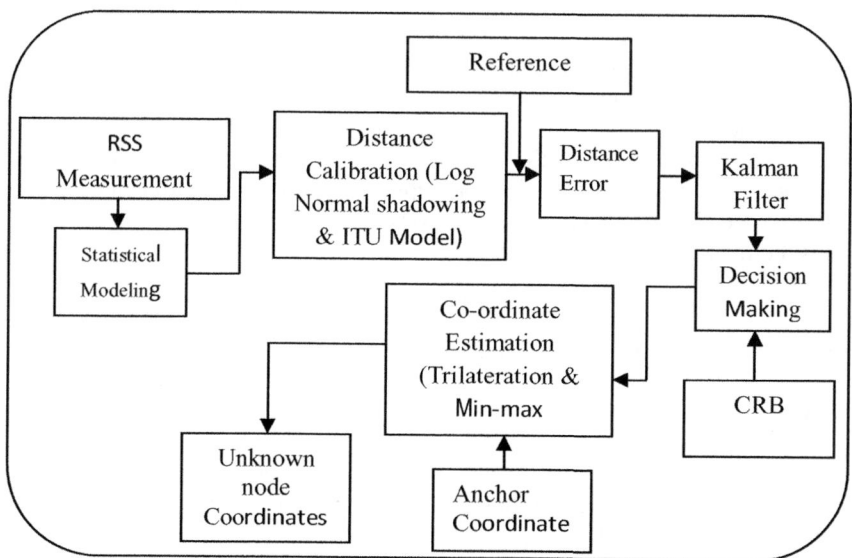

Fig. 1. Block diagram of proposed RSS based localization algorithm

Once every unknown nodes compute their relative co-ordinates through the proposed RSS based localization algorithm, they are termed as anchor nodes. Now, each node in the network performs neighbor discovery process to obtain the distance information of its neighboring nodes. With this information, the transmission power level at which each node need to be operated to reach their neighboring nodes with acceptable RSS is adaptively controlled by the proposed Distance Based Transmission Power Control (DBTPC) scheme. The block diagram of the proposed DBTPC scheme is presented in Figure 2.

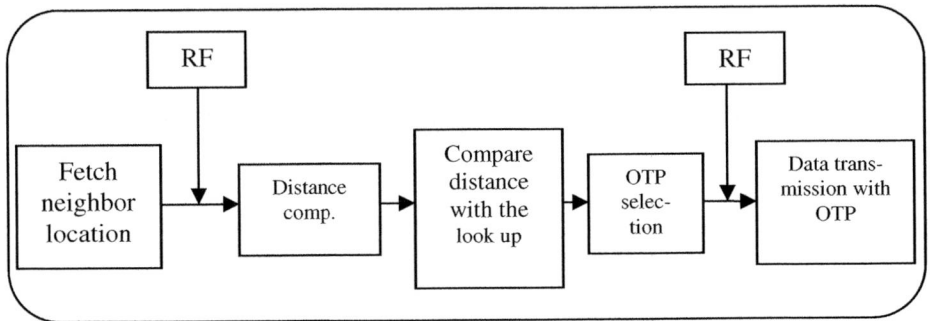

Fig. 2. Block diagram of proposed DBTPC scheme

3.1   RSS Based Localization Algorithm

The anchor node equipped with GPS receiver transmits signal with a particular transmission power to the unknown node, whose location is to be determined. The Received Signal Strength (RSS) value of the transmitted signal is measured in the unknown node. As the RSS values are fluctuating, more samples of RSS values are measured for different time instances. The above said procedure is repeated for different channels. The link qualities of all the channels are analyzed through statistical modeling. The channel possessing less Standard Deviation (*SD*) is concluded as the best channel. However, when the anchor node is operated at different transmission power level, then best channel selection depends on two other parameters like Packet Reception Rate (*PRR*) and Transmission Power (*P_t*).

   Thus the Best Channel of Transmission (BCT) is expressed as,

$$BCT = fn(PRR, P_t, SD)$$

(1)

Where *PRR* - Packet Reception Rate.
   P_t -Transmission Power
   SD - Standard Deviation

The distance of the unknown node with respect to anchor node is computed from the ensemble mean RSS value of the best channel using two models namely path loss log normal shadowing and ITU indoor attenuation models.

   Let d_{mn} be the distance between anchor node *m* and unknown node *n* where *m* =1,2,....*M* and *n* =1,2,.....*N*.

The distance *dmn'* computed from log-normal shadowing model [15] is given below,

$$d_{mn}' = d_{ref}(10^{[(Pt(dBm)-\overline{PL}(d_{ref})-X_\sigma-Pr(d_{mn}')[dBm])/(10n_p)]})$$ (2)

where $\overline{PL}(d_{ref})$ is the ensemble path loss at a short reference distance d_{ref}, X_σ is the zero mean Gaussian random variable with standard deviation σ and n_p is path-loss exponent, typically lies between 2 and 4.

The distance *dmn"* computed from ITU indoor attenuation model [16] is given below,

$$d_{mn}'' = d_{ref}(10^{[(Pt(dBm)-20\log f-L\log d_{ref}-P_f(f_{rmn})+28-X_\sigma-Pr(d_{mn}')[dBm])/(10n_p)]})$$ (3)

Where L is the distance power loss coefficient, f is the frequency, f_{max} is the number of floors between the node m and node n, $P_f(f_{mn})$ is floor loss penetration factor.

The distance errors of the calculated distance by the two models are minimized by one-dimensional Kalman estimator [17].The convergence rate of Kalman filter estimator is decided based on the CRB (Cramer Rao Bound) [18].

Thus through one-dimensional Kalman filter estimator, the estimated distance $\hat{d}_{mn}$ is computed as given in equations below:

$$\hat{d}_{mn} = \left(\frac{d_{mn}'}{\sigma_1}+\frac{d_{mn}''}{\sigma_2}\right) \bigg/ \left(\frac{1}{\sigma_1^2}+\frac{1}{\sigma_2^2}\right)$$ (4)

(Or)

$$\hat{d}_{mn} = (d_{mn}'\sigma_2^2 + d_{mn}''\sigma_1^2)\bigg/(\sigma_1^2+\sigma_2^2)$$ (5)

where $\sigma 1$ = standard deviation for d_{mn}' about the mean

$\sigma 2$ = standard deviation for d_{mn}'' about the mean

The above estimated distance can also be rewritten as

$$\hat{d}_{mn} = d_{mn}'+ K(d_{mn}''-d_{mn}')$$ (6)

where $K = \sigma_1^2\bigg/(\sigma_1^2+\sigma_2^2)$ is defined as the Kalman gain.

The CRB value is given below,

$$CRB_{bound} = \frac{\sigma^2}{n_d}$$ (7)

where $\sigma = \dfrac{(\sigma_1 + \sigma_2)}{2}$

n_d =number of samples

Lateration is one of the most popular techniques for node positioning in Wireless Sensor Network. The lateration technique focused in this paper is tri-lateration. Assume that there are three anchor nodes with known positions (x_a, y_a) where a=1, 2, 3, a node at unknown position (x_u, y_u) and estimated distance value $\hat{d}_{mn}$.

Using the model of multi-lateration [2], the 2-D coordinates of the unknown node can be determined as below:

$$\begin{bmatrix} x_u \\ y_u \end{bmatrix} = 2 \begin{pmatrix} y_m - y_{m-1} & \cdots & y_1 - y_m \\ \vdots & \ddots & \vdots \\ x_{m-1} - x_m & \cdots & x_m - x_1 \end{pmatrix} \begin{bmatrix} (d_{1n}^2 - d_{mn}^2) & - & (x_1^2 - x_m^2) & - & (y_1^2 - y_m^2) \\ \cdot & & \cdot & & \cdot \\ \cdot & & \cdot & & \cdot \\ (d_{mn-1}^2 - d_{mn}^2) & - & (x_{m-1}^2 - x_m^2) & - & (y_{m-1}^2 - y_m^2) \end{bmatrix} \tag{8}$$

where (x_u, y_u) be the coordinates of the unknown node, (x_a, y_a), a=1,2,....m be the co-ordinates of the known nodes, d_{in}, i=1,2,...m be the distance between the node i and node n. Then from the point of intersection of all three anchors node's connectivity circles, the relative coordinates of unknown node is computed. But in real time scenario, overlapping circles may exist due to wireless medium constraints, this result in inaccurate relative coordinate estimation of unknown node. To resolve this problem an efficient min-max bounding box concept is used in the proposed algorithm. The connectivity circles of anchor nodes are modeled as square positioning Cells (PC). Then the Final Bounding Box (FBB) which is the overlapping box of all the square boxes is determined as below:

FBB is given by

$$FBB = \bigcap PC_n \tag{9}$$

3.2   Distance Based Transmission Power Control (DBTPC) Scheme

Let D represent the maximum coverage of (S=M+N) sensor nodes that are randomly deployed in indoor Wireless Sensor Network. After the Localization phase, each node is aware of its relative coordinates. Now each node broadcast a request message to obtain its neighbor's topological information. After receiving the network topological information, the proposed Distance Based Transmission Power Control (DBTPC) scheme placed in the transceiver module of the sensor node is switched ON. This DBTPC scheme, selects the Optimal Transmission Power (OTP) required to reach its neighboring node. This scheme minimizes the energy utilization in the node, thereby enhanced the lifetime of the network. The model used in the proposed DBTPC scheme is presented in the following section.

The DBTPC scheme present in the transceiver module of each sensor node controls its transmission power based on two factors namely distance information of its neighbors (d_i) where i=1, 2,..., (S-1) and its own residual energy (E_k) at K^{th} instant.

Thus the Optimal Transmission Power (OTP) can be expressed as,

$$OTP = \begin{cases} fn(E_k, d_i) & ; \quad E_k > e_t \\ 0 & otherwise \end{cases}$$

(10)

Where e_t the total amount of energy is consumed per packet transmission and is given by [11],

$$e_t = P_t \times \frac{L}{R_b} (Joules)$$

(11)

where P_t - Transmission power, L - Packet size, Rb- the bit rate.

The proposed DBTPC scheme is validated by two models namely (i) Connectivity model and (ii) Energy model.

Connectivity model

Let d_k represent the connectivity information of a sensor node at instant. If neighboring nodes are within the connectivity of a sensor node, then $d_k = 1$ else $d_k = 0$. Let x_k be a variable that indicates the communication status of a sensor node such that,

$$x_k = \begin{cases} 1, & transmission\,occurs \\ 0, & no\,transmission \end{cases}$$

(12)

The necessary condition at which the proposed DBTPC scheme can be enabled in the sensor node is when both $d_k = x_k = 1$.

Energy model

The energy consumption in nodes can be the sum of energy spent in the sensing module, processing module and transceiver module. It is found from the literature [14] that energy consumption in sensing and processing is negligible compared to that of transceiver module. In transceiver module, the energy consumption can be controlled by adaptively selecting the transmission power to reach the neighboring nodes based on their distance. The proposed DBTPC scheme is designed to perform this task.

Let E_k be the residual energy in a node at k^{th} instant that can be expressed as,

$$E_k = E_{k-1} - x_k E_1(d_k) + (1 - x_k)e_i$$

(13)

Where E_{k-1} is the residual energy in the node at $(k-1)^{th}$ instant, e_i is the energy spent by the nodes when they are in idle state, $E_1(d_k)$ is the energy consumed when the node decides to transmit.

$$E_1(d_k) = pe_t$$

(14)

Where p - number of transmitted packets.

The lifetime of the node can be defined as the time taken to drain out of initial battery energy and is given by [3],

$$\tau = \frac{E_{batt}}{\lambda_t e_t} = \frac{E_{batt} R_b}{\lambda_t L P_t} (sec)$$

(15)

where λ_t - average transmission rate, E_{batt} - initial battery energy.

The expected power profile obtained by the proposed DBTPC scheme is given in Figure 3. Transmitting with a lower power reduces the communication range of the node. Hence the profile justifies that the communication between nodes that are closer to each other can be achieved with lower transmission power level of operation.

Consider each sensor node consists of an ordered set of transmission power levels, $P_t = \{P_1, P_2, \ldots\ldots\ldots, P_l)$ where l represents the number of power levels such that $P_1 < P_2 < \ldots\ldots\ldots < P_l$. Let d_{ij} be the distance between any two nodes that are within the maximum coverage D, Then the Optimal Transmission Power (OTP) is given in equation (16).

If each node has the maximum coverage D, it has the privilege to communicate with the maximum power, P_l. However, the proposed DBTPC scheme adaptively controls the transmission power P_l in accordance with its distance information. The Energy reduction is further ensured by enabling the DBTPC scheme in the transceiver module which in turn wakeup the RF module from idle to active state. Thus the proposed DBTPC scheme provides the energy saving mode of operation.

$$OTP = \begin{cases} P_1, & 0 \le d_{ij} < D - \dfrac{(l-1)}{l}D \\[2mm] P_2, & D - \dfrac{(l-1)}{l}D \le d_{ij} < D - \dfrac{(l-2)}{l}D \\ \cdot & \cdot \\ \cdot & \cdot \\ \cdot & \cdot \\ P_l, & D - \dfrac{(l-(l-1))}{l}D \le d_{ij} \le D \end{cases} \qquad (16)$$

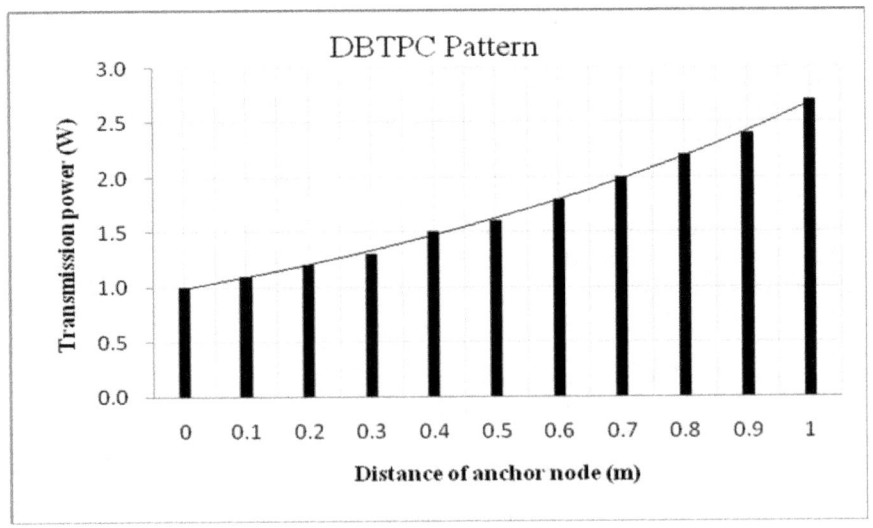

Fig. 3. Power Profile of the Proposed DBTPC scheme

4   Results and Discussions

The proposed RSS based Localization algorithm is analyzed using Matlab version 7.0. The results of analysis are given below.

4.1   RSS Analysis

The experimentation is done in the indoor environment using Zigbee series 1 RF module and the associated X-CTU software of MAXSTREAM. The experiment has been repeated for five different channels (B, C, D, E and F) with five different frequencies. The experimental observations recorded in indoor environment are presented in Table 1. The table shows 20 samples of RSS (Received Signal Strength) values measured at 20 different time instances for a specific distance. Figure 4 illustrate the relationship between distance and RSS measurement for channels B with frequency values shown in Table 1. It is seen that the Received Signal Strength of the unknown node decreases as distance between the anchor and unknown node increases. Similar relationship between Distance and RSS measurement is also found for remaining four channels (C, D, E, and F).

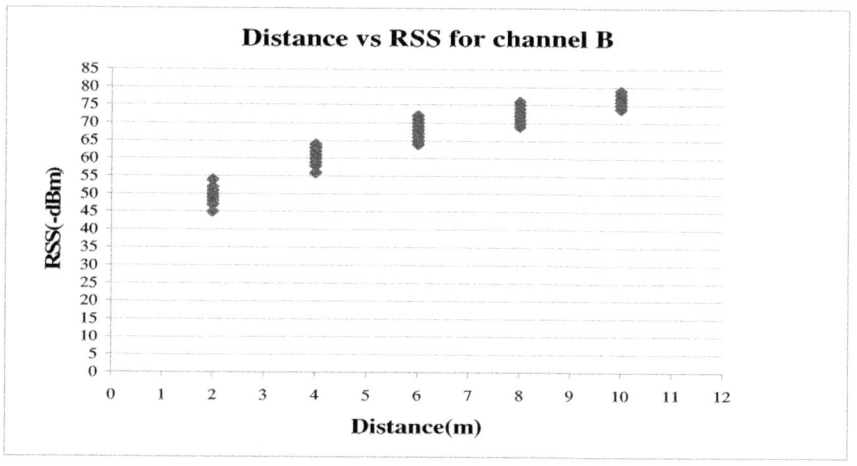

Fig. 4. Distance vs RSS for channel B

The statistical modeling is done to find the best channel. The results are presented in Table 2. It is found that the channel E is selected as the best channel for distances 2m, 4m and 10m as it possesses low standard deviation. For distances 6m and 8m, channel D is selected as the best channel of transmission. The average RSS value of channels D and E are computed.

From log normal shadowing path loss model and ITU attenuation model, the distance between the anchor node and unknown node is calculated as illustrated in Figure 5. It is seen that the calculated distance is closer to the actual distance in case of log normal shadowing path loss model than ITU attenuation model.

Table 1. Experimental results in indoor environment

Dist. (m)	RSS Measurements (dBm)				
	Channel B (2.404-2.406) GHz	Channel C (2.409-2.411) GHz	Channel D (2.414- 2.416) GHz	Channel E (2.419-2.421) GHz	Channel F (2.424-2.426) GHz
22	-50,-48, 47, -45,-45,-51, -53,-54,-52, -49,-49,-51, -52,-50,-52, -51,-50,-49, -50,-48	-50,-50,-47, -49,-51,-52, -45,-55,-50, -50,-49,-48, -49,-47,-49, -50,-51,-51, -52,-50	-50,-50,-49, -48,-47,-50, -51,-53,-54, -55,-47,-55, -49,-48,-47, -51,-52,-47, -48,-50	-49,-51,-49, -50, -50,-49, -49,-51, -49, -48,-51,-50, -51,-50,-51, -50,-49,-49, -49,-50	-49,-51,-50, -52, -53,-50, -48,-50, -49, -50,-50,-51, -52,-54,-52, -49,-47,-49, -50,-51
4	-59,-60,-62, -60,-58,-61, -62,-63,-64, -58,-56,-60, -61,-62,-61, -60,-61,-62, -59,-60	-59,-60,-61, -62,-60,-60, -61,-60,-56, -61,-59,-60, -60,-58,-56, -61,-62,-61, -60,-59	-60,-59,-61, -58,-59,-60, -60,-61,-60, -59,-59,-58, -61,-62,-61, -63,-61,-64, -61,-61	-59,-60,-61, -60,-59,-61, -60,-61,-59, -60,-61,-62, -60,-59,-60, -61,-62,-63, -60,-61	-60,-61,-63, -60,-59,-58, -61,-60,-58, -60,-61,-59, -61,-62,-59, -63,-60,-61, -60,-59
6	-67,-69,-70, -71,-69,-70, -69,-67,-66, -66,-65,-64, -67,-69,-70, -71,-72,-69, -67,-68	-66,-67,-70, -69,-71,-71, -72,-70,-69, -70,-71,-70, -72,-71,-70, -71,-72,-73, -69,-70	-66,-68,-69, -68,-66,-66, -68,-66,-69, -66,-68,-69, -66,-66,-68, -66,-68,-69, -66,-65	-66,-68,-66, -66,-66,-69, -68,-66,-66, -70,-66,-65, -66,-66,-66, -68,-66,-69, -66,-65	-67,-69,-70, -71,-66,-67, -69,-70,-66, -72,-71,-72, -70,-66,-66, -69,-70,-67, -67,-68
8	-70,-71,-72, -70,-73,-75, -76,-69,-72, -73,-72,-71, -70,-69,-71, -72,-73,-74, -75,-74	-70,-73,-74, -73,-75,-72, -73,-74,-73, -72,-71,-72, -73,-73,-74, -75,-73,-73, -74,-74	-72,-73,-71, -73,-73,-74, -75,-73,-73, -72,-74,-73, -72,-74,-75, -73,-72,-74, -74,-72	-72,-73,-72, -73,-71,-74, -74,-73,-73, -75,-72,-70, -71,-70,-73, -72,-73,-74, -72,-72	-72,-73,-71, -73,-70,-74, -70,-73,-75, -76,-71,-73, -77,-70,-71, -72,-73,-74, -73,-73
10	-76,-75,-77, -77,-75,-76, -77,-77,-77, -76,-77,-77, -79,-77,-76, -75,-74,-77, -76,-78	-76,-77,-77, -76,-77,-77, -77,-79,-76, -77,-77,-79, -80,-79,-77, -77,-73,-77, -77,-78	-77,-76,-77, -79,-79,-77, -77,-77,-76, -77,-77,-79, -77,-77,-77, -79,-77,-77, -77,-78	-77,-76,-77, -77,-77,-77, -77,-77,-77, -77,-77,-77, -77,-79,-77, -77,-77,-79, -77,-76	-76,-77,-77, -77,-75,-79, -77,-77,-77, -76,-75,-76, -77,-77,-77, -79,-79,-75, -77,-78

4.2 Kalman Analysis

Figure 6 shows the relationship between actual distance and estimated distance estimation obtained with and without one-dimensional Kalman filter. It is seen that with kalman, the estimated distance is very close to the actual distance. The percentage of accuracy improved by kalman and the number of iteration taken to achieve them is

Table 2. Standard Deviation for five channels

DIST. (m)	B (2.404-2.406) GHz	C (2.409-2.411) GHz	D (2.414-2.416) GHz	E (2.419-2.421) GHz	F (2.424-2.426) GHz
2	2.39	2.13	2.62	0.92	1.69
4	1.87	1.65	1.53	1.23	1.38
6	2.13	1.67	1.26	1.37	2.05
8	1.98	1.29	1.04	1.31	1.92
10	1.11	1.44	0.96	0.61	1.18

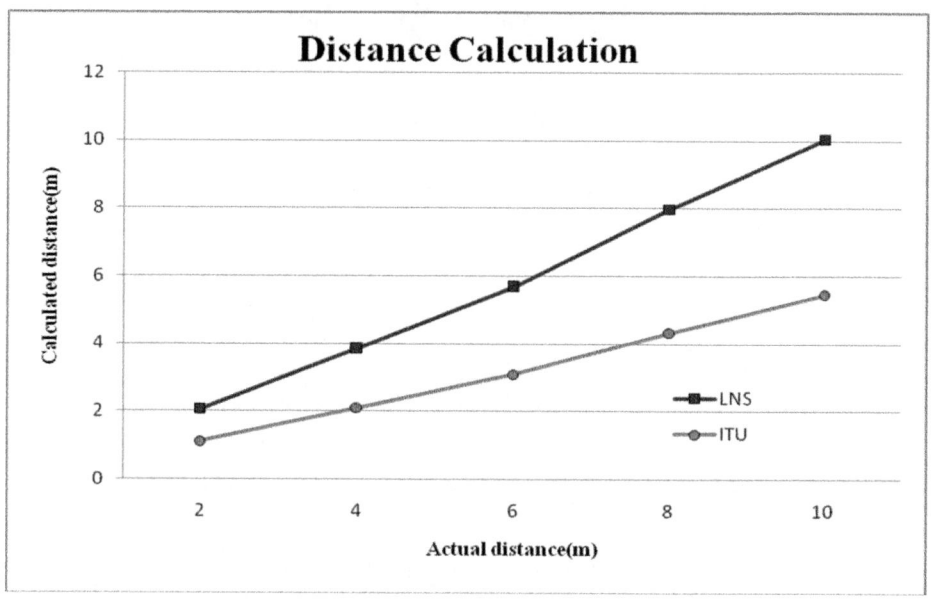

Fig. 5. Distance Calculation

presented in Table 3. It is inferred that percentage of accuracy improvement is gradually increased as the distance of separation between nodes is increased with the cost of number of iteration.

4.3 Trilateration Analysis

The relative coordinate of the unknown node is determined by trilateration and the accuracy is improved using min-max algorithm. Figure 7 shows the results of trilateration with min-max algorithm.

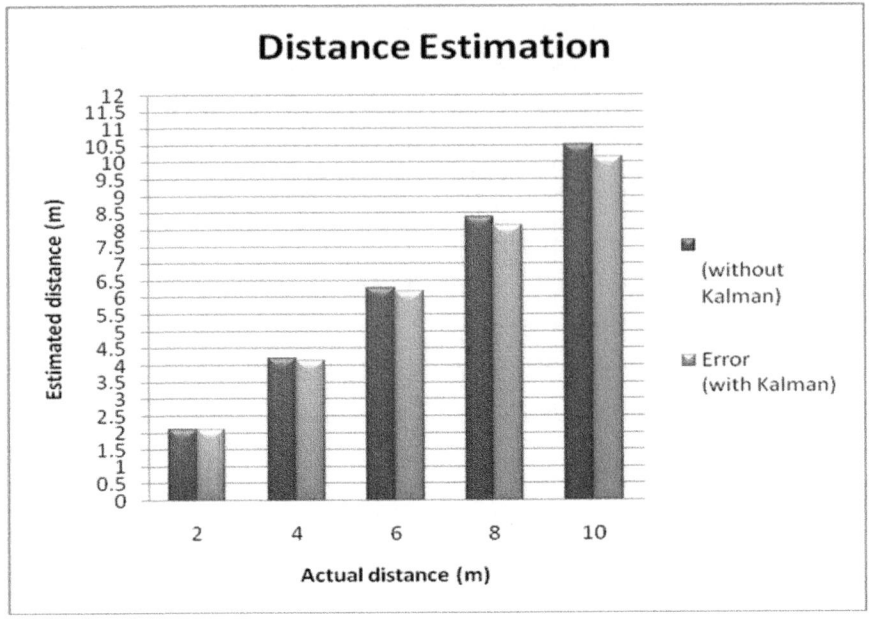

Fig. 6. Distance Estimation

Table 3. Kalman Filter Analysis

Act. Dist(m)	Error (without Kalman)	% of error With out Kalman	% of acc. With out Kalman	Error (with Kalman)	% of error With Kalman	%of acc. With Kalman	% of Impro-ved acc.	No. of itera-tion
2	0.10	5.06	94.93	0.08	4.47	95.52	0.59	1
4	0.19	4.84	95.15	0.13	3.30	96.69	1.55	2
6	0.27	4.61	95.38	0.19	3.21	96.78	1.40	2
8	0.37	4.62	95.37	0.14	1.77	98.22	2.85	3
10	0.52	5.28	94.72	0.16	1.68	98.32	3.6	3

Table 4 shows the result obtained with and without min-max bounding box algorithm. It is found that accuracy is improved through min-max bounding box algorithm.

4.4   Real-Time Experimentation Transmission Power Analysis

In real time, RSS values are measured between pairs of nodes for five different transmission power levels and the distance estimation is done using the proposed RSS

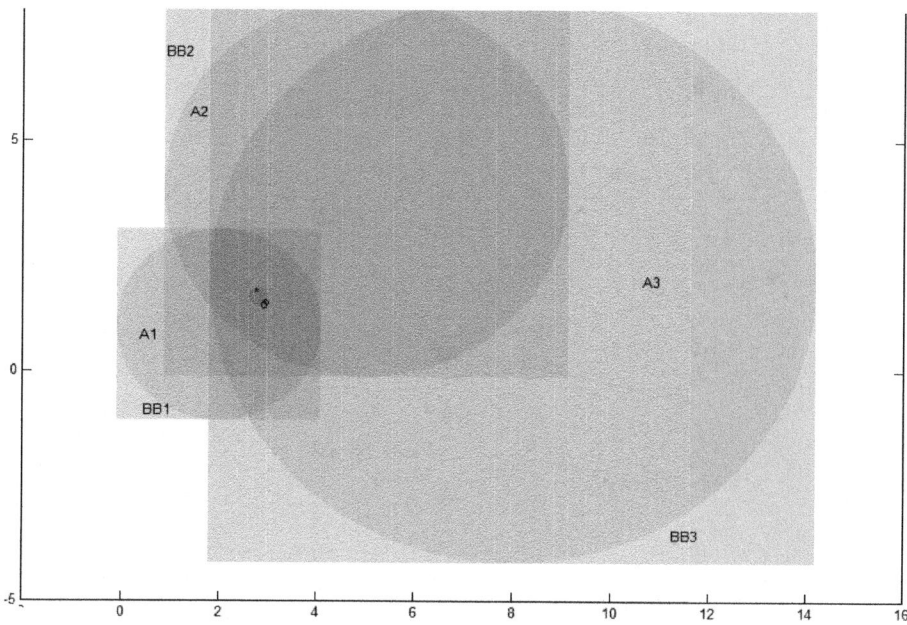

Fig. 7. Coordinates of unknown node estimated using min-max algorithm

Table 4. Co-ordinates Estimation

Anchor coordinates (x,y)		Distance between anchor node and unknown node(m)	Coordinates of the unknown node without min-max algorithm (x,y)	Coordinates of the unknown node with min-max algorithm (x,y)	Actual Coordinate
1	(2,1)	2.089	(2.12,1.75)	(2.94,1.47)	(3.3, 1.7)
2	(5,4)	4.132			
3	(8,2)	6.192			

localization algorithm. To analyze the energy consumption per packet transmission, a simple experimental setup is developed using zigbee series 1 RF module as illustrated in Figure 8. The various transmission power levels supported by zigbee series 1 RF module are lowest (0.16mW), low (0.25mW), medium (0.39mW), high (0.63mW) and highest (1mW). One experimental scenario is shown in Figure 8, in which two nodes A and B are kept at 8m distances apart.

The maximum distance (D) that can be covered by node A when operated at the highest transmission power in indoor environment is found to be 15m. Node A can communicate with node B as it is within its coverage, but it will definitely leads in reduction of its lifetime if it is continuously operated at the highest transmission

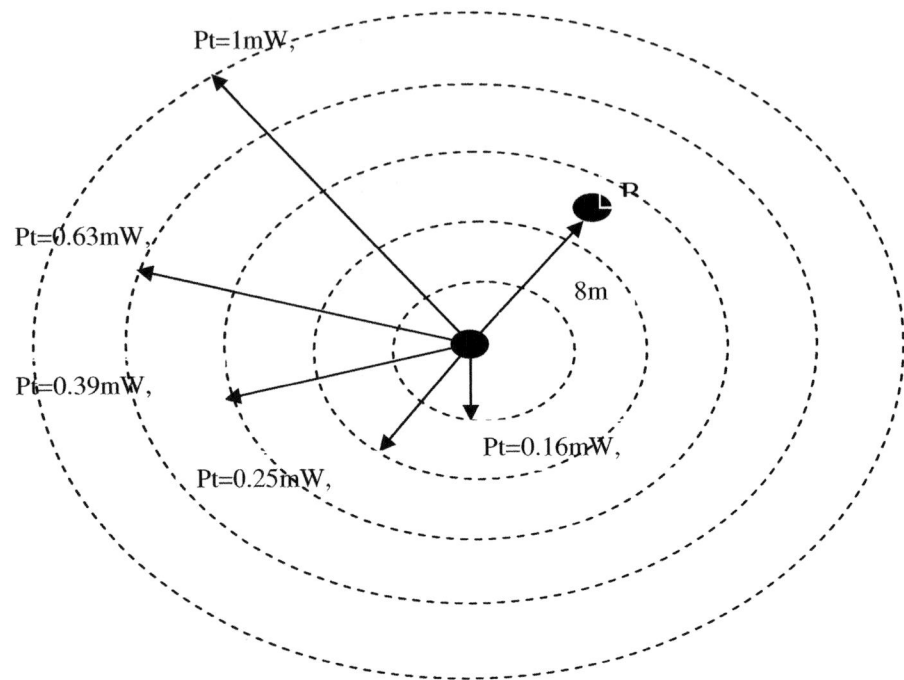

Fig. 8. Experimental setup to analyze the energy consumption

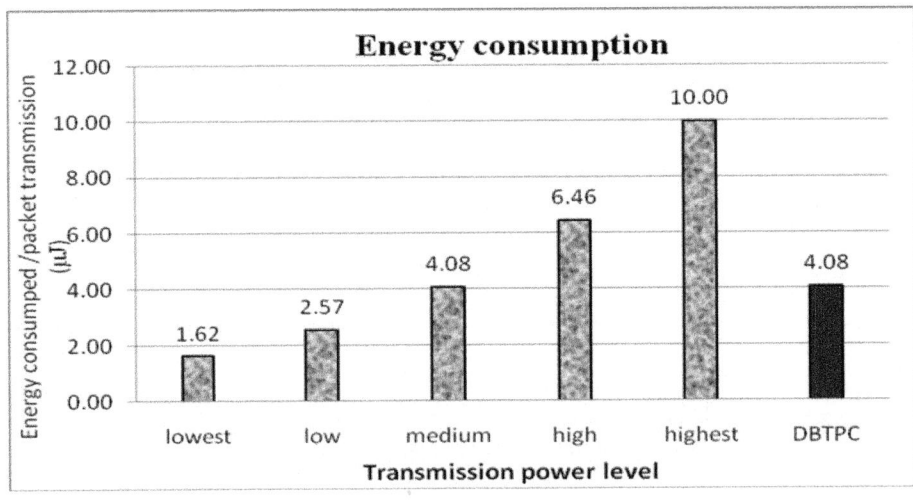

Fig. 9. Energy consumption vs Transmission Power level

power. By applying the proposed DBTPC scheme, node A dynamically adjusts its power level to medium power level to reach node B. Using equation 11, the energy consumed per packet transmission is computed for all the five power levels and plotted as shown in Figure 9. From the experimental result, it is found that by applying the proposed scheme, energy consumed per packet transmission is reduced to around 145%, when node A is operated in medium power level than in highest power level.

5 Conclusion and Future Work

In this paper, an efficient RSS based distributed localization algorithm and DBTPC scheme are proposed. The inaccuracies incurred in the RSS based localization algorithm are refined using Kalman filter estimator and trilateration with min-max algorithm. The energy consumption in the node for data transmission among neighboring nodes are reduced by the proposed DBTPC scheme which selects the Optimal Transmission Power based the distance information of the neighboring nodes. Hardware design and embedding the proposed DBTPC scheme in the transceiver module of the sensor node are under progress.

Acknowledgement

The authors would like to thank Tata Consultancy Services (TCS) for funding this project.

References

1. Yick, J., Mukherjee, B., Ghosal, D.: Wireless Sensor Network survey. ELSEVIER Journal on Computer Networks 52, 2292–2330 (2008)
2. Karl, H., Willig, A.: Protocols and Architectures for Wireless Sensor Networks. Wiley Publications, Chichester (2005)
3. Yedavalli, K., Krishnamachari, B., Ravula, S., Srinivasan, B.: Ecolocation: A Sequence Based Technique for RF Localization in Wireless Sensor Networks. IEEE Transactions on Mobile Computing 7(1), 81–94 (2005)
4. Zanca, G., Zorzi, F., Zanella, A., Zorzi, M.: Experimental comparison of RSSI-based localization algorithms for indoor Wireless Sensor Networks. In: REALWSN 2008, pp. 1–5 (2008)
5. Zhao, D., Men, Y., Zhao, L.: A hop-distance algorithm for self-localization of Wireless Sensor Networks. In: International Conference on Software Engineering, Artificial Intelligence and Parallel/Distributed Computing, vol. 2, pp. 108–112. IEEE computer society, Los Alamitos (2007)
6. Patwari, N., Ash, J.N., Kyperountas, S., Hero III, A.O., Moses, R.L., Correal, N.S.: Locating the nodes. IEEE Signal Processing Magazine, 54–69 (2005)
7. Sugano, M., Kawazoe, T., Ohta, Y., Murata, M.: Indoor localization system using RSSI measurement of Wireless Sensor Network based on ZIGBEE standard. In: International Conference on Wireless Sensor Networks, pp. 6–12 (2006)
8. Srbinovska, M., Gavrovski, C., Dimcev, V.: Localization Estimation System Using Measurement of RSSI Based on ZIGBEE Standard. In: Electronics 2008, pp. 45–51 (2008)

9. Reichenbach, F., Timmermann, D.: Indoor localization with low complexity in Wireless Sensor Networks. In: IEEE International Conference on Industrial Informatics, pp. 1018–1023 (2006)
10. Patwari, N., Hero, A.O.: Demonstrating Distributed Signal Strength Location Estimation. In: SenSys 2006, pp.353–354 (2006)
11. Onel, T., Ersoy, C., Delic, H.: Information Content-Based Sensor Selection and Transmission Power Adjustment for Collaborative Target Tracking. IEEE Transactions on Mobile Computing 8(8), 1103–1116 (2009)
12. Paschalidis, I.C., Lai, W., Starobinski, D.: Asymptotically Optimal Transmission Policies for Large-Scale Low-Power Wireless Sensor Networks. IEEE Transactions on Mobile Computing 15(1), 105–118 (2007)
13. Panichpapiboon, S., Ferrari, G., Tonguz, O.K.: Optimal Transmit Power in Wireless Sensor Networks. IEEE Transactions on Mobile Computing 5(10), 1432–1447 (2006)
14. Arroyo-Valles, R., Marques, A.G., Cid-Suerio, J.: Optimal Selective Transmission under Energy Constraints in Sensor Networks. IEEE Transactions on Mobile Computing 8(11), 1524–1538 (2009)
15. Log-distance path loss model, http://www.wikipedia.com
16. ITU Model for Indoor Attenuation, http://www.wikipedia.com
17. Rojas, R.: The Kalman Filter. Technical report, Freie University of Berlin (2003)
18. Larsson, E.G.: Cramer-Rao bound analysis of distributed positioning in sensor networks. IEEE Signal Processing Letters 11(3), 334–337 (2004)
19. Kay-I.: Cramer-Rao Bound and Minimum Variance Unbiased Estimator, ch. 3

A Novel Feature Vectors Construction Approach for Face Recognition

Paul Nicholl[1], Afandi Ahmad[2,3], and Abbes Amira[4]

[1] School of Electronics, Electrical Engineering and Computer Science
The Queens University, Belfast, Northern Ireland
P.Nicholl@qub@ac.uk
[2] Department of Electronic and Computer Engineering,
School of Engineering and Design, Brunel University, Uxbridge, United Kingdom
[3] Department of Computer Engineering,
Faculty of Electrical and Electronic Engineering,
Universiti Tun Hussein Onn Malaysia, Batu Pahat, Johor, Malaysia
afandia@uthm.edu.my
[4] Nanotechnology and Integrated Bio-Engineering Centre (NIBEC),
Faculty of Computing and Engineering, University of Ulster,
Jordanstown Campus, Co Antrim, Northern Ireland
A.Amira@ulster.ac.uk

Abstract. This paper discusses a novel feature vectors construction approach for face recognition using discrete wavelet transform (DWT). Four experiments have been carried out focusing on: DWT feature selection, DWT filter choice, features optimization by coefficients selection as well as feature threshold. In order to explore the most suitable method of feature extraction, different wavelet quadrant and scales have been studied. It then followed with an evaluation of different wavelet filter choices and their impact on recognition accuracy. An approach for face recognition based on coefficient selection for DWT is the presented and analyzed. Moreover, a study has been deployed to investigate ways of selecting the DWT coefficient threshold. The results obtained using the AT&T database have shown a significant achievement over existing DWT/PCA coefficient selection techniques and the approach presented increases recognition accuracy from 94% to 97% when the Coiflet 3 wavelet is used.

Keywords: Face recognition, discrete wavelet transform, coefficient selection, feature selection, feature optimization.

1 Introduction

In recent years, the demand for sophisticated security systems has risen significantly. Both commercial and governmental organisations require methods of protecting people and property. These often involve identifying people; to control access to resources or to detect individuals on a watch list. Solutions employing biometric techniques are being used widely to facilitate these needs [1].

M.L. Gavrilova et al. (Eds.): Trans. on Comput. Sci. XI, LNCS 6480, pp. 223–248, 2010.
© Springer-Verlag Berlin Heidelberg 2010

A variety of biometric approaches have been investigated or adopted. For example, fingerprint recognition [2] has been used in crime solving for many years and is being increasingly installed in consumer devices, such as laptop computers. It is generally accurate and can be deployed with minimal cost. It does, however, suffers from the requirement of many biometric methods that an individual being identified must be compliant in the process – two fingerprints (or sets of fingerprints) must be supplied in order to create a match. Some biometrics are less intrusive – voice scans [3] can be taken without user compliance, although accuracy is currently low.

Face recognition has received a large amount of attention from researchers in recent years [4]. It has the potential to provide a robust biometric which, although unlikely to exceed the accuracy of techniques like iris or fingerprint scanning, could fulfill the needs of many scenarios. Much of the interest in face recognition has been prompted by humans' own remarkable ability to recognize faces [5]. This ability encompasses recognition of faces from thousands of known individuals, even in cases where there is partial occlusion of the face, poor illumination, or there has been a change in appearance. Automatic face recognition also requires less compliance by the individual being identified. A face image for matching can be taken without the individual posing or even knowing that the image is being captured.

A multitude of techniques have been applied to face recognition and they can be separated into two categories: geometric feature matching and template matching. Geometric feature matching involves segmenting the distinctive features of the face, for examples eyes, nose, mouth, etc., and extracting descriptive information about them such as their widths and heights. Ratios between these measures can then be stored for each person and compared with those from known individuals [6].

On the other hand, template matching is a holistic approach to face recognition. Each face is treated as a two-dimensional (2-D) array of intensity values, which is compared with other facial arrays. Techniques of this type include principal component analysis (PCA) [7], where the variance among a set of face images is represented by a number of eigenfaces. The face images, encoded as weight vectors of the eigenfaces, can be compared using a suitable distance measure [8]. In independent component analysis (ICA), faces are assumed to be linear mixtures of some unknown latent variables. The latent variables are assumed non-gaussian and mutually independent, and they are called the independent components of the observed data [9]. In neural network models (NNM), the system is supplied with a set of training images along with correct classification, thus allowing the neural network to ascertain a weighting system to determine which areas of an image are deemed most important [10].

Hybrid multiresolution approaches have received much attention in recent years. The discrete wavelet transform (DWT) [11] has been used along with a number of techniques, including PCA [12], ICA [13] and support vector machines (SVM) [14]. DWT is able to extract features that are localized in both space and frequency by convolving a bank of filters with an image at various locations. However, to date, no systematic examination has been performed which determines how to best employ DWT for face recognition. The effect of employing different filters and scales has not been examined.

This research study attempts to investigate these issues. Initially, experimentation is performed using the Haar [15] and biorthogonal 4.4 [16] wavelets, in order to determine the most appropriate wavelet quadrants and scales. The study is then widened to cover a range of wavelets, with filters examined from the Daubechies [17], symlet [17], Coiflet [17] and biorthogonal [18] families. Results are analysed in order to ascertain whether scales and wavelet filters can be intelligently chosen for face recognition applications. In addition, an approach for face recognition based on DWT coefficient selection is presented and analysed. This operates by attempting to optimize the feature vectors produces by DWT, thereby improving results. An added benefit of the process is that it can automatically segment the face image, eliminating the need to manually crop images and possibly removing useful information.

The remainder of this paper is organized as follows. Section 2 investigates which DWT features should be utilized for face recognition. Section 3 analyses which wavelet filters perform best in this domain. Section 4 addresses the optimization of feature vectors using coefficient selection. Section 5 investigates how to choose a threshold for coefficient selection. Section 6 provides concluding remarks.

2   DWT Feature Selection

2.1   Concepts

In order to assess whether DWT can enhance face recognition system performance, a study is performed which attempts to determine how to employ it for this purpose. A number of variables are assessed, including: quadrant – which DWT quadrant(s) should be used for feature extraction?; scale – which scale(s) should be used for feature extraction?; and filter – which wavelet filters produce the best results?. This section attempts to address the first two points and experiments are conducted on the AT&T database. Each experiment is performed on coefficients taken from a specific wavelet scale and quadrant. A high-level overview of the recognition approach adopted is given in Figure 1.

2.2   Experiments

The experiments start with system training. For this stage, each training image undergoes wavelet transformation to the xth scale. DWT coefficients from the specified quadrant at the xth scale undergo PCA, producing a set of principal components. The training images are then projected onto the set of principal components, producing a weight vector for each image, which represents the features for the image. Probe images are processed in a similar manner, with each image decomposed to the same scale and coefficients from the same quadrant

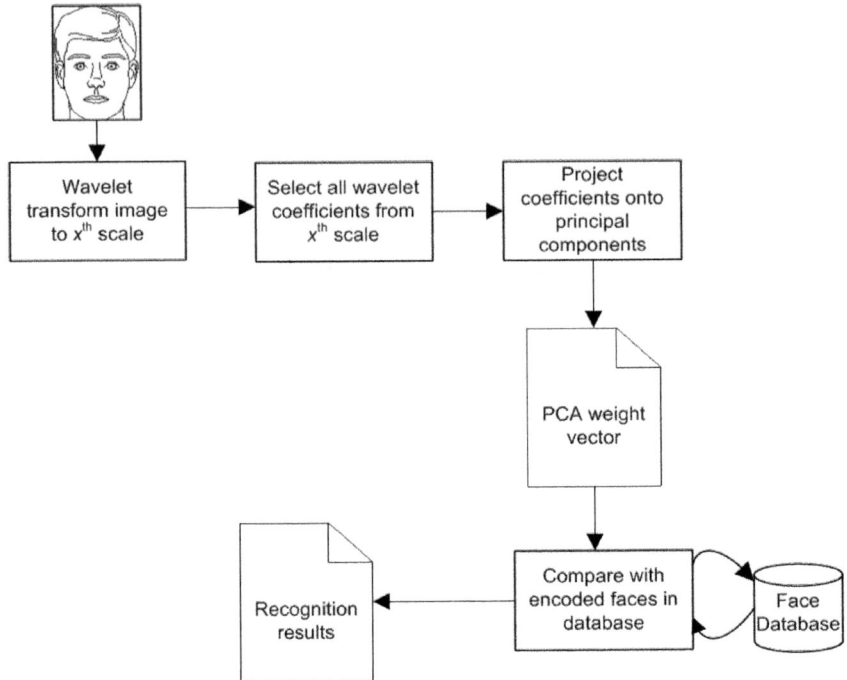

Fig. 1. Overview of recognition approach

extracted. This image is projected onto the same principal components and a weight vector is produced. This vector can be compared with those of training images. For the experiments described here, the Euclidean distance measure is employed.

For this study, five randomly-selected training images are used for each individual, with the remaining five being used as probe images. Other than minor re-scaling, the images undergo no preprocessing. As assessing wavelet filters is not the object of the experiments in this section, only two filters are adopted: Haar and biorthogonal 4.4. PCA is then employed, reducing the feature set further. The images used for system training also from the gallery set. As there are 200 training images, up to 200 principal components can be used to encode each face. However, when there are fewer than 200 features (pixels or wavelet coefficients) per training image (as is the case for higher wavelet scales), using more than 200 principal components is redundant. Encoded gallery images are compared with probe images using the Euclidean distance measure.

2.3   Results

This section presents recognition results for the Haar and biorthogonal 4.4 wavelets. The Haar wavelet has been chosen for its simplicity. The biorthogonal 4.4 wavelet has been chosen to represent a more sophisticated filter. Results are presented for the first five scales. It is worth mentioning that recognition accuracies for the sixth scale are significantly lower, due to the reduced number of coefficients at this scale.

The results for the Haar wavelet are shown in Figures 2 to 6. As can be seen from the graphs, the choice of scale does have a significant effect on recognition rate. For example, in Figure 2, the first and second scales perform better initially, however, the recognition rates fall sharply as more eigenvectors are used. This would suggest that, for HH quadrants, most of the useful information in these scales is encoded within the first 20 eigenvectors. The results for the third scale are more consistent, although they do not match the second scale peak recognition rate of 66.5%. The results for the fourth scale and fifth scale are lower, with the recognition rates leveling off at 64 and 16 eigenvectors respectively, due to the number of coefficients per quadrant at these scales being 16 and 64, respectively.

Fig. 2. Recognition results for Haar wavelet and HH quadrant

Figures 3 and 4 show results for LH and HL quadrants. In both cases, the fourth scale produces the best results, followed by the third. In addition, recognition rates for the first two scales significantly deteriorate as eigenvectors increase, whereas this does not occur for the third, fourth and fifth scales. Peak recognition rates are higher than for HH, with 74% of faces recognized correctly using the LH quadrant and 78% with the HL quadrant. Figure 5 provides results for the LL quadrant. By a significant margin, the best results with the Haar wavelet are achieved using this quadrant. The best scale for LL is the third, producing recognition rates up to 95%. Scales 2, 1 and 4 produce similar performance, with maximum recognition rates of 94%, 93% and 93%,

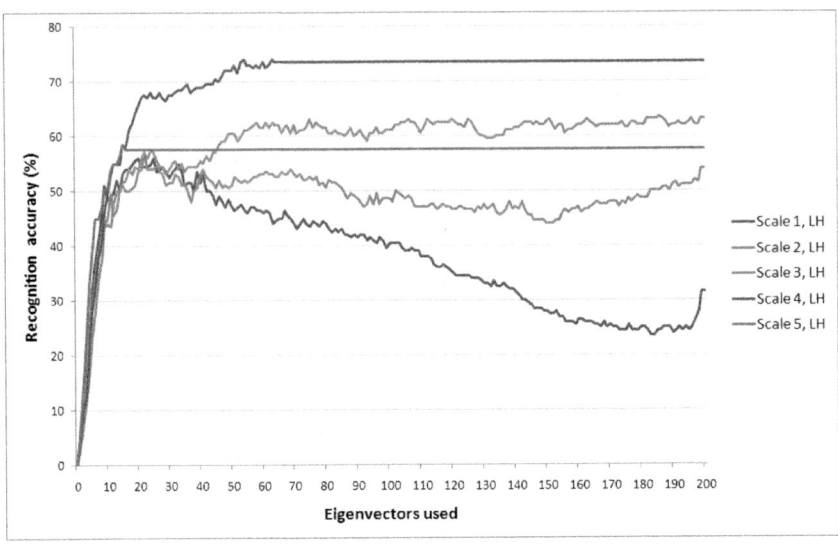

Fig. 3. Recognition results for Haar wavelet and LH quadrant

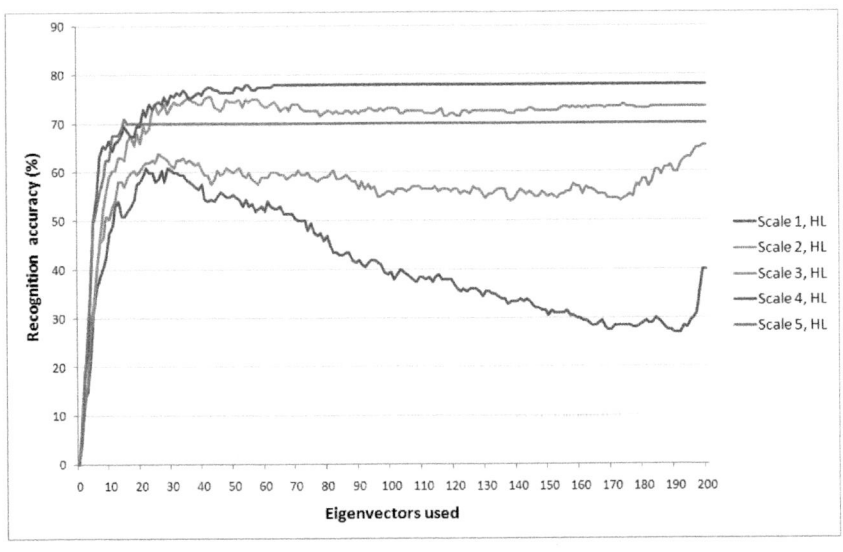

Fig. 4. Recognition results for Haar wavelet and HL quadrant

respectively. Scale 5 matches up to 87.5% of faces correctly. Figure 6 illustrates the extent to which the LL quadrant outperforms other quadrants.

Figures 7 and 8 show performance for the biorthogonal 4.4 wavelet. For coefficients from the LL quadrant, recognition rates are similar to those for Haar: scales 2 and 3 correctly recognize up to 94.5% of faces, with scales 1 and 4 recognizing 94% and 93%, respectively. From Figure 8, it can be seen that

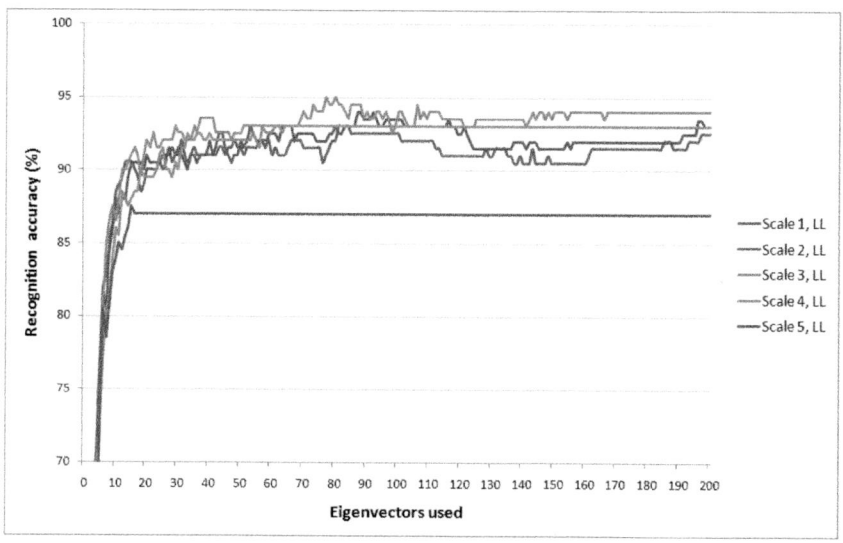

Fig. 5. Recognition results for Haar wavelet and LL quadrant

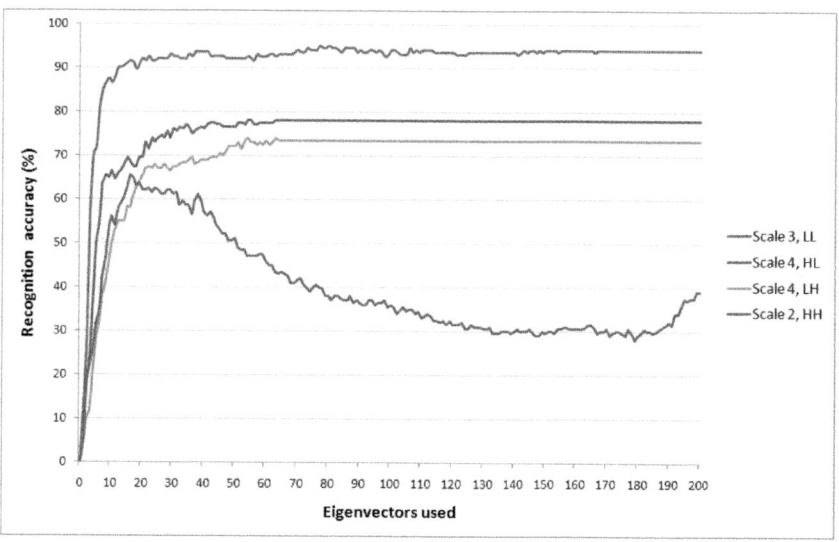

Fig. 6. Best recognition results for each quadrant using the Haar wavelet

the LL quadrant significantly outperforms the other quadrants. When compared with the Haar results, the most significant difference is that the best-performing scales for the LH and HL quadrants are the second and fifth respectively, as opposed to the fourth. However, the significance of this is minimal, due to the wide margin between these quadrants' results and those for LL.

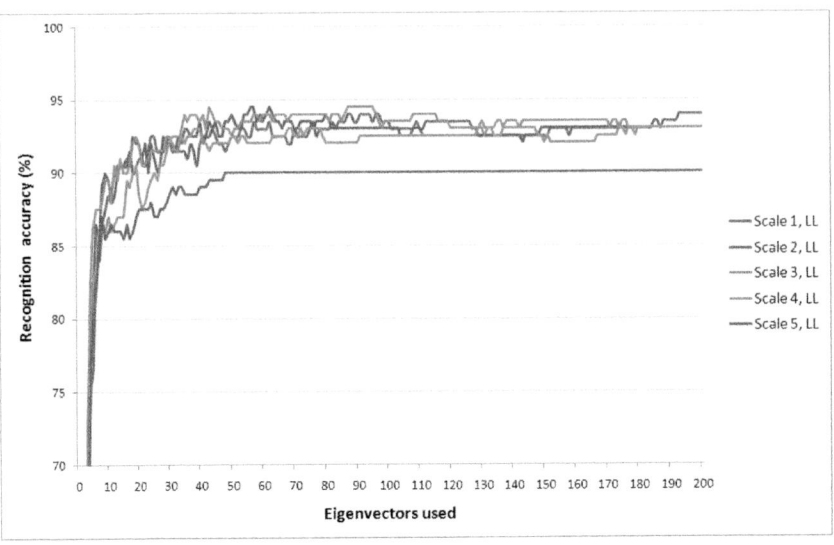

Fig. 7. Recognition results for biorthogonal 4.4 wavelet and LL quadrant

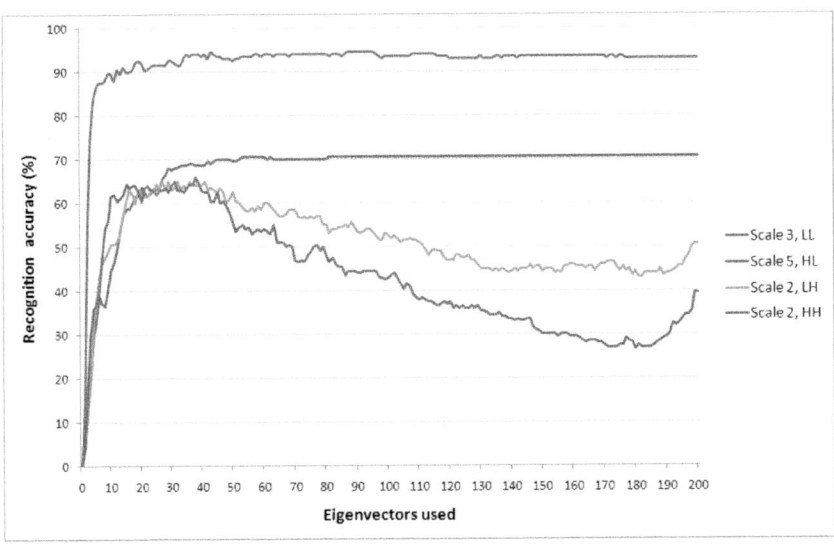

Fig. 8. Best recognition results for each quadrant using the biorthogonal 4.4 wavelet

The results achieved for these experiments help to guide decisions regarding the experiments that are still to be performed. When DWT coefficients are used for training a PCA-based recognition system, those from the LL quadrant appears to be much more discriminative in the process of face classification. As other quadrants isolate high-frequency features such as edges, small errors in alignment or facial expression between the images will significantly detract from accuracy. Conversely, the LL quadrant benefits from the removal (or reduction in impact) of high-frequency features. The consequence from these conclusions is that quadrants other than LL need not be investigated further. Remaining experiments will focus on observations from the LL quadrant. The effect of scale in the LL quadrant is less clear. Although the third scale produced best results for both wavelet filters tested, there was less variation between results for different scales than there was for different quadrants. It would therefore be appropriate to investigate the effect of scale further in remaining experiments.

3   DWT Filter Choice

3.1   Concepts

In this section, a study is performed to determine whether the choice of wavelet filter has a significant effect on recognition accuracy. Various wavelet families exist, each providing a different compromise between compactness and smoothness. Within a family, individual wavelets vary in the number of vanishing moments they contain. A vanishing moment limits a wavelet's ability to represent polynomial behavior or information in a signal. For example, a wavelet with one moment easily encodes polynomials of one coefficient, or constant signal components. A two moment wavelet encodes polynomials with two coefficients, i.e. constant and linear signal components; and three moment wavelets encode 3-polynomials, i.e. constant, linear and quadratic signal components.

Most wavelets can be described as orthonormal, meaning that they have a unit magnitude and are orthogonal. The consequence of having a unit magnitude is that convolution of a signal with a wavelet does not change the total energy of the signal. Orthogonality indicates that the inner product of the wavelet basis functions at different scales is zero. A signal can therefore be completely represented using a finite number of wavelet basis functions. The same wavelet filters are generally used for decomposition and reconstruction.

3.2   Experiments

Four wavelets are tested from each of the following wavelet families shown in Table 1. Matlab is used for experimentation and the filters are provided by the Matlab wavelet toolbox. As before, the AT&T database is used for experimentation, with five training images and five testing images used for each individual. Only the LL quadrant is used for feature extraction, at scales 1 to 5.

3.3   Results

This section presents recognition results for the examined wavelet filters. Figures 9 to 12 provide the results. Choice of wavelet family seems to have little effect on the maximum possible recognition rate – filters from the Daubechies and biorthogonal wavelet families matched up to 96.5% of faces correctly, whereas

Table 1. Wavelets filters descriptions

Wavelets	Descriptions
Daubechies	Designed to have the maximum possible number of vanishing moments for a given support size. Daubechies wavelets are widely used for solving a broad range of problems, such as identifying self-similarity properties of a signal, signal discontinuities, etc.
Symlet	Designed to be more symmetrical than Daubechies wavelets, yet compactly supported. Their increased symmetry makes them more adept at analysis of a signal than Daubechies wavelets.
Coiflet	Designed to have the maximum number of vanishing moments for both the wavelet and scaling filters, they are efficient at extracting the low frequency information in a signal.
Biorthogonal	Use separate filters for decomposing (or analyzing) and reconstructing signals. This allows each filter to be optimized for its specific purpose. Although the decomposition filters are not orthogonal with respect to the reconstruction filters, the two sets of filters are orthogonal with respect to themselves. Biorthogonal wavelets have been found to be effective in image processing applications such as fingerprint image compression [19].

filters from the symlet and coiflet families recognized 97%. The choice of filter within a wavelet family seems to be more significant. For example, although the biorthogonal 5.5 wavelet matches up to 96.5% of faces correctly, the biorthogonal 3.3 wavelet only reaches 93%. The exact nature of the relationship between wavelet and recognition performance however is unclear.

The number of non-zero coefficients in a wavelet filter (known as support size) has a number of effects on the performance of the wavelet. Filters with a larger support size are more adept at analyzing and representing complex features contained within the signal/image, however, they are more likely to be affected by artifacts at the edge of the image. Computational complexity of the wavelet transform is also increased when filters with larger support sizes are used.

Figure 13 illustrates the relationship between the support size of the low-pass filter for each wavelet with the maximum recognition rate for the filter. As all coefficients are taken from LL quadrants, only the low-pass filter is employed to calculate them. The graph shows all the maximum recognition rates achieved, for all wavelets and scales tested. The graph reveals little correlation between the two parameters. Although accuracy is highest for wavelets with a support size of 30 (Daubechies and symlet 15) and 40 (Daubechies and symlet 20), it drops again for wavelets with a support size of 50 (Daubechies and symlet 25). However, as larger filters are known to be more affected by boundary conditions, and the images used for experimentation are relatively small, this is not unexpected. Figure 14 presents the maximum recognition rates for each scale. Accuracy for

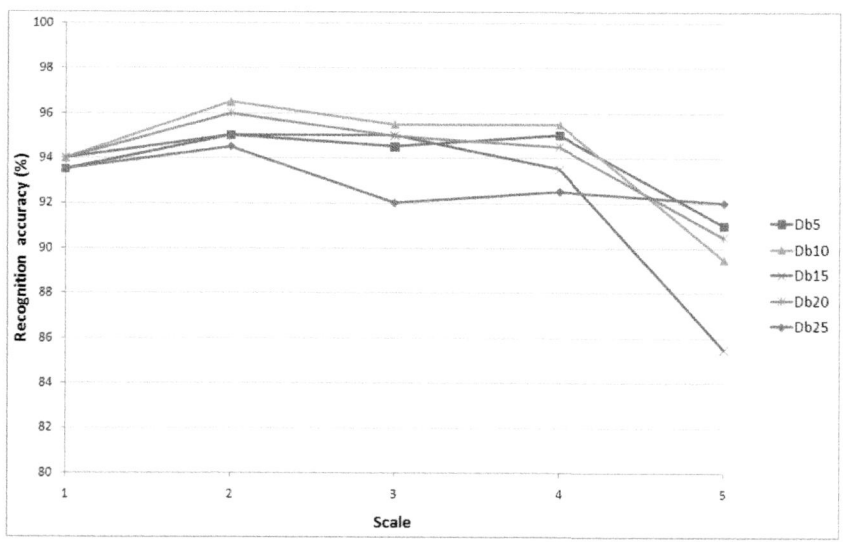

Fig. 9. Maximum recognition rates for selected Daubechies wavelets

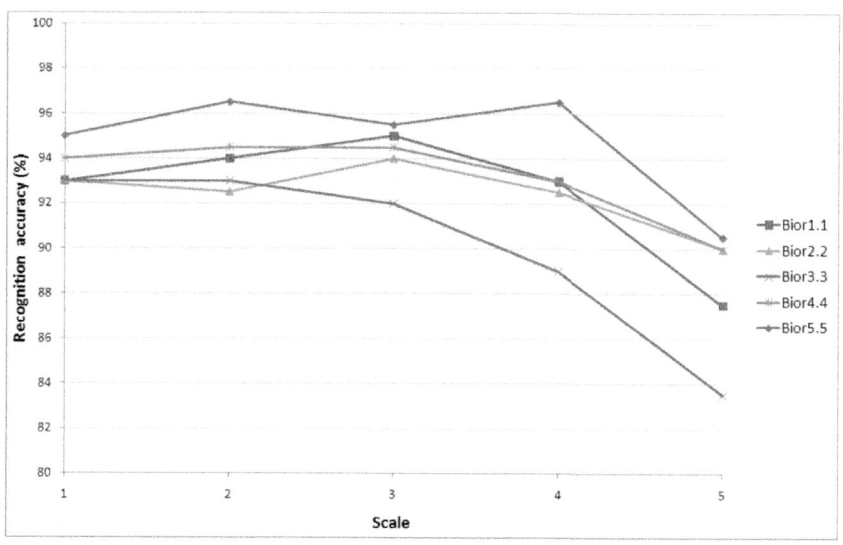

Fig. 10. Maximum recognition rates for selected biorthogonal wavelets

scales 2, 3 and 4 are similar to each other, with scale 1 providing slightly lower accuracy and scale 5 significantly lower than the best three. Scale 2 appears to provide somewhat more consistent accuracy than scales 3 or 4.

Figure 15 compares one of the best-performing wavelet filters (biorthogonal 5.5, 4th scale) against recognition in the spatial domain. As can be seen form the graph, recognition accuracy is increased significantly, with maximum recognition

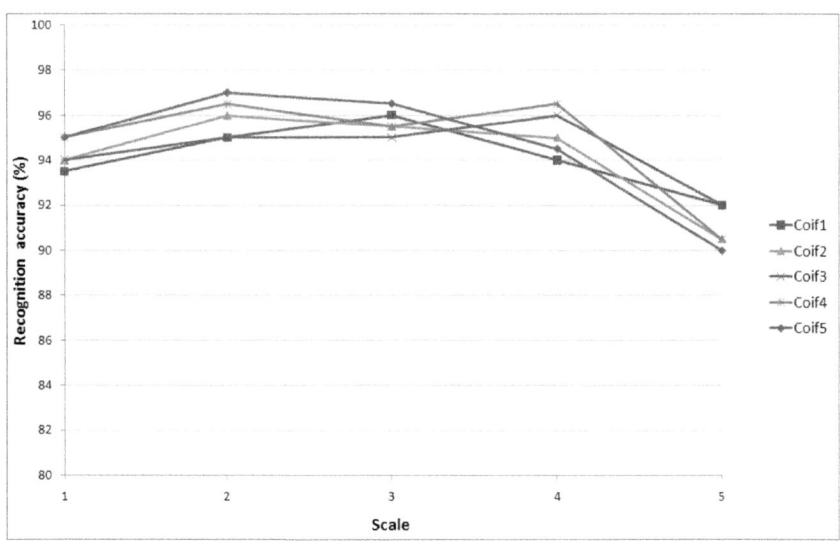

Fig. 11. Maximum recognition rates for selected coiflet wavelets

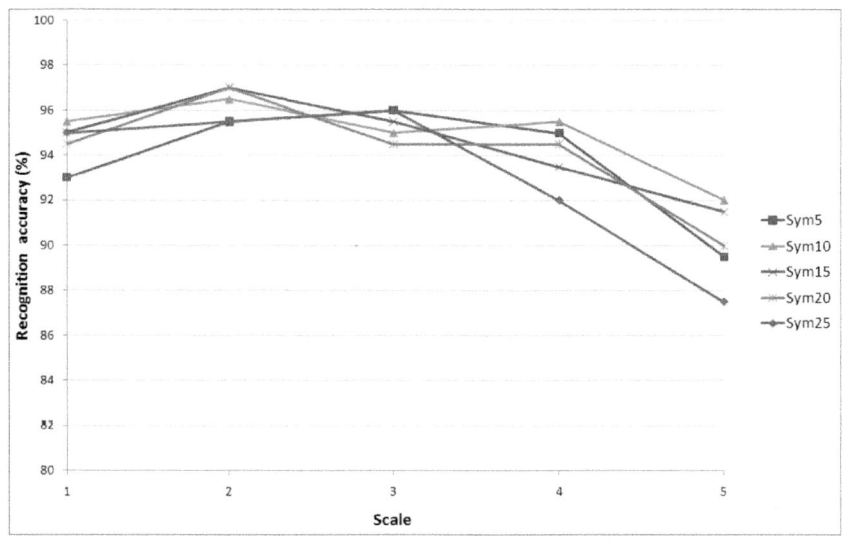

Fig. 12. Maximum recognition rates for selected symlet wavelets

rates increasing from 93% to 96.5%. This corresponds to 50% decrease in the number of incorrectly classified images.

Sample execution times are provided in Table 2. Training and classification times are given for images in the spatial and wavelet domains (biorthogonal 5.5 wavelet). Training time decreases from 0.122 seconds per image in the spatial domain to between 0.0480 and 0.0610 seconds in the wavelet domain. Although

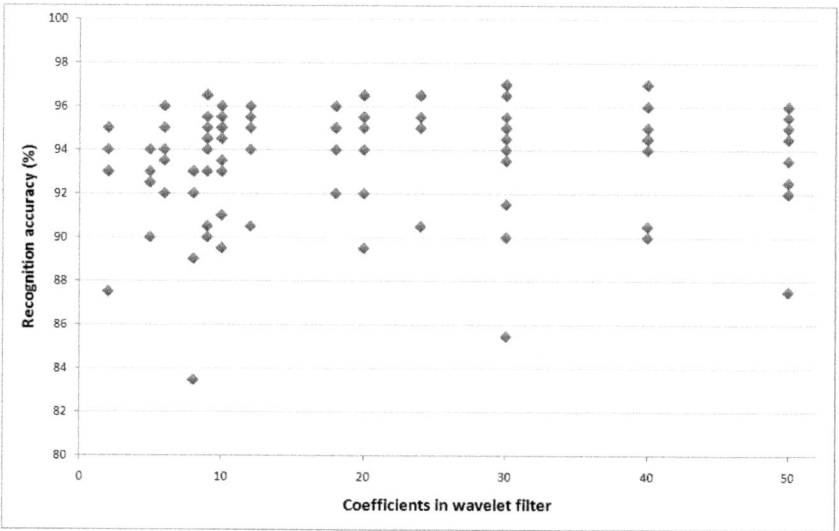

Fig. 13. Maximum recognition rate and support size of low-pass filter for all tested wavelets and scales

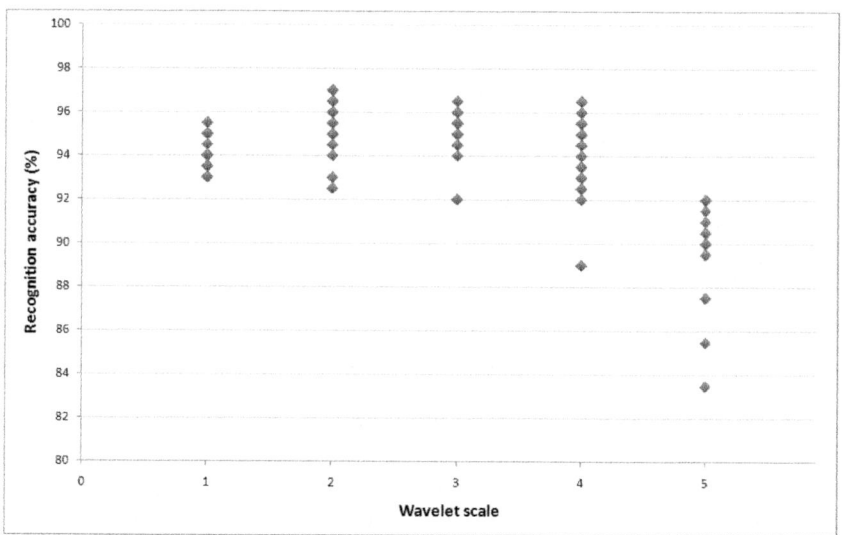

Fig. 14. Maximum recognition rate and scale for all tested wavelets and scales

there is a time penalty involved in performing DWT, this is offset by the resulting reduction in coefficients during PCA training. Classification time per spatial domain image is 0.0375 seconds and ranges from 0.0293 to 0.0479 seconds for images in the wavelet domain. The execution times were obtained using a single processor 2.4 GHz Pentium 4, with 512 MB of RAM. (Although the hardware

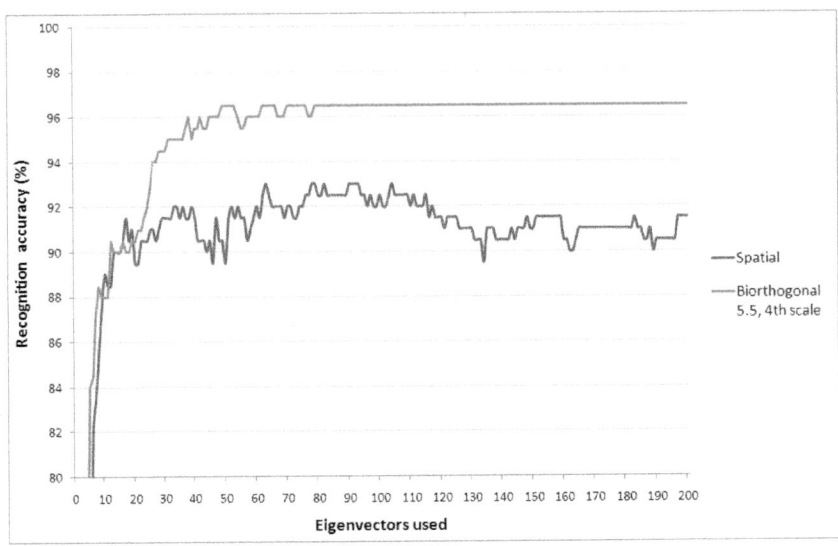

Fig. 15. Comparison of recognition results for biorhogonal 5.5 wavelet, 4th scale with results for recognition in the spatial domain

Table 2. Comparison of training and classification times for AT&T database images in the spatial and wavelet domains (biorthogonal 5.5 wavelet)

Scales	Training time per image (s)	Classification time per image (s)
Spatial	0.163	0.0375
DWT, 1st scale	0.0610	0.0293
DWT, 2nd scale	0.0509	0.0373
DWT, 3rd scale	0.0482	0.0402
DWT, 4th scale	0.0480	0.0420
DWT, 5th scale	0.0517	0.0479

used is not of a high specification, the execution times would differ only in magnitude if more powerful equipment had been used.)

To summarize, it is clear that DWT has the potential to significantly enhance recognition rates for PCA-based face recognition. For the AT&T database, maximum recognition rates increase from 93% for recognition in the spatial domain to 97% in the wavelet domain. There is not a substantial difference between recognition rates for the wavelet families tested, although coiflet filters produced slightly more consistent results. Across all the tested wavelet filters, there was no strong correlation between the support size of the low-pass filter and the results. Scale did appear to have an effect on results, with the 2nd scale slightly outperforming the 3rd and 4th scales. The 1st scale produced slightly lower results, with the 5th scale performing significantly worse.

4   Optimizing Features by Coefficient Selection

4.1   Concepts

In this section, an approach for face recognition based on coefficient selection for DWT is presented and analyzed. One problem with many face recognition techniques is that the areas of the face images to be used for recognition have to be chosen. Images are often cropped by creating an arbitrary bounding box around the face and discarding the information outside the box [20]. There is often a trade-off between ensuring that the most relevant parts of a face image are selected for recognition and removing information that is not useful or may detract from the process.

Similarly, with PCA, eigenvectors are ordered by their corresponding eigenvalues, with the vectors with the highest eigenvalues being used to encode the face images [7]. A number of variations of this approach include excluding the initial eigenvector, or choosing eigenvectors based on their energy values. However, the number of eigenvectors chosen and the number discarded are often arbitrary choices.

The recognition approach is based on standard DWT/PCA face recognition. Figure 16 provides a general overview of the system. As can be seen from the diagram, face images firstly undergo DWT coefficient selection, followed by PCA coefficient selection. The output from this stage is a coefficient vector, which is compared with those of the gallery face images. Recognition results are returned as the identities of the most likely matches in the database.

The purpose of DWT coefficient selection is to select the most discriminative DWT coefficients. Each training image undergoes wavelet decomposition to

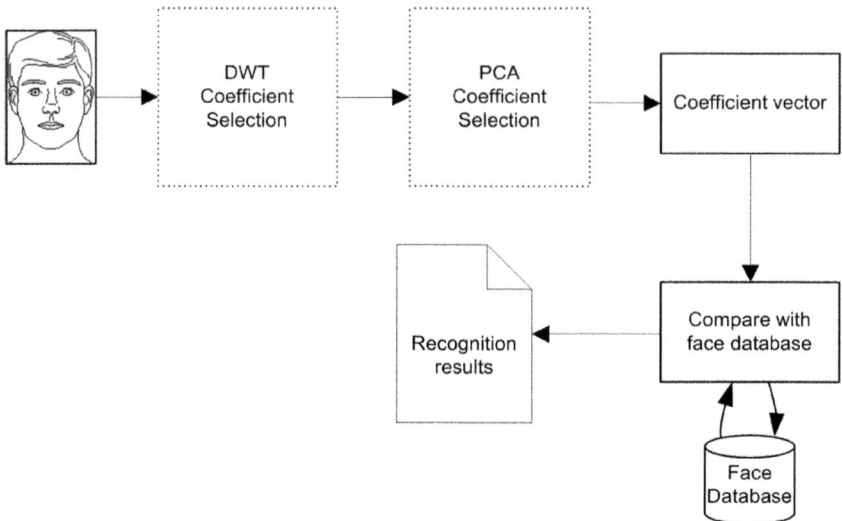

Fig. 16. System overview

a specified scale, with the low-pass coefficients being selected to form the image's observation vector. The distribution of these coefficient values is then examined to determine each coefficient's discriminative power. The inter-class and intra-class standard deviations for each coefficient are calculated and the ratio of these two values is determined. This ratio indicates how tightly the coefficient's values are clustered within each class, compared to the spread within the complete training dataset. The selection of DWT coefficients is therefore based on the maximisation of the following criterion:

$$J = \frac{\sigma_{inter}\left(A_m\right)}{\sigma_{intra}\left(A_m\right)} \tag{1}$$

where $\sigma_{inter}\left(A_m\right)$ and $\sigma_{intra}\left(A_m\right)$ represent inter-subject and intra-subject standard deviation spanned by DWT coefficients in the feature space A_m respectively. The DWT coefficients with the highest ratios are the most discriminative and chosen for recognition.

Figure 17 shows the steps involved in DWT coefficient selection. Figure 18 provides an illustration of the ratios calculated for a set of faces. Brighter areas in the image represent DWT coefficients with higher inter-class to intra-class ratios. As can be seen, brighter areas include the eyes, nose, mouth and the outline of the face. As would be expected, the image background and to a lesser extent, areas such as the forehead have lower values and have therefore been deemed to be less discriminative.

The second component of the approach is PCA coefficient selection. This initialises by performing PCA on the selected DWT coefficients, creating a set of eigenvectors and associated eigenvalues. Each training face's DWT coefficients are then projected onto the eigenvectors, producing a projection vector for each image. Eigenvectors are generally ordered by descending corresponding eigenvalues and selected using one of a number of approaches:

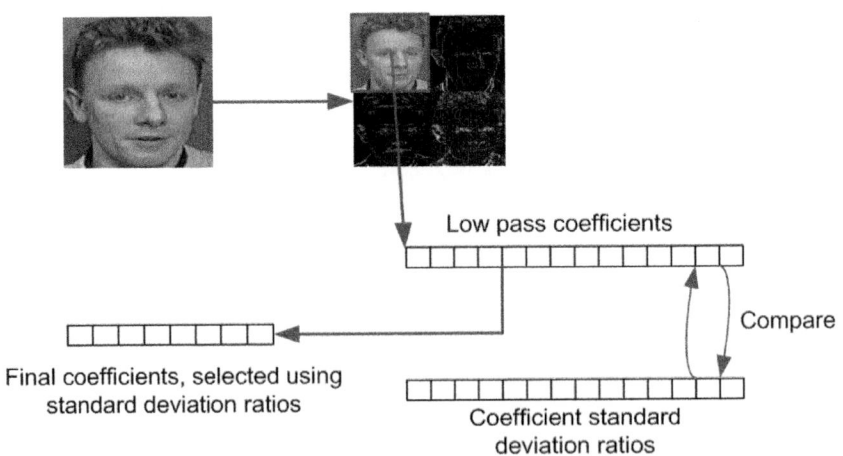

Fig. 17. DWT coefficient selection

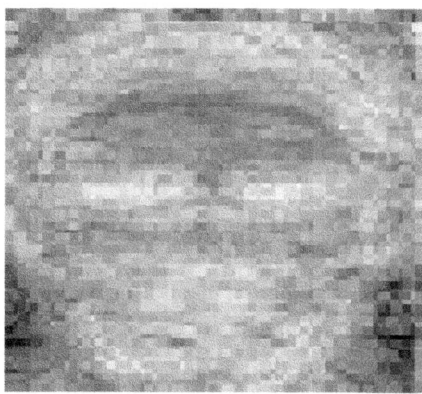

Fig. 18. Inter to intra-class ratios of DWT coefficients

- All eigenvectors corresponding to non-zero eigenvalues are used to create the eigenspace;
- The first x eigenvectors are chosen, where x often corresponds to 60% of the total eigenvector set [8];
- All eigenvectors are used apart from the first, which usually represents mostly variation in illumination [8];
- Eigenvectors are chosen based on energy values, with the first y being selected so that their cumulative energy exceeds a predetermined percentage of the total energy of all eigenvectors [21];
- Eigenvectors can be chosen based on their stretch values, where the stretch of an eigenvector is the ratio of its eigenvalue over the maximum eigenvalue [21]; and
- Eigenvectors can be chosen based on the ratios of inter-class to intra-class variance values, where those with the highest values are deemed most discriminative and selected [22].

The approach adopted for this study is based on the inter-class to intra-class standard deviation ratios. As with DWT coefficient selection, the ratios of inter-class to intra-class standard deviations are calculated. Projection coefficients with the highest ratios indicate that the associated eigenvector is highly discriminative and may contribute to better recognition accuracy. This method eliminates the need to guess which eigenvectors represent mostly variation in image illumination. Once training is complete and the most discriminative eigenvectors have been selected, classification can be performed using a simple distance measure, such as Euclidean. The adoption of this approach brings together similar coefficient selection strategies for both stages of the feature vector selection – DWT coefficient selection and PCA eigenvector selection.

4.2   Experiments

Experiments are performed which determine the benefits of DWT coefficient selection and PCA eigenvector selection separately, as well as in a combined

Table 3. Comparison of DWT coefficient selection recognition rates with those of standard DWT/PCA approach, along with percentages of DWT coefficients required to achieve maximum rate

| Wavelet | Scale | Recognition Rate (%) | | | Coefficients Required (%) |
		Standard Approach	Coefficient Selection	Increase (%)	
Haar	1	93	95	2	66
	2	94	95	1	50
	3	95	96	1	58
	4	93	95.5	2.5	68
Biorthogonal 4.4	1	94	96	2	69
	2	94.5	96	1.5	78
	3	94.5	96.5	2	83
	4	93	94	1	95
Coiflet 3	1	94	97	3	73
	2	95	97	2	85
	3	95	97	2	98
	4	96	96	0	95
Daubechies 10	1	94	95.5	1.5	66
	2	96.5	97.5	1	99
	3	94	96.5	2.5	75
	4	95.5	97	1.5	98
Symlet 10	1	95.5	96.5	1	99
	2	96.5	96.5	0	90
	3	95	95	0	90
	4	95.5	95.5	0	92
				Average Increase (%):	1.37

framework. As the technique is more suited to face data sets with little variation in pose/location, the AT&T database of faces is used for experimentation. The images contain variation in lighting, expression and facial details (for example, glasses/no glasses). For the experiments described in this study, five images for each individual are used for system training, with the other five used for testing.

4.3   Results

A number of wavelet filters are investigated, and decomposition is performed to between one and four levels. Selection percentages from 1% to 100% are tested and PCA is used for classification. Where the selection percentage is 100%, this is equivalent to no coefficient selection being applied. Results are shown in Table 3.

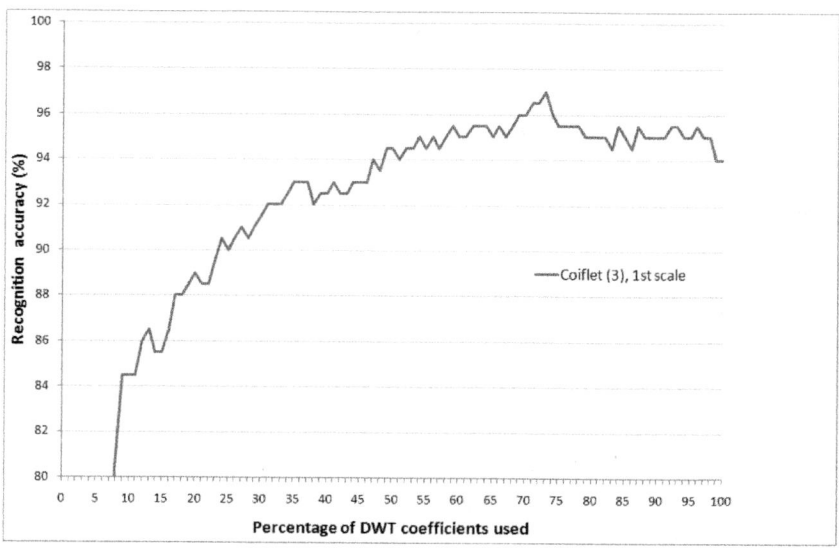

Fig. 19. Recognition rates for various DWT coefficient selection percentages, using Coiflet 3 wavelet, 1st scale

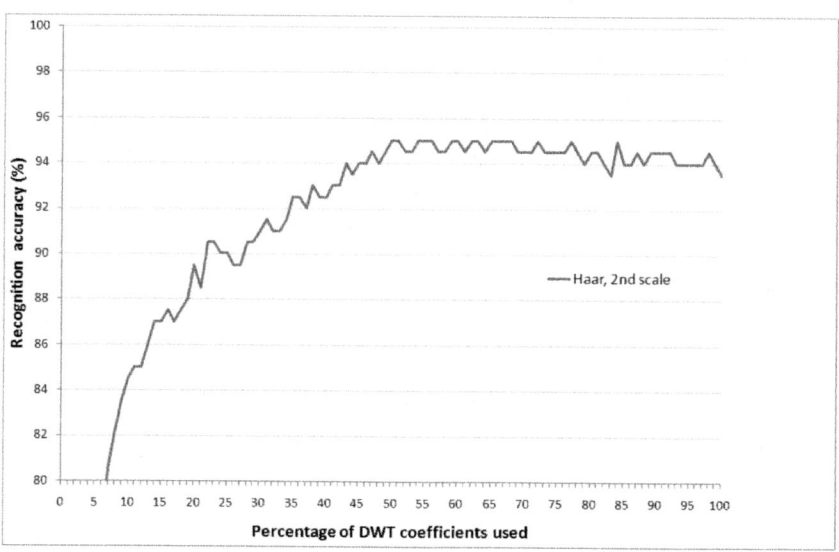

Fig. 20. Recognition rates for various DWT coefficient selection percentages, using Haar wavelet, 2nd scale

The results show that DWT coefficient selection has increased maximum recognition rate in 16 out of the 20 cases tested. The percentages of coefficients required to achieve the new maximum are also shown. In one case: Coiflet 3, 1st

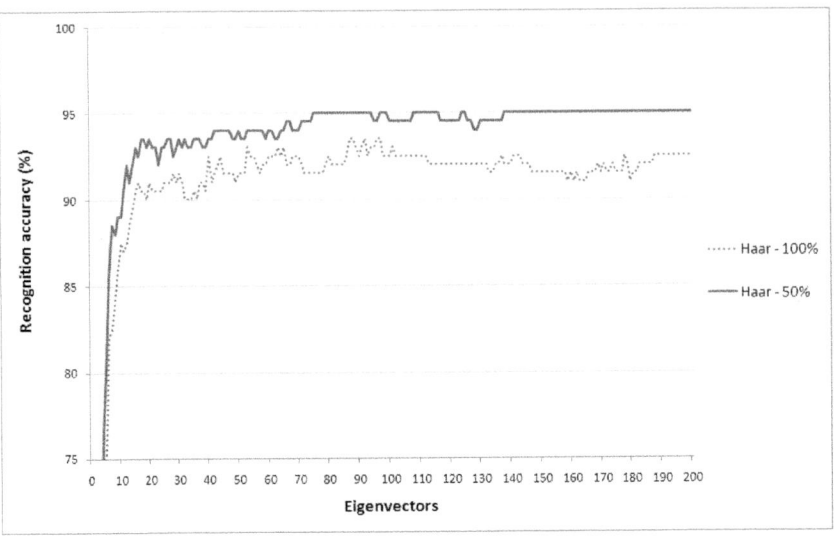

Fig. 21. Haar 2nd scale, all coefficients vs. top 50% coefficients

scale, the recognition rate has risen from 94% to 97%, which corresponds to 43% reduction in incorrectly-classified faces. The graph in Figure 19 provides more detail for this case. As the percentage of DWT coefficients increases, recognition accuracy also increases until 73% of coefficients are used. The trend then changes, with the recognition rate generally decreasing as the remaining coefficients are added. Another example: Haar, 2nd scale as can be seen in Figure 20, where the maximum recognition rate of 95% is reached with 50% of coefficients. This case is shown in more detail in Figure 21, which compares recognition accuracies for each of the two cases (50% vs. 100% of coefficients) for varying numbers of eigenvectors. The graph illustrates the full benefit of the approach, as the recognition rate for 50% of coefficients is consistently better than that for 100%.

Table 4. Comparative results on AT&T database

Method	Accuracy (%)	References
DCT/HMM	84	[23]
ICA	85	[24]
Weighted PCA	88	[25]
Gabor filters & rank correlation	91.5	[26]
2D-PHMM	94.5	[27]
NMF	96	[28]
TNPDP	96.5	[29]
LFA	97	[30]
DWT/PCA with coefficient selection	97.5	Proposed

The average increase in accuracy for this case is 2.7%. Similar improvements were seen across other wavelets and scales.

As Table 4 shows, the approach described compares well with other techniques from the literature that have used this training set. It should be noted that although the AT&T database is relatively small, the technique could be extended to other face databases. However, the coefficient selection approach is particularly suited to data sets with little variation in pose and alignment, therefore, images would have to undergo a normalization step prior to recognition. If this was performed, it is expected that results for other databases would be similar to those for the AT&T database.

5   Feature Threshold

In this section, a study is performed to investigate ways of choosing the DWT coefficient selection threshold. Although the recognition increases offered by DWT coefficient selection are significant, they are only achievable through a judicious choice of threshold. The maximum possible increases in accuracy offered by DWT coefficient selection can be seen in Table 3. Increases in recognition accuracy range from 0% to 3%, with the average increase being 1.37%. However, the results presented are the best for each wavelet and scale, found after tests employing varying numbers of DWT coefficients. For coefficient selection to be viable, the number of DWT coefficients to use as features must be chosen automatically. Two approaches are investigated for choosing this threshold.

5.1   Percentage Midpoint Average (PMA)

The first approach is referred to as percentage midpoint average (PMA). PMA assumes that a number of tests runs have been carried out with appropriate wavelets and scales, and full accuracy data obtained. For each test set, the minimum percentage of DWT coefficients required to produce the maximum recognition accuracy is recorded. The highest percentage of DWT coefficients producing the same maximum accuracy is also noted. The average of these two figures is then calculated, as the percentage midpoint for the current test set. The average of the percentage midpoints for all the test runs is calculated, with this percentage being chosen as the selection threshold.

Tests are performed on the AT&T database to determine the effectiveness of this approach. The PMA value is calculated from recognition results obtained previously, and determined to be 81.36%. DWT coefficient selection results, using 81.36% of coefficients, are shown in Table 5. The results indicate that this approach is not effective, with recognition accuracy decreasing by an average of 0.025% from the results obtained without DWT coefficient selection. This is not unexpected, as the approach is not sophisticated. It assumes that the same percentage of coefficients should be chosen in each case, regardless of the choice of wavelet filter and scale or the individual characteristics of the data set, such as the amount of background (non-face) in the image.

Table 5. Maximum recognition rates using DWT coefficient selection with PMA threshold

Wavelet	Scale	Recognition Rate (%)		
		All coefficients	PMA	Increase (%)
Haar	1	93	94	1
	2	94	94	0
	3	95	94.5	-0.5
	4	93	94.5	1.5
Biorthogonal 4.4	1	94	95	1
	2	94.5	95	0.5
	3	94.5	96.5	2
	4	93	92	-1
Coiflet 3	1	94	94.5	0.5
	2	95	97	2
	3	95	94	-1
	4	96	94	-2
Daubechies 10	1	94	94.5	0.5
	2	96.5	96	-0.5
	3	94	96.5	2.5
	4	95.5	93	-2.5
Symlet 10	1	95.5	95.5	0
	2	96.5	95	-1.5
	3	95	93.5	-1.5
	4	95.5	94	-1.5
		Average Increase (%):		-0.025

5.2   Optimal Ratio Average (ORA)

The second approach is referred to as optimal ratio average (ORA). As with PMA, ORA assumes that a number of tests runs have been carried out with appropriate wavelets and scales, and full accuracy data obtained. As explained in previously, DWT coefficient selection operates by calculating the ratios of inter-class to intra-class standard deviations for each coefficient: this value is used to select the most discriminative coefficients. In ORA, the cut-off ratio that produces the highest recognition rate for each test run is recorded. The average of the cut-off ratios for all test runs is chosen as the selection threshold.

Tests are performed on the AT&T database to determine the effectiveness of this approach. The ratio threshold value is calculated from the DWT coefficient selection results obtained previously. Unlike with PMA, a different percentage of DWT coefficients may be chosen for each wavelet and scale, depending on how discriminative its coefficients are. Results are provided in Table 6 and indicate

Table 6. Maximum recognition rates using DWT coefficient selection with ORA threshold

| Wavelet | Scale | Recognition Rate (%) | | |
		All coefficients	ORA	Increase (%)
Haar	1	93	94.5	1.5
	2	94	94.5	0.5
	3	95	94.5	-0.5
	4	93	94	1
Biorthogonal 4.4	1	94	95.5	1.5
	2	94.5	95	0.5
	3	94.5	95.5	1
	4	93	93.5	0.5
Coiflet 3	1	94	95.5	1.5
	2	95	96.5	1.5
	3	95	96	1
	4	96	96	0
Daubechies 10	1	94	96	2
	2	96.5	96	-0.5
	3	94	96.5	2.5
	4	95.5	96	0.5
Symlet 10	1	95.5	94.5	-1
	2	96.5	95	-1.5
	3	95	95	0
	4	95.5	95.5	0
		Average Increase (%):		0.6

that the approach is effective, increasing recognition accuracy by an average of 0.6% over recognition without DWT coefficient selection. However, this is less than 50% of the maximum possible increase of 1.37% that DWT coefficient selection could provide. Although ORA is more flexible than PMA in handling varying datasets, it is likely that an optimized system would utilize one specific wavelet and scale for both system training and identification. This would allow a more relevant threshold ratio to be chosen, which would increase recognition accuracy.

6   Conclusions

In this paper, a novel feature vectors construction approach for face recognition using DWT has been discussed. The first set of experiments performed focused on the choice of DWT features. It is reveals that, where direct coefficient values were used for recognition, the LL quadrant provided the best results. For the wavelet filters tested, the highest recognition rate achieved for this quadrant was 95%. The highest accuracies for the HL, LH and HH quadrants were 78%, 74% and 66%, respectively. However, these tests did not provide enough information to indicate whether particular scales perform consistently better than others.

The second set of tests has been designed to identify which wavelet filters were the most effective at extracting features for face recognition with the specified database. The maximum recognition rates were compared for five wavelet filters each from the Daubechies, symlet, Coiflet and biorthogonal wavelet families. LL coefficients were used as features, with the first five scales investigated. The results indicated that there was no strong link between choice of wavelet family and recognition rate, although Coiflet wavelets produced the most consistent performance, across various filters and scales. When the results from all wavelet families and filters were examined together, there was no obvious correlation between the support size of the scaling filter and the maximum recognition rates.

The choice of scale did appear to have some effect, with the second, third and fourth scales outperforming the first scale by a small margin and the fifth scale by a significant margin. In case of feature optimisation by coefficient selections, the results show that DWT coefficient selection has increased maximum recognition rate in 16 out of the 20 cases tested. For instance, recognition accuracy increased from 94% to 97% for the Coiflet 3 wavelet, 1st scale.

Finally, for the feature threshold, two approaches have been investigated which are PMA and ORA. Results obtained shown that the PMA is ineffective approach, with recognition accuracy decreasing by an average of 0.025% from the results obtained without DWT coefficient selection. Unlikely, results for ORA approaches indicate better recognition accuracy by an average of 0.6%.

References

1. Jain, A., Bolle, R., Pankanti, S.: Biometrics: Personal Identification in Networked Society. Kluwer Academic Publishers, Dordrecht (1999)
2. Maltoni, D., Maio, D., Jain, A., Prabhakar, S.: Handbook of Fingerprint Recognition. Springer, New York (2003)
3. Campbell, W., Sturim, D., Reynolds, D.: Support Vector Machines Using GMM Supervectors for Speaker Verification. IEEE Signal Processing Letters 13(5), 308 (2006)
4. Zhao, W., Chellappa, R., Phillips, P.J., Rosenfeld, A.: Face Recognition: A Literature Survey. ACM Comput. Surv. 35(4), 399–458 (2003)
5. Kanwisher, N., Moscovitch, M.: The Cognitive Neuroscience of Face Processing: an Introduction. Cognitive Neuropsychology 17(1), 1–11 (2000)
6. Amira, A., Farrell, P.: An Automatic Face Recognition System Based on Wavelet Transforms. In: IEEE International Symposium on Circuits and Systems, ISCAS 2005, vol. 6, pp. 6252–6255 (May 2005)
7. Turk, M., Pentland, A.: Eigenfaces for Recognition. Journal of Cognitive Neuroscience 3, 71–86 (1991)

8. Moon, H., Phillips, P.: Computational and Performance Aspects of PCA-based Face Recognition Algorithms. Perception 30, 303–321 (2001)
9. Yuen, P.C., Lai, J.H.: Face Representation Using Independent Component Analysis. Pattern Recognition 35(6), 1247–1257 (2002)
10. Kussul, E., Baidyk, T., Kussul, M.: Neural Network System For Face Recognition. In: ISCAS, vol. (5), pp. 768–771 (2004)
11. Mallat, S.: A Wavelet Tour of Signal Processing, 3rd edn. The Sparse Way. Academic Press, London (2008)
12. Feng, G., Yuen, P., Dai, D.: Human Face Recognition Using PCA on Wavelet Subband. Journal of Electronic Imaging 9, 226–233 (2000)
13. Harandi, M.T., Ahmadabadi, M.N., Araabi, B.N.: Face Recognition Using Reinforcement Learning. In: ICIP, pp. 2709–2712 (2004)
14. Kemal Ekenel, H., Sankur, B.: Multiresolution face recognition. Image and Vision Computing 23(5), 469–477 (2005)
15. Haar, A.: Zur Theorie der orthogonalen Funktionensysteme. Mathematische Annalen 69(3), 331–371 (1910)
16. Daubechies, I., et al.: Orthonormal Bases Of Compactly Supported Wavelets. Comm. Pure Appl. Math. 41(7), 909–996 (1988)
17. Daubechies, I.: Ten Lectures on Wavelets. Society for Industrial Mathematics (1992)
18. Dahmen, W., Aachen, R., Micchelli, C.: Biorthogonal Wavelet Expansions. IGPM Preprint (114) (May 1995)
19. Hopper, T., Brislawn, C., Bradley, J.: WSQ Gray-Scale Fingerprint Image Compression Specification. Federal Bureau of Investigation Tech. Rep. IAFIS-IC-0110-V2 (Criminal Justice Information Services, Washington, DC) (1993)
20. Bartlett, M., Movellan, J., Sejnowski, T.: Face recognition by Independent Component Analysis. IEEE Transactions on Neural Networks 13(6), 1450–1464 (2002)
21. Kirby, M.: Geometric Data Analysis: An Empirical Approach to Dimensionality Reduction and the Study of Patterns. John Wiley & Sons, Inc., New York (2000)
22. Wang, J., Plataniotis, K., Venetsanopoulos, A.: Selecting Discriminant Eigenfaces For Face Recognition. Pattern Recognition Letters 26(10), 1470–1482 (2005)
23. Nefian, A., Hayes, M.: Hidden Markov Models For Face Recognition. In: ICASSP 1998, pp. 2721–2724 (1998)
24. Kim, J., Choi, J., Yi, J., Turk, M.: Effective Representation Using ICA for Face Recognition Robust to Local Distortion and Partial Occlusion. IEEE Trans. Pattern Anal. Mach. Intell. 27(12), 1977–1981 (2005)
25. Wang, H.Y., Wu, X.J.: Weighted PCA Space And Its Application In Face Recognition. In: Proceedings of 2005 International Conference on Machine Learning and Cybernetics, Washington, DC, USA, pp. 4522–4527. IEEE Computer Society, Los Alamitos (2005)
26. Ayinde, O., Yang, Y.H.: Face Recognition Approach Based On Rank Correlation Of Gabor-Filtered Images. Pattern Recognition 35(6), 1275–1289 (2002)
27. Samaria, F.: Face Recognition using Hidden Markov Models. PhD thesis, Cambridge University Engineering Department (1994)
28. Xue, Y., Tong, C.S., Chen, W.S., Zhang, W.: A Modified Non-negative Matrix Factorization Algorithm for Face Recognition. In: ICPR 2006: Proceedings of the 18th International Conference on Pattern Recognition, Washington, DC, USA, pp. 495–498. IEEE Computer Society, Los Alamitos (2006)

29. Lu, J., Tan, Y.P.: Enhanced Face Recognition Using Tensor Neighborhood Preserving Discriminant Projections. In: 15th IEEE International Conference on, Image Processing, ICIP 2008. pp. 1916–1919 (October 2008)
30. Ersi, E.F., Zelek, J.S.: Local Feature Matching For Face Recognition. In: CRV 2006: Proceedings of the The 3rd Canadian Conference on Computer and Robot Vision (CRV 2006), Washington, DC, USA, p. 4. IEEE Computer Society, Los Alamitos (2006)

An Extended Proof-Carrying Code Framework for Security Enforcement

Heidar Pirzadeh[1], Danny Dubé[2], and Abdelwahab Hamou-Lhadj[1]

[1] Department of Electrical and Computer Engineering
Concordia University
Montreal, QC, Canada
{s_pirzad, abdelw}@ece.concordia.ca
[2] Department of Computer Science
Laval University
Quebec City, QC, Canada
danny.dube@ift.ulaval.ca

Abstract. The rapid growth of the Internet has resulted in increased attention to security to protect users from being victims of security threats. In this paper, we focus on security mechanisms that are based on Proof-Carrying Code (PCC) techniques. In a PCC system, a code producer sends a code along with its safety proof to the consumer. The consumer executes the code only if the proof is valid. Although PCC has been shown to be a useful security framework, it suffers from the sheer size of typical proofs -proofs of even small programs can be considerably large. In this paper, we propose an extended PCC framework (EPCC) in which, instead of the proof, a proof generator for the program in question is transmitted. This framework enables the execution of the proof generator and the recovery of the proof on the consumer's side in a secure manner using a newly created virtual machine called the VEP (Virtual Machine for Extended PCC).

Keywords: Software Security, Proof-Carrying Code, Virtual Machine.

1 Introduction

Modern computer systems have become so complex that traditional security mechanisms built around anti-viruses and intrusion detection mechanisms can no longer sustain the severity of today's ever-increasing security threats. One can claim that, except perhaps for security experts and professionals, it is too big a burden, or even unrealistic, for users to bear sole responsibility for adequate security and protection of their computing systems. Proof-Carrying Code (PCC) techniques have been introduced to reduce the impact of this problem by allowing a consumer of a computer program to verify a proof of its general safety properties, sent by the code producer, before executing it [8].

In a PCC system, there are typically two main parties, (1) a code producer, who builds machine code along with its safety proof (expressed typically in a formal

M.L. Gavrilova et al. (Eds.): Trans. on Comput. Sci. XI, LNCS 6480, pp. 249–269, 2010.
© Springer-Verlag Berlin Heidelberg 2010

logic), and (2) a code consumer, who wishes to run the compiled code as long as it satisfies predetermined safety policies.

A typical interaction between the producer and consumer encompasses several steps. In the first step, the producer sends the consumer a program, which consists of the code and additional annotations such as loop invariants and function pre- and post-conditions. The consumer provides the received code to the Verification Condition Generator (VCGen), which generates a verification condition based on a set of safety policies that need to be satisfied. A verification condition is a logical formula that, if satisfied, implies that the code satisfies the safety policies.

The consumer, then, sends the generated verification condition to the producer. The producer runs a theorem prover (in many cases along with necessary human intervention) to obtain a proof that corresponds to the received verification condition. Next, the producer submits the proof to the consumer. The consumer uses a proof checker to verify that the received proof is indeed a proof of the verification condition that was initially generated. If the check succeeds the code is considered trustworthy and can be executed.

It should be noted that it is very common to have a copy of the VCGen on the producer's side to simplify the interaction between the code producer and the code consumer. In this way, the code consumer receives the annotated code as well as the safety proof during the first step of the interaction. Fig. 1 shows by the components according to the order by which they are executed in the PCC process. The steps involved in a typical interaction between producer and consumer as discussed above. The ovals are the artifacts that are generated/sent, the arrows represent the flow of the artifacts, and the rectangles show the components that perform computations. The starting point of the interaction is represented by a closed circle (•). At the end of the interaction, a switch (symbolically shown as a triangular tri-state buffer) checks the result of the proof checking; if the proof checking succeeds the code is considered trustworthy and can be executed (on the CPU shown as a rhomboid) if not the switch remains off and the code will not be executed.

One of the key properties of a PCC framework is that the Trusted Computing Base (TCB) (specified by the orange curved rectangle in Fig. 1) contains relatively small and simple components such as VCGen and a proof checker while the theorem prover is on the producer's side and therefore out of the consumer's TCB. The reason for that is twofold: performance and security. That is, in general, proving the verification condition is a resource consuming task which can result in low performance. Furthermore, considering that the theorem prover is a large and complex program, it could not be placed on the consumer's side as it could hardly be trusted. Another important property of the PCC framework is that PCC programs are tamper-proof. An intruder cannot modify the code or the proof in a way that results in execution of a malicious code on the consumer's side. Any attempt to tamper with either the code or the proof results in a validation error during the proof checking process.

Despite the fact that PCC can be a powerful security mechanism, it is still not widely accepted in practice due to two keys issues. First, it is usually difficult to write proofs for large programs. Although with the recent advances in Certifying compilation [3, 28] some safety properties of programs can automatically be proved as certificates, this is limited to basic safety properties and only possible for a restricted class of programs. For example, it may not be possible to prove automatically safety properties if the software system is complex or the policies are sophisticated [29]. The

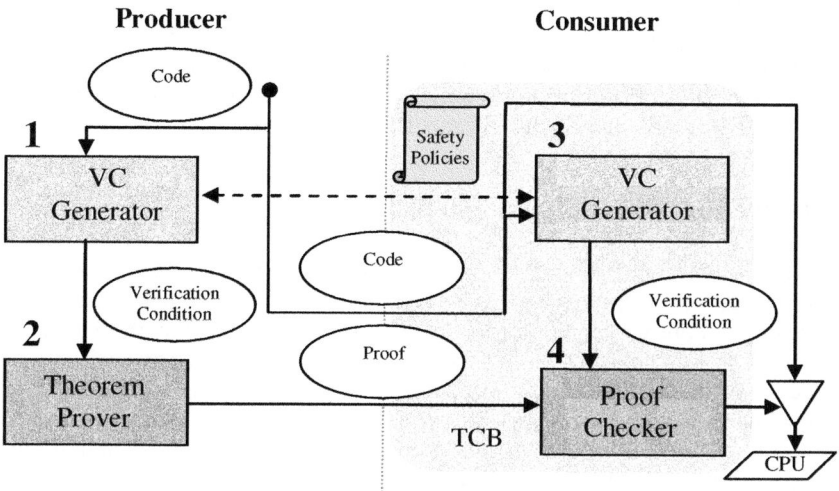

Fig. 1. Conventional PCC framework: typical steps and involved components

second limitation, which is the topic of this paper, is concerned with the difficulty in communicating and storing the proofs which are inherently large [11]. It is common to have proofs that are 1000 times larger than the associated code, which renders the use of PCC impractical for all but the tiniest examples [11]. This is further complicated when dealing with systems with limited storage and processing resources such as mobile and handheld devices, and networks with low survivability and scarce resources.

Clearly, there is a need for efficient techniques to reduce the size of proofs. The approaches proposed to alleviate this issue which include the use of data compression techniques [8, 10, 11, 25] suffer from drawbacks of their own, among which the most important one is the enlargement of the TCB, A large TCB increases the chance of defects which may cause an unsafe program to be accepted.

In this paper, we propose a novel approach to solving the proof size problem while avoiding to increase significantly the TCB. Our approach is based on the innovative idea of sending a program that generates the proof instead of the proof itself. This is inspired by the concept of Kolmogorov complexity [16], where the complexity of a string x is the shortest computer program that produces x on a so-called universal computer, i.e., a machine that computes the string, prints it, and then halts. One important observation is that the ideal compressed form for a given proof is the shortest program that outputs that proof.

To allow the proof generator program to execute on the consumer's side, we have developed a virtual machine that we call VEP (the Virtual Machine for Extended PCC). VEP is written in C and has less than 300 lines of code, which is an acceptable addition to the consumer's TCB. The design of VEP is relatively simple to be able to easily verify that is safe. It has also been developed with security in mind so as the running programs do not access unauthorized resources. Using the VEP, we believe that proofs, which are represented as programs, can be executed safely on the consumer's side while keeping the consumer's TCB reasonably small.

Organization of the paper: In the next section, we provide background information about PCC, and discuss studies related to our work. In Section 3, we present the extended PCC framework, followed by the VEP and its components. We show the effectiveness of our approach by applying it to several benchmark proofs in Section 4. We conclude the paper and discuss future directions in Section 5.

2 Background and Related Work

It is desirable that proofs be represented in a compact format. One way to reach this goal is through proof optimization in which the proofs are rewritten in a more compact form which preserves the meaning of the original form of the proof [13, 2]. This could be done by replacing all the occurrences of a given term t with a smaller equivalent term s in the proof (e.g., in the arithmetic system, there could be a rule $x *$ $1 \rightarrow x$ which always reduces the size of a term). Necula et al. experimented with proof optimization in an approach called lemma extraction and were able to obtain a minor reduction gain of 15% in the size of the proofs [2].

Another way of compacting the proofs is through data compression. Data compression techniques compress data by searching for more efficient encodings that take advantage of repetition in the data. These techniques are not well exploited in PCC framework due to the following reasons. The consumer of compressed data must first decompress it, which requires a safe decompressor on the consumer's side. Generating the proof of safety for a normal decompressor (a relatively large program with about 7000 lines of code) can be a difficult task not worth performing because one would only obtain a specific decompressor that cannot work with a proof compressed by an appropriate but different compressor. In other words, each time a new decompressor is used, a proof of its safety is required. The objective of the VEP is to tackle this problem by tailoring it to the needs of executing proof generators that could be, as shown in our case study, a compressed file along with a decompression tool.

Necula et al. proposed a new strategy called Oracle-based Proof-Carrying Code (OPCC) [11]. In OPCC, the handling of the proofs on the consumer's side is changed. As shown in Fig. 2, this change in strategy, led to a change in the framework, namely, they assumed that the consumer uses a non-deterministic proof checker.

The untrusted theorem prover on the producer's side records a sequence of bits that shows which sub-goals failed and needed backtracking. Then, the producer sends to the consumer this bit-stream that serves as a proof witness. On the consumer's side, the received bit stream works as an "oracle" which can guide the trusted non-deterministic proof checker to avoid back-tracking. Every time the checker must make a choice between the possible ways to proceed, it consults some bits from the oracle. In this approach, the trusted non-deterministic proof checker is, in fact, a non-deterministic theorem prover having the task of proving the verification condition. The oracle is used to drive the theorem prover to a final proof without search.

Experimental evidence shows that oracle strings, as suggested by Necula et al., can be about 1/8 of the code size and about 30 times smaller than proofs in traditional PCC [11]. However, Wu et al. [14] suggested that the code size relation might be deceptive as the size Java class files, that are necessary to be sent along with the proof witness in a SpecialJ proof-carrying Java system, is not included in calculation.

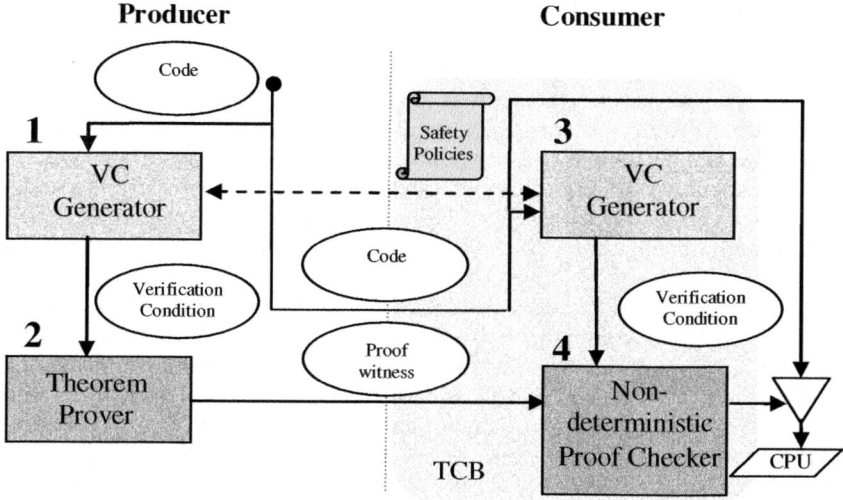

Fig. 2. OPCC framework: typical steps and involved components

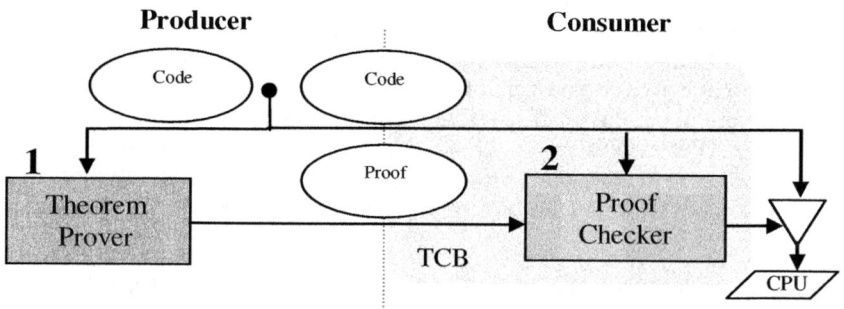

Fig. 3. FPCC framework: typical steps and involved components

One of the most important downsides of the OPCC is that it involves complex trusted components, such as a non-deterministic proof checker plus the usual PCC components. The TCB in OPCC is about 26000 lines of C code which is larger than the TCB size in traditional PCC (15000-20000 LOC). Any flaw in the implementation of these components can compromise safety of the system. As a matter of fact, the Special-J system [3], used in Necula et al.'s approach, showed a critical leak in its type axioms found by League [5].

Although the above approach has resulted in proofs which were smaller than the original proofs, they had to significantly enlarge the TCB. In fact, Appel points out that the VCGen (and consequently the TCB size) even in traditional PCC is too large [27] and it needs to be verified. As shown on Fig. 3, Foundational Proof-Carrying (FPCC) [27] Code aims to further reduce the TCB size by removing the VCGen from the consumer's side.

FPCC uses a foundational mathematical logic for defining the semantics of the machine instructions and the proof rules. In this way, Appel et al. avoid using the VCGen by defining the operational semantics of machine instructions and the safety policies in a higher-order logic. This is done by modeling the machine instruction with a transition from one machine state (set of memory and registers) to another machine state and defining the safety policy accordingly. Similar to the PCC, a theorem prover should produce a proof of safety to be accompanied by the code. The proof checker verifies the safety proof before the program is executed. FPCC is concerned with minimizing the TCB of the system, by not including the VCGen as shown in Fig. 3.

While the original FPCC uses deductive reasoning to encode proof rules, some variants of FPCC use computational reflection to replace deduction by computation [31]. FPCC is likely to be more secure than traditional PCC because it has a smaller TCB. However, the proofs in FPCC, in comparison with traditional PCC, are more complex to produce and, as stated by Appel et al., can explode exponentially [27]. According to Necula, the proof size in FPCC is 20% bigger than the proof size in traditional PCC [11]. Therefore, even though in FPCC, it is only necessary to send a proof generator, the complexity of producing the proofs, in the first place, renders FPCC hard to use in practice.

Wu et al. [14] proposed submitting annotated programs that can be checked for safety by a verified logic program. The program logic clauses are derived as lemmas from the (trusted) axioms about safe execution of the machine. This way, it is not necessary to build and check a large proof at the code consumer's side. However, according to [32], there exist issues about scalability of the results, as reported by Wu et al. [14], and effective engineering of their verifiers.

While we are not in favor of possible compromises to the security of the system due to a large TCB expansion (as we have in OPCC), we also like to overcome the difficulty in communicating and storing the proofs which are inherently large (as we have in traditional PCC and more severely in conventional FPCC) in a practical way.

3   The Extended Proof-Carrying Code Framework (EPCC)

3.1   Overview

Fig. 4 describes the steps involved in the proposed Extended Proof-Carrying Code (EPCC) framework [17]. In an EPCC system, there are two main parties, a code producer, on the left-hand side, who sends a code along with its safety proof generator program[1], and a consumer, on the right-hand side, who wishes to run the code provided that it is proven safe by the system.

The interaction between these two parties consists of the following steps. In the first step, the producer runs a theorem prover to obtain a safety proof of the code he intends to send. Similar to what is done in FPCC [27], the producer is not constrained to generate the safety proof in the logic that the consumer imposes. The producer can

---

[1] A proof generator is a program whose sole function is to output the proof. This program aims to be a more compact representation of its resulting proof and does not necessarily rediscover the proof.

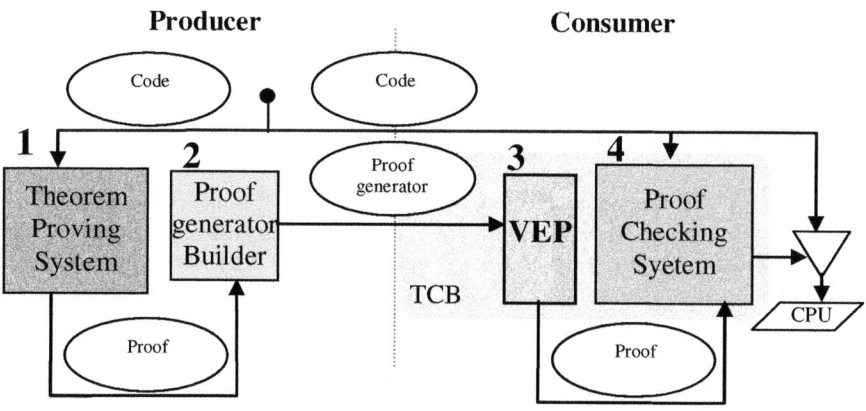

Fig. 4. EPCC framework: typical steps and involved components

use this opportunity to build the proof in a logic (e.g., a higher-order logic) that results in a smaller proof. In other words, the producer has the possibility of reducing the size of the safety proof by using a custom logic which can be later converted (translated) to the logic set by the consumer. In the second step, the producer writes a proof generator program, which outputs the safety proof in the format which is acceptable by the consumer.

Next, the producer submits the code accompanied by its safety proof generator program to the consumer. At this point, the proof generator program is yet another program that the producer sends to the consumer. It is as untrustworthy as the payload code itself. So it seems we are in a kind of chicken-and-egg situation: before running the untrustworthy payload code, the consumer needs to verify its attached proof, which requires execution of the proof generator program, which is also untrustworthy. One possible way to overcome this issue is to simply verify the safety of the proof generator program using traditional PCC. This solution has the obvious drawback of necessitating a proof for each proof generator program, which could hinder the practical aspect of our approach due to the complexity of writing proofs. We propose, instead, to run the proof generator in a tightly sandboxed environment: our carefully designed virtual machine, the VEP (the Virtual machine for Extended Proof-carrying code). The design of the VEP is discussed in more details in Section 4.

Upon receiving the code and the corresponding proof generator program, the consumer runs the proof generator (only for a single time) on the VEP and obtains the safety proof. The next steps are similar to the traditional PCC: The consumer runs the proof checker; after the proof check succeeds the consumer can safely execute the code. As one can easily observe, the EPCC framework is tamper proof, just like PCC.

The EPCC framework not only makes PCC more scalable and practical by reducing the proof size but also provides the code consumer with the possibility to use a safe environment in which a large class of proof generators that can be executed in a secure manner, regardless of the original logic in which the proofs were represented. In this way, EPCC leaves the easiest tasks to the consumer and gives adequate means to the producer to do the hard tasks. This major flexibility for the consumer and producer is gained through the VEP, a minor TCB extension, which can be verified easily. Technically, except for the VEP, the security of EPCC is as strong as the traditional PCC. Currently, a verified VEP is being developed (using conventional PCC). A verified VEP would potentially make EPCC exactly as secure as PCC.

4   Virtual Machine for Extended Proof-Carrying Code (The VEP)

The VEP [12] is intended to be a sandbox interpreter for the proof generator programs. Any defect in the VEP might give an opportunity to an attacker to write a malicious proof generator such that its execution on the VEP turns the VEP into an attacker against the consumer. Therefore, the safe execution of the proof generator depends greatly on the safety of the VEP and the way it imposes the security requirements. In this section, we present the design of the VEP starting from the general requirements that the VEP needs to satisfy to be deemed secure.

4.1   Requirements

The virtual machine design process starts by capturing the requirements. In the case of the VEP, we dealt with the following requirements.

1. The VEP should run as a virtual machine, deployed on different platforms to allow portability of proofs. This is similar in principle to the concept of universal computing proposed by Kolmogorov when describing the characteristics of the ideal decompressor [16].
2. It should enable the execution of the proof generator at the consumer's side in a secure manner. It should provide a tightly controlled set of resources for proof generation. Network access, the ability to inspect the host system, or read from input devices and write into file streams should be disallowed. Moreover, the VEP should be able to perform some sort of execution monitoring to verify that these constraints are maintained.
3. As indicated in EPCC framework, the VEP is a part of the TCB of the consumer. Knowing that any bug in TCB can compromise the security of the whole system, we need the VEP to be small and simple such that it is relatively easy to check for its safety. This would give the VEP the potential to be proved safe by the PCC itself.
4. The proof generators are sent in the VEP language. Consequently, this language should be flexible enough so that it allows compact proof generators to be written.
5. The VEP should be designed with performance in mind since it adds an overhead to the processing of proofs on the consumer's side.
6. The design of the VEP should be based on proven practices and common technologies to facilitate its adoption.

It should be noted that the above requirements are not equally important. The three first requirements are the most important ones in case trade-offs need to be made. For example, the low complexity and small code size both depend on the number of instructions in the VEP instruction set. On one hand, having a small set of instructions results in a virtual machine with low complexity, on the other hand, a large list of instructions makes the code smaller. Although these two factors are contradictory, there can be a good balance between them. Therefore, finding good trade-off has been one of the guiding principles in designing the VEP.

4.2 Machine Type

Conventionally, a virtual machine (VM) can either be stack-based or register-based. Implementing a universal computer can be achieved with a stack machine which has more than one stack or has one stack with random access. Nevertheless, register machines can be universal computers; therefore, both approaches can satisfy Requirement 1.

The most popular virtual machines, however, such as the Java Virtual Machine [6] and the Common Language Runtime [7], use a stack machine type rather than the register-oriented architectures due to the simplicity of their implementation. Hence, a stack-based machine helps us to better fulfill Requirement 3 (simplicity of the design). The simple stack operations can be used to implement the evaluation of any arithmetic or logical expression and any program written in any programming language (for execution on register machines) can be translated into an equivalent stack machine program. Moreover, the stack machines are easier to compile to, which could potentially help the adoption of the VEP (Requirement 6).

Finally, we chose the stack machine type over the register one because a compiled code for a stack machine has more density than the one for the register machine. In an experiment, Davis et al. [4] showed that the corresponding register format code after eliminating unnecessary instructions was around 45% larger than the stack code needed to perform the same computation. This can especially affect the size of the proof generator written for the VEP as mentioned earlier.

4.3 Instruction Set Architecture

The Instruction Set Architecture (ISA) of a virtual machine is the VM interface to the programmer. In the case of the VEP, available data types and the set of memory spaces are defined by ISA. The ISA definition also includes the specification of the set of opcodes (machine language) and the VEP's instruction set. Next, we discuss each of these parts and their design choices.

Data Types
On the VEP, we have two distinct types of values: *numbers* and *pairs*. Considering that the VEP is implemented using 32-bits machine words, the least significant bit of the word shows the data type of the stored value. This bit is not visible to the programmer while the remaining 31 bits are visible. If we have a word that references a *pair*, the content of the word represents the address of a *pair* in memory. For a word with its type *number*, the content of the word is a signed integer.

Memory
The VEP uses three blocks of memory: a code space (an array whose elements are bytes), a heap (an array whose elements are pairs, see below), and a stack (an array whose elements are bytes).

A *pair* is an ordered sequence of two values; the representation includes two words, for both values, and a third word which stores the reference counter for the simple garbage collection system of the VEP.

The stack grows towards the high addresses (the first item pushed on the stack is stored at address zero) and the stack pointer points at the topmost element. The heap provides the programmers with additional flexibility by supplying the VEP with memory for objects of arbitrary lifespan.

Fig. 5 shows the schemata of the stack and the heap in the VEP. For each of these two schemata, sample binary contents are shown on the right-hand side and the human readable format of the same content on the left-hand side. The second stack element from the top has the type pair (the type bit is one) and rest of the bits show the address of the pair in the heap which is 1 (1p in human-readable format). The pair 1p in the heap is a pair of the two values 34 and 0p which are respectively the *car* and the *cdr*[2] of 1p, where car returns the first item of the pair and *cdr* returns the second one. It should be mentioned that the values in the pairs follow the same typing convention as we have in the stack.

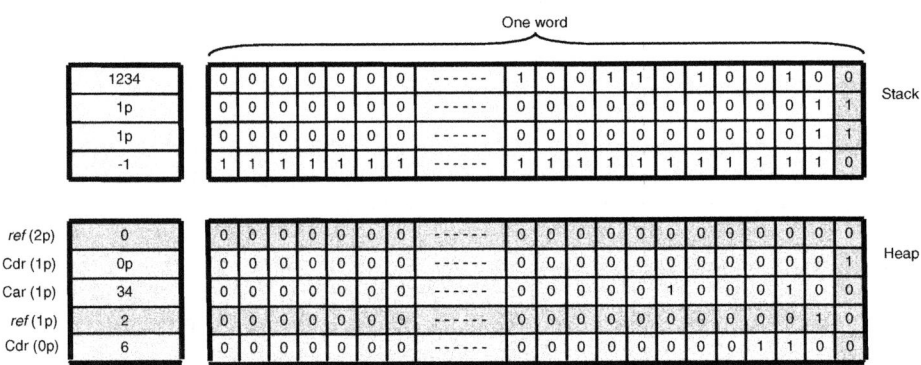

Fig. 5. Schemata of the stack and the heap

Memory Management

The VEP provides automatic memory management of the heap, thus there can be no dangling reference or memory leak due to manual memory management errors and the programmer can put more time on productivity instead of managing low-level memory operations.

The VEP relies on the reference counting [23] to automatically detect unused objects and collect them from memory. A major drawback of reference counting is its failure in reclaiming cyclic garbage data structures. We took the care of designing the VEP so that it does not have the ability to perform *destructive updates* on the pairs. Every value in the VEP is built up out of existing values; hence, it is impossible to create a cycle of references, resulting in a reference graph (a graph which has edges from objects to the objects they reference) that is a directed acyclic graph. This way,

---

[2] Analogous to the LISP operations on binary tree structures, where *cdr* returns a list consisting of all but the first element of its argument and *car* returns the first element.

the weakness of the reference counting garbage collector is avoided due to the lack of circular structures in the VEP heap.

Instruction Set
The design of the instruction set is one of the most interesting and important aspects of the VEP design. The code space, being made of bytes, naturally leads to an instruction set of 256 instructions. The VEP has a RISC-like instruction set which provides random access to stack, many arithmetic, logical, comparison, data transfer, and control instructions and restricted access to the pair-based heap.

This gives application developers a good flexibility in implementing their ideas and innovations when developing VEP-enabled proof generators. It also guarantees an acceptable execution performance (that is in line with Requirement 5). We provide the VEP with a rich set of data transfer instructions which might help to execute the proof generators on the VEP more efficiently. The distribution of the instructions is based on an interpretation of the work of Hennessy et al [18] where they found the 10 simple instructions that account for 96% of the instructions executed for a collection of integer programs running on the popular Intel 80x86. We used Table 1 as a reasonable guide for determining an appropriate distribution of instructions (in line with requirement 6).

Table 1. Distribution of instructions interpreted from [18]

Rank	80x86 instructions	% Execution
1	Data transfer instructions	38.00%
2	Control instructions	22.00%
3	Comparison instructions	16.00%
4	Arithmetical instructions	13.00%
5	Logical instructions	6.00%
	Total	96.00%

The VEP instructions can be classified into the following categories.

- Data transfer instructions (POP, PEEK, POKE, LOAD1, LOAD2, LOAD3, LOAD4, PEEKI n, POKEI n, LOADI n, PUSH-PC, READC): These instructions move data from one location in memory to another. These instructions come in a variety of ranges and density of operations, for instance, PEEKI n, POKEI n have shorter range (i.e., they can perform their operations only on the top eight elements of the stack), while PEEK and POKE have broader range (they can perform their operations only on all elements of the stack) and less density of operations (e.g. a LOAD1 -1 followed by a PEEK, is equivalent to PEEKI -1).
- Control instructions (HALT, NOP, JUMP, JMPR, JMPRF, JMPRT): Machines and processors, by default, process instructions sequentially. Redirection from this sequence is possible through control instructions. The most basic and common kinds of program control are the unconditional jump and the conditional jumps (branches). Control instructions also include instructions which directly affect the entire machine such as HALT or *no operation* (NOP).

- Comparison instructions (EQU, LEQ, LTH, NEQ): These instructions compare values by using a specific comparison operation. Typical comparison instructions include "equal" and "not equal".
- Arithmetic instructions (ADD, SUB, MUL, DIV, MOD): The basic four integer arithmetic operations are addition, subtraction, multiplication, and division.
- Logical instructions (BSHIFT, BAND, BNOT, BOR): These instructions usually work on a bit by bit basis. Typical logical operations include "logical negation" or "logical complement", "logical and", "logical or".
- Heap related instructions (CONS, CAR, CDR, ISPAIR): These instructions whether perform their action on a pair (CAR and CDR, respectively return the first and the second item of a pair), constructs a pair (CONS), or verify if a value is a pair (ISPAIR).
- Input/Output instructions (OUTPUT): The VEP provides a tightly-controlled set of resources for proof generators to run in. In order to be able to output the resulting proof, a proof generator is allowed to print characters onto the standard output. This is the sole way provided by the VEP for a proof generator to communicate with the outside world. Other than that, network access, the ability to inspect the host system, or reading from input devices and writing into file streams are disallowed.

Almost all of the instructions take their arguments from the stack and have no (immediate) operands. In particular, PEEK, POKE, and jump instructions are intended to be used along with "LOAD* val;" instruction[3]. This keeps almost all of the instructions to a single variant (no need to handle various addressing modes). Prevalence of LOAD* explains the existence of the 1-byte instruction LOADI for constants close to zero. These choices achieve simplicity of the VEP and compactness of the byte-code. The only instructions with immediate operands (other than LOAD*) are POKEI and PEEKI, which are extremely frequent as they are the typical means to implement the write/read of the local variables on the stack.

A note-worthy point about the VEP instructions set is the absence of instructions which operate on network or gives the ability to inspect the host system. Furthermore, there are no instructions which can read from input devices and write into file streams. These are to enable the execution of the proof generator at the consumer side in a secure manner. That is, we tried to enforce security policies such as no access to files or no access to the network on instruction set design level. Thus, the selected instructions provide the VEP with a tightly-controlled environment for proof generator to run in.

4.4   Security Enforcement by the VEP

We designed the VEP such that it guarantees a certain number of fundamental safety properties in order to execute the untrusted code in a secure manner. Memory safety is one of these properties which prevents reading and writing to illegal memory locations. The code space is read-only and the legal code space locations are

---

[3] LOAD* val pushes the numeric value encoded by the next * byte(s) in the code space onto the stack.

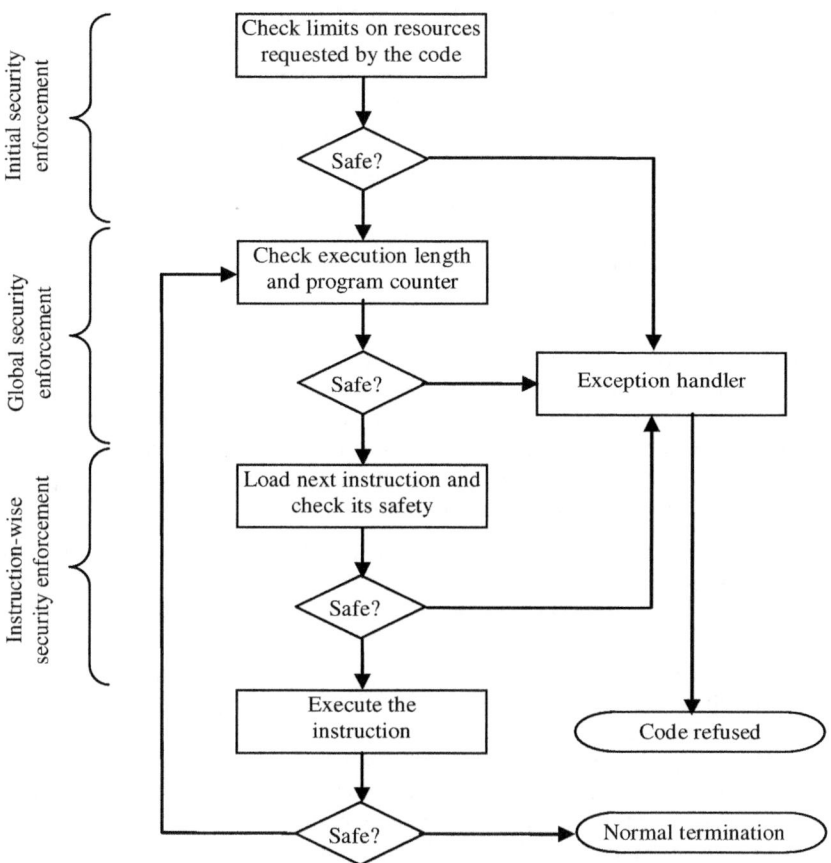

Fig. 6. The flowchart of Security Enforcement of the VEP

$0, \ldots, Nc - 1$, where Nc is the code size. Even the instruction loading must be performed as legal reads from the code space.

In the case of the stack, reads and writes are permitted. Any read or write to the stack is preceded by a memory check which ensures that the read and write are going to be performed on valid stack locations as their destination. What is a valid destination varies from instruction to instruction. Generally, the valid read and write destinations are stack locations ranging from the bottom to the top of the stack.

In the case of the heap, reads and writes are very restricted. Since the construction of the pairs is governed by the VEP, the programmer has no means to modify the type bit to forge a new pair and he has no means to read and write in the heap other than to use CONS, CAR, CDR. Furthermore, *memory safety* in the VEP asserts that each operation has a sufficient amount of required memory (stack and/or heap) to perform the instruction (e.g., the VEP raises an error if an attempt is made to pop when the stack is empty or to push an item onto a full stack). *Control-flow safety* prevents jumps outside of the code space, and *resource bound check* enforces limitations on the size of the code space, the size of the stack, the size of the heap, and the number of instructions the VEP may execute. There are other security requirements such as *type safety* and *numeric safety* which will be explained in following subsections.

The security enforcement by the VEP is simple and straightforward. The VEP enforces these security requirements at different levels. Categorizing the security checks according to their enforcement level shows better how easy the VEP security enforcement is to perform and understand. Fig. 6 shows a complete schema of the security enforcement mechanism and its different levels.

Initial Security Enforcement

A proof generator makes requests for resources. These requests are made using a declaration in the header of the proof generator. Each time, the VEP verifies whether the requested amount of resources is no greater than the maximum value settled in an agreement between the producer and the consumer. The requested code size and stack size are, respectively, denoted by Nc and Ns. The amount of needed heap size of the proof generator is represented in number of pairs Nh.

- *Code size*: If the VEP refuses or fails to allocate the requested block of memory, the VEP refuses the proof generator. Otherwise, the VEP allocates a block of Nc bytes of memory as the code space and inserts the code into the code space.
- *Stack size*: If the VEP refuses or fails to allocate the requested block of memory above agreed-upon limit, the VEP refuses the proof generator. Otherwise, the VEP allocates a block of Ns words of memory as the stack memory.
- *Heap size*: If the VEP refuses or fails to allocate the requested block of memory, the VEP refuses the proof generator. Otherwise, the VEP allocates a block of $3*Nh$ words of memory as the heap memory.
- *Length of Execution*: The proof generator should finish its task within a definite number of operations No. In the case where the No is more than the limit the VEP refuses the proof generator.

When the proof generator is not refused during the initial security enforcement, it is ready to be executed by the VEP.

Global Security Enforcement

Throughout the execution, the VEP enforces two security checks globally, which are independent of the next instruction that is about to be executed. The global security enforcement consists of checking the following aspects:

- *Length of execution*: Before fetching the next instruction, the VEP makes sure that the elapsed time of the execution of the proof generator (measured as the number of executed operations) has not exceeded the number of operations (i.e., No). If the number of executed operations is less than the approved number, then the check is passed, otherwise the code is refused for having run for too long.
- *Program counter*: The VEP should check if the program counter points inside the code space (i.e., non-negative and less than the code size).

Instruction-Wise Security Enforcement

The third level of security enforcement by the VEP is the fine-grained level and is done per instruction. This level of security prevents the proof generator from performing any unsafe operation.

Generally, after fetching each instruction and before the execution of the instruction, the VEP performs a combination of the following checks.

- *Number of operands*: The number of operands of an instruction can vary from zero to two implicit operands on the stack, depending on the instruction. For an instruction that requires one or more operands on the stack, the existence of a sufficient number of operands must be checked before execution of the instruction. If insufficient operands lie on the stack, the execution is discontinued and the proof generator is not considered safe.
- *Type of operands*: The VEP checks if the type of the operands conforms to the operation. As mentioned earlier, the values in the VEP can be numbers or pairs. The VEP can distinguish the type of an operand according to its type bit. Depending on the instruction and the operand, the latter may have to be a number, it may have to be a pair, or it may be free to be of either types. Checking the type of operands ensures that a code is free of type-mismatches according to the VEP's type system.
- *Legal range of operands*: The arithmetic instructions should have legal arguments. The VEP checks the operand legality to prevent potential errors of using partial operators with arguments outside their defined domain (e.g., division by zero).
- *Legal code destination*: Before changing the program counter to the jump destination, the VEP checks if the destination is within the code space. It should be mentioned that the VEP does not enforce the concept of instruction boundaries.
- *Legal stack destination*: For any instruction which results in a read or a write to the stack, the VEP ensures that the reads and writes have legal stack locations as their destination.
- *Stack overflow*: The VEP verifies whether there is enough stack space to perform an instruction which works with stack memory.
- *Heap overflow*: The VEP verifies whether there is enough free space on the heap to perform an instruction which works with the heap memory.

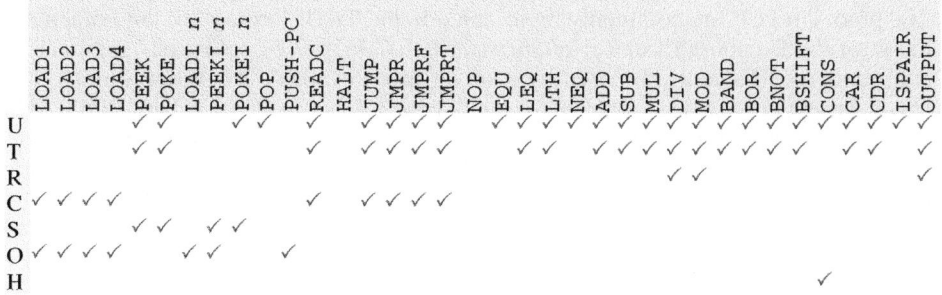

Fig. 7. Instruction-wise security enforcement

As shown in Fig. 7, the complete set of instructions[4] with their safety checks can be simply put into a table. In this way it would be an easy task to verify the safety of the VEP.

4.5 The VEP versus Other VMs

There are many systems that execute untrusted codes in virtual machines to limit their access to system resources. Therefore, a question one could ask is "why not use another existing virtual machine instead of the VEP?" Here, we highlight the main reasons of choosing the VEP over two popular virtual machines, which are the Java virtual machine (JVM) [6] and the .NET platform (CLR) [7].

Any virtual machine that we choose would be a part of the TCB in EPCC framework. Knowing that any bug in the TCB can compromise the security of the whole system, we should choose a virtual machine which increases the size of the TCB the least. Using either JVM or .NET results in a large TCB (these large TCBs were the motivations for introducing the PCC approach in the first place). Appel et al. [1] measured the TCBs of various Java virtual machines at between 50,000 and 200,000 lines of code. The TCB size in these VMs is even larger than the TCB size of the traditional PCC. Therefore, using these virtual machines to extend the PCC framework would result in an undesirably large TCB and hence an ineffective PCC framework.

For EPCC, we need a virtual machine so simple that, it is feasible for a human to inspect and verify it. None of the mentioned virtual machines or any other ones that we are aware of has been developed with this goal in mind. JVM, .NET, and other well-known virtual machines focus essentially on performance, portability, etc. Similar to other components of the TCB in traditional PCC and OPCC, the VEP is implemented in C language. However, unlike the OPCC that extends the TCB for about 9000 lines of C code, the implementation of the VEP is less than 300 lines of code which makes it possible to be easily verifiable by humans and gives it the potential of being proven safe in the future. Therefore, we have shown that the VEP is orders of magnitude smaller and it is simpler than popular virtual machines.

5 Application of EPCC

The proofs in PCC are commonly represented in the Twelf format [26] (an implementation of the Edinburgh Logical Framework (LF) [36]). We applied our approach to six proofs (see Table 2) produced by a solver made available by Aaron Stump[5]. The solver accepts quantified Boolean formulas benchmarks in the standard QDIMACS format, and emits proof terms showing whether the formula evaluates to true or to false. These proofs are the same as the ones considered in Stump's work [15], where easy benchmark formulas from [21] different domains (formal verification, planning, etc)[6] were solved to generate the proof terms. All proofs use a form of implicit LF [9]

---

[4] All 256 available opcodes are assigned to these 36 instructions; few instructions with immediate argument cover more than one opcode as the argument is encoded in the opcode itself.

[5] http://www.cs.uiowa.edu/~astump/software.html

[6] Interested readers can see [30] for a complete description of the domains and families of the proved formulae.

and can be as large as 7.4 megabytes. Although these proofs were not specifically designed for PCC, we believe that they can be fair representatives of large proofs, and be used in the absence of large PCC proofs due to the complexity of building them.

5.1 Building a Proof Generator

We created a proof generator for each of Stump's proofs. Our proof generator consists of a package that comprises a compressed version of the proof and a VEP machine executable decompressor. That is, we built a self-decompressing executable program which will generate the original proof as a result of being executed on the VEP. For this purpose, we reused an existing off the shelf compression tool, Gzip [22], which we modified to make it VEP-enabled. We could have also created our own program that generates the proof by looking at patterns in the proofs and creating programs that would explore these patterns forming a compact representation of a proof as a running program. We deliberately chose not to proceed this way to show that our framework can be equally used with existing programs, relieving the users of our framework from creating proof generators from scratch. However, we recognize that one of the main drawbacks of our approach lies in the need to adapt any program used to represent a proof to the VEP, a task that may turn to be difficult and time consuming. There is definitely a need to further investigate this issue as a key future direction.

Fig. 8 shows the steps involved in EPCC. The first and second steps are similar to traditional FCC. Given a proof, in the 3rd step a component called "proof generator builder" indicated by as a box with upward diagonal pattern in is responsible for building a proof generator. As shown in Fig. 8, the proofs are compressed using Gzip. To decompress the proofs on the consumer's side, we needed to send the decompression tool that can run on the VEP along the compressed proofs. For this purpose, we modified Gzip component that performs the decompression task (called gunzip). This involved using static allocation, removing all preprocessor commands and function prototypes, in-lining functions, etc. In order to in-line the functions without causing an increase in the code size, we used the `computed goto` construct [24], which is a `goto` statement for which the address of the target is computed by an expression of type `void*`.

The modified decompressor fetches its input (compressed data) from a literal string (array of compressed data) and outputs the decompressed data on the standard output. For the decompressor to fetch its input from a literal string, and to print a character, respectively, `readcmp` and `putchar` were developed as two special functions.

The modified gunzip C code (which now contains about 2000 LOC) is re-compiled to generate the assembly code of the gunzip (see Fig. 8). For this, we developed our own C compiler that supports a subset of C constructs that map to the VEP instruction set. The C compiler is based on the C89 open source complier [20]. Since the computed `goto` is not supported by the ANSI C89 grammar, we added it to the C89 grammar.

The assembly code generated by the compiler is then given to the assembler as input which results in having the VEP-executable gunzip machine code as its output (see Fig. 8). Our assembler implemented in C, permits *assembly-time* arithmetic operations to take place in order to compute constants to include in the assembled program. Thus, the expressions are evaluated during the assembly and the results become permanent parts of the code.

Gunzip machine code and the compressed proof are packaged to form a proof generator sent to the consumer. This packaging is performed manually by allocating the compressed stream statically in the code space. This saves us a lot on stack space in comparison with the case we dynamically allocation the compressed data in a global variable. The compressed stream is then read by the decompressor using the auxiliary function `readcmp`. This is the only function that we add to the existing decompressor code so that it can read the compressed data from within the decompression code.

Before sending the proof generator, the producer needs to add the request in code size, heap size, stack size, and execution time to the proof generator program header. For this, he has the option of running the proof generator on a copy of the VEP installed on his side. The VEP contains a feature that can add automatically the actual amount of the consumed resources to the proof generator program header.

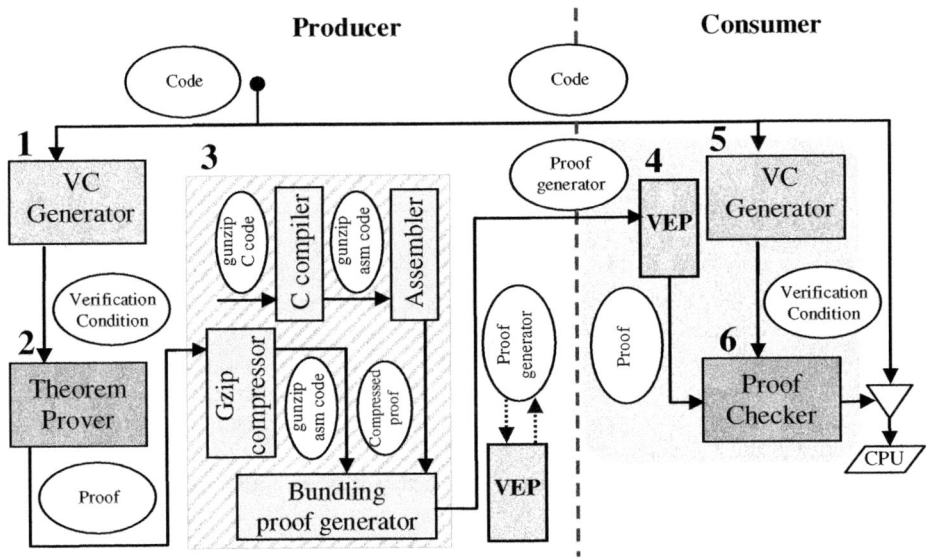

Fig. 8. Detailed diagram of our sample implementation of EPCC

5.2   Results of Applying the Approach

Table 2 shows the results of applying our approach to the proofs selected for this study. For each proof, the original proof size (N) and the size of the proof generator (NPG) are represented.

The size of the proof generator excluding a compressed proof is about 15KB (which is the size of gunzip machine code and is constant for all of our proof generators). The proof generators average 2.9% the original proofs which is about 34 times smaller than before, which constitutes a significant gain in size reduction. The proof generator reduction in size relative to the original size of the proof is represented as the percentage of space savings (*SS*):

$$Space\ Savings = 1 - (NPG\ /\ N)$$

The space saving ratio of proof generators to the size of the original proofs ranges from 87.19% up to 96.77%. The table also shows the elapsed times of the execution of proof generators on the VEP. All times are reported in seconds on an Intel Core Duo CPU 2.00GHz, 2MB cache, 1GB main memory, running Windows XP. We can see that the VEP performed in less than a second for processing the proof generators.

Table 2. The size effect of representing proofs as programs

Experiment	N	NPG	SS %	Elapsed time	Domain
cnt01e	164 KB	21 KB	87.19	< 1s	Formal verification
tree-exa2-10	337 KB	25 KB	92.58	< 1s	Pattern matching
toilet 02 01.2	917 KB	45 KB	95.09	< 1s	Planning
1qbf-160cl.0	1407 KB	59 KB	95.80	< 1s	Formal verification
tree-exa2-15	3847 KB	115 KB	97.01	< 1s	Pattern matching
toilet 02 01.3	7377 KB	238 KB	96.77	< 1s	Planning

6   Conclusion and Future Work

In this paper, we presented an extension to a traditional proof-carrying code framework in which proofs tend to be considerably large to transmit. Our extended framework is based on the idea of representing proofs and programs that are sent to the consumer. As such, the consumer runs the program and generates the original proof. The proof generator program should be the shortest possible to maximize the size reduction gain.

We developed a virtual machine called the VEP that runs on the consumer's side and which is responsible of running the proof generator program. The implementation of the VEP contains less than 300 lines of code which is a minor extension to the consumer's TCB.

The VEP enables the proposed extended PCC framework to make the PCC idea more scalable and practical by providing the code consumer with the possibility of using a safe environment in which a large class of proof generators can be executed in a secure manner, regardless of the original logic in which the proofs were represented.

In the future, a first practical step will be to obtain a VEP that has been proven safe using the conventional PCC framework. In this way, the VEP would not increase the size of the TCB at all. Writing an oracle-based proof generator could be another possible direction to explore. This proof generator could be one which uses the proof witness in order to rebuild the original proof. Therefore, there would be no need to use any non-deterministic proof checker on the consumer side and the verification could be done with the original PCC proof checker. In this way, we would not force the consumer to change the PCC structure to gain the benefit of small proofs in OPCC and there will be no need for compromises in the size of the TCB. When both the proof generator and the proof checker can work incrementally, the whole proof need not be rebuilt at any one time on the consumer side. Instead, the output of the proof generator can be piped into the input of the proof checker, which consumes (and verifies) parts of the proof as soon as they are output.

In addition, we intend to continue experimenting with the proposed approach using larger proofs. This can be hard to achieve due to the unavailability of proofs for large systems.

Finally, we intend to compare the results of our approach with existing approaches such as the oracle PCC [11], although the size reduction gain should not be the only criterion that needs to be used in the comparison since, again, any approach that increases considerably the TCB poses risks to security no matter the size compression ratio achieved.

References

1. Appel, W., Wang, D. C.: JVM TCB: Measurements of the trusted computing base of Java virtual machines, Tech. Rep. CS-TR-647-02, Princeton University (2002)
2. Cheney, J. R.: First-order term compression: techniques and applications, Master's thesis, Carnegie Mellon University (August 1998)
3. Colby, C., Lee, P., Necula, G.C., Blau, F., Plesko, M., Cline, K.: A certifying compiler for Java. SIGPLAN Not. 35(5), 95–107 (2000)
4. Davis, B., Beatty, A., Casey, K., Gregg, D., Waldron, J.: The case for virtual register machines. In: Proceedings of the 2003 Workshop on interpreters, Virtual Machines and Emulators, IVME 2003. San Diego, California, June 12 - 12, pp. 41–49. ACM, New York (2003)
5. League, C., Shao, Z., Trifonov, V.: Precision in practice: a type-preserving java compiler. In: Hedin, G. (ed.) Proceedings of the 12th International Conference on Compiler Construction, Warsaw, Poland, April 07-11. Lecture Notes In Computer Science, pp. 106–120. Springer, Heidelberg (2003)
6. Lindholm, T., Yellin, F.: Java Virtual Machine Specification, 2nd edn. Addison-Wesley Longman Publishing Co., Inc. (1999)
7. Meijer, E., Gough, J.: Technical Overview of the Common Language Runtime (2000)
8. Necula, G.C.: Proof-carrying code. In: Proceedings of the 24th ACM SIGPLAN-SIGACT Symposium on Principles of Programming Languages, POPL 1997. Paris, France, January 15-17, pp. 106–119. ACM, New York (1997)
9. Necula, G.C.: A Scalable Architecture for Proof-Carrying Code. In: Kuchen, H., Ueda, K. (eds.) FLOPS 2001. LNCS, vol. 2024, pp. 21–39. Springer, Heidelberg (2001)
10. Necula, G.C., Lee, P.: Safe kernel extensions without run-time checking. SIGOPS Oper. Syst. Rev. 30, 229–243 (1996)
11. Necula, G.C., Rahul, S.P.: Oracle-based checking of untrusted software. In: Proceedings of the 28th ACM SIGPLAN-SIGACT Symposium on Principles of Programming Languages, POPL 2001. London, United Kingdom, pp. 142–154. ACM, New York (2001)
12. Pirzadeh, H., Dubé, D.: VEP: a virtual machine for extended proof-carrying code. In: Proceedings of the 1st ACM Workshop on Virtual Machine Security, VMSec 2008. Alexandria, Virginia, USA, October 27-27, pp. 9–18. ACM, New York (2008)
13. Rahul, S.P., Necula, G.C.: Proof Optimization Using Lemma Extraction. Technical Report. UMI Order Number: CSD-01-1143., University of California at Berkeley (2001)
14. Wu, D., Appel, A.W., Stump, A.: Foundational proof checkers with small witnesses. In: Proceedings of the 5th ACM SIGPLAN International Conference on Principles and Practice of Declaritive Programming, PPDP 2003. Uppsala, Sweden, August 27-29, pp. 264–274. ACM, New York (2003)

15. Stump, A.: Proof Checking Technology for Satisfiability Modulo Theories. Electron. Notes Theor. Comput. Sci. 228, 121–133 (2009)
16. Li, M., Vitnyi, P.: An Introduction to Kolmogorov Complexity and its Applications, vol. 3. Springer Publishing Company, Heidelberg (2008) (incorporated)
17. Pirzadeh, H., Dubé, D.: Encoding the Program Correctness Proofs as Programs in PCC Technology. In: Proceedings of the 2008 Sixth Annual Conference on Privacy, Security and Trust, October 01-03, pp. 121–132. PST. IEEE Computer Society, Washington (2008)
18. Hennessy, J.L., Patterson, D.A.: Computer Architecture: a Quantitative Approach, vol. 3. Morgan Kaufmann Publishers Inc., San Francisco (2003)
19. Jansen, W., Karygiannis, T.: NIST special publication 800-19 – mobile agent security. Technical report, National Institute of Standards and Technology, Computer Security Division, Gaithersburg, MD 20899. U.S. (2000)
20. American National Standards Institute, "Programming Language C," Document ANSI X3.159-1989
21. Giunchiglia, E., Narizzano, M., Tacchella, A.: Quantified boolean formulas satisfiability library (qbflib) (2001), http://www.qbflib.org
22. Deutsch, P.: GZIP File Format Specification Version 4.3. RFC. RFC Editor (1996)
23. Christopher, T.W.: Reference count garbage collection. Software – Practice and Experience 14(6), 503–507 (1984)
24. Griffith, A.: GCC: the complete reference. McGraw-Hill/Osborne (2002)
25. Ireland, A.: On the Scalability of Proof Carrying Code for Software Certification. In: Proc. Workshop on Software Certificate Management, November 8, pp. 31–34 (2005)
26. Pfenning, F., Schürmann, C.: System Description: Twelf - A Meta-Logical Framework for Deductive Systems. In: Ganzinger, H. (ed.) CADE 1999. LNCS (LNAI), vol. 1632, pp. 202–206. Springer, Heidelberg (1999)
27. Appel, A.W.: Foundational proof-carrying code. In: 16th Annual IEEE Symposium on Logic in Computer Science (LICS 2001), pp. 247–258 (2001)
28. Necula, G.C., Lee, P.: The design and implementation of a certifying compiler. SIGPLAN Not. 33(5), 333–344 (1998)
29. Mobius, Public, Deliverable D4. 1: Scenarios for Proof-Carrying Code, FP6-015905, Information Society Technologies (2006)
30. Narizzano, M., Pulina, L., Tacchella, A.: Report of the third QBF solvers evaluation, Journal of Satisfiability. Boolean Modeling and Computation 2, 145–164 (2006)
31. Barthe, G., Crégut, P., Grégoire, B., Jensen, T., Pichardie, D.: The MOBIUS Proof Carrying Code Infrastructure. In: de Boer, F.S., Bonsangue, M.M., Graf, S., de Roever, W.-P. (eds.) FMCO 2007. LNCS, vol. 5382, pp. 1–24. Springer, Heidelberg (2008)
32. Chlipala, A.J.: Implementing Certified Programming Language Tools in Dependent Type Theory. Doctoral Thesis. UMI Order Number: AAI3311660, University of California at Berkeley (2007)

NPT Based Video Watermarking with Non-overlapping Block Matching

S.S. Bedi[1], Shekhar Verma[2], and Geetam S. Tomar[3,*]

[1] MJP, Rohilkhand (U.P.), India
erbedi@yahoo.com
[2] Indian Institute of Information Technology-Allahabad
Deoghat, Jhalwa, Allahabad – 211012, India
sverma@iiita.ac.in
[3] M.I.R. Labs, 223 New Jiwaji Nagar, Gwalior 474011 India
Ph. No.: +91-9450965336; +91-9425129523
gstomar@ieee.org

Abstract. The paper presents a naturalness preserving transform (NPT) based collusion and compression resistant watermarking technique for video. An image that is statistically similar to a video frame is chosen as the watermark and this image is embedded independently in consecutive frames of the video. To enhance the resistance to inter frame collusion based attacks, a non-overlapping block matching is used to determine the region for placing the watermark in consecutive frames. Only a trace of the watermark image is embedded which enhances the robustness of the watermark to different attacks. When a frame and the image become substantially different, another image is chosen as the watermark. The size of the watermark determines the quality of the watermarked video frames. Watermark extraction is blind and requires only the region where the watermark was originally placed. The reconstruction process is iterative and bestows immunity the watermark against noise and lossy compression. Analysis indicates that the watermark is sufficiently immune to second order inter-frame statistical attacks and is quite robust to image level compression. Experimental results confirm these theoretical findings and demonstrate the resistance of the technique to temporal frame averaging, additive noise and JPEG based compression. However, the technique is limited by the fact that the original video sequence (frames) is required for reconstruction based recovery of the watermark from the watermarked video sequence.

Keywords: Watermarking, Video, naturalness preserving transform, attacks.

1 Introduction

Video Watermarking can be used for copyright protection or for tracking its distribution. In addition, it can be used for protection against duplication and to monitor broadcast [1]. A watermark can be embedded in video by considering it as a sequence of image frames or by recognizing the temporal dependency in addition to the inherent spatial redundancy [2]. Watermarking can also be done in the raw video or in

---

* Corresponding author.

M.L. Gavrilova et al. (Eds.): Trans. on Comput. Sci. XI, LNCS 6480, pp. 270–292, 2010.
© Springer-Verlag Berlin Heidelberg 2010

different stages of compression process [3]. This can be done the pre-quantized transform stage [4], in the post-quantization stage [5], prior to or posterior to rounding [6] or in the VLC/CABAC codes [7]. The temporal dimension can be exploited in both the uncompressed and compressed by embedding fixed or varying watermarks, content dependent [8] or independent watermarks in static or dynamic regions of the frames [9]. Any process that alters or makes watermark extraction difficult is an attack [10]. Watermarked video can be subject to attacks similar to watermarked image. Filtering, lossy frame compression, noise addition and geometrical manipulations, video editing constitutes such attacks [10]. The twin requirements to be fulfilled are imperceptibility and robustness. Different watermarking schemes proposed address the requirements of invisibility and robustness in the image domain. These schemes can be divided into two broad categories: spatial domain and transform domain watermarking. The transform domain schemes [11] are more robust to noise, geometrical manipulations, cropping and lossy compression. Different types of transform domain watermarking are more robust to different types of manipulations. DCT based watermarking is more robust to lossy compression [12], DWT based watermarking to noise [13], DFT based watermarking to geometrical changes [14]. Video watermarking introduces a number of new issues that are not present in image watermarking due to the addition of the temporal dimension. The large amount of data and intraframe redundancy makes video susceptible to temporal attacks like frame averaging, frame dropping, and statistical attacks like collusion I and II [15] for the detection or removal of the watermark. In addition, motion estimation and compensation in video compression process is highly detrimental to watermarking in the predictive frames. These specific attacks render image or any frame-by-frame watermarking schemes vulnerable to video specific attacks. Non-hostile video processing like photometric manipulations like gamma correction, spatial filtering, trans coding that modify the pixel values in all the frames; spatial and temporal de-synchronization like changes in aspect ratio, spatial resolution; and video editing constitute non malicious attacks unique to video domain [16]. Desynchronization can also be as a malicious process to defeat watermarking [17]. It is the process of identifying the association between spatial and temporal coordinates of the watermarked signal and that of an embedded watermark. A spatial desynchronization attack [18] may perturb the synchronization between the signal and the embedded watermark by geometrical variations like rescaling, reorientation etc. so that detector is not able to locate the position of the watermark. Temporal synchronization attack [19] exploits the high degree of temporal similarity allowing frame dropping, insertion of arbitrary or averaged frames etc. This makes watermark extraction extremely difficult. Real time watermarking [20] can be additional need that puts a limit to the complexity of the watermarking algorithm and requires watermarking in the compressed domain itself.

The remaining paper is organized as follows. Section 2 discusses approaches specific to robust video watermarking techniques. Section 3 enumerates a set of design guidelines for embedding an imperceptible watermark immune to collusion attacks. In section 4, a watermarking technique based on non-overlapping block matching and NPT has been proposed for embedding watermarks in video. In section 5, a theoretical analysis has been presented along with implementation results to evaluate and establish the efficacy of the proposed technique. The conclusions are given in section 6.

2 Video Watermarking Approaches

Video watermarking endeavors to hide a watermark in a manner such that the watermark survive discovery through statistical estimation or obliteration by desynchronization in addition to immunity to noise and compression. Various methods have been proposed that range from methods that are just an extension of image watermarking to video specific techniques where both the watermark sequence and the embedding process are determined by the characteristics of host video sequences. Some of the representative approaches are as follows. In [21], a PN sequence is spread over the video frames in the spatial domain and repeated over the entire video in a sequential manner such that the watermark frame correlation was 0 or 1. This rendered the watermark collusion resistant to some extent. In Just Another Watermarking System (JAWS), higher dimensional PN sequences were utilized to mark the video frames such that inter frame watermark correlation was close to unity, Bit plane decomposition of the video frames was performed in [22] for watermark insertion. The video sequence is watermarked by replacing one of the least significant planes of each frame by a PN binary sequence. These result in low pair wise correlation in watermarked signal for independent host frames and high pair wise correlation for identical host frames. In transform domain approaches, noise like sequences is added to the transform coefficients with a key support for collusion resistance. All these spatial or transform domain approaches are vulnerable to desynchronization and require generation and storage of keys. Temporal partitioning for multiresolution scene dependent watermarking is proposed in [23]. The watermark is embedded such that the correlation of watermarks is high for static and low for dynamic scenes to attain collusion resistance. Average luminance in sub-regions is modified for watermarking [24]. The average luminance regions are categorized on the basis of partitioning using a secret key. Local correlations render the watermark collusion resistant. In [25], sets of video hash functions have been utilized to generate video content dependent noises like watermark sequences for achieving statistical invisibility and collusion resistance. A method to improve imperceptibility by reduction of flickering effect is proposed in [26]. The work also proposes methods for determination of video scene characteristics for robust watermarking. It is shown in [27] that frame-by-frame watermarking embedding, which is oblivious to content characteristics is highly susceptible to collusion attacks. A content dependent watermark with time dependent embedding strength has been proposed as an effective deterrent to collusion based attacks. In-depth analysis and design guidelines for collusion-immune watermarking are presented in [28]. The work indicates that statistical invisibility can be achieved through content dependent watermarking. Such watermarks exhibit high correlation for similar frames and negligible correlation for independent frames. A spatially localized water marking technique with random footprints has been proposed along these guidelines [29]. Apart from these, watermark embedding has been proposed in the motion vectors [30], post quantized transform coefficients [31]. Watermark embedding in motion vectors makes the watermark fragile [32] while compressed domain is mostly tied to the compression standard. Both achieve the requirements of real time watermarking but are intolerable to transcoding, scaling etc. Intermediate transform domains like NPT have been used for image watermarking [33]. NPT based image watermarking is resistant to noise addition and tolerates high degree of compression. The process embeds only traces of the watermark in the host, not the complete watermark and the perturbed part of the original signal that becomes the part of the final signal is tightly coupled to the original host. This makes a potential candidate for collusion resistant video watermarking. In this study, NPT based watermarking of image has been explored and extended for efficient watermarking of video signals.

3 Design Guidelines

A transparent and robust watermark should be such that the watermark is present and detectable in every frame of the video sequence. Moreover, the watermarked video frames should also have high fidelity so that the watermark is not discernable to the human eye. These requirements need to be satisfied along with the collusion resistance, compression tolerance, noise and synchronization immunity. These requirements and constraints can be met through the design guideline that is intuitively developed as follows. First, to tolerate compression, the transform coefficients of the watermarked signal should be large enough to survive the quantization. In block-based compression, quantization is applied on each coefficient of the block to a different degree. These coefficients are almost decorrelated. If the watermark is applied in the spatial domain and is an independent broadband signal, then, it shall lie in the mid to high range AC coefficients where the high quantization level can obliterate these low amplitude watermarks. Moreover, a watermark that is statistically similar to the block it is embedded in and also exhibits high pair wise correlation with the target block in the reference frame resists both intra and inter-frame compression [36]. Thus, a compression robust watermark requires that the watermark signal be a narrow band signal with coefficients that are large enough to withstand quantization but low enough not induce visible distortions. The embedding process or the watermark should be such that the watermark coefficients are correlated to one another and to the host signal so that they map towards the DC coefficients in the DCT domain. Transforms that have poor energy compaction like Hartley transform can be employed for data hiding [34]. To be immune to noise, it is required that different watermark coefficients are correlated and the watermark is able to withstand a variance reducing integrate and dump filter type operations [35]. Multi-frame collusion attacks manifests in two forms [36]. Type I collusion arises when a number of uncorrelated frames of a video sequence are marked with highly correlated watermarks. This case arises when a fixed watermark is used to watermark all the frames and that watermark is independent of the video sequence. Type II collusion arises when large number of statistically similar frames of a video sequence is marked with linear combination of uncorrelated independent watermarks [15]. This happens when a noise like pattern in embedded in the video. Type I Collusion can be evaded if the watermark sequence is independent in each frame because the linear frame combining will not enhance the watermark. The requirements for Type II collusion are opposite to that of Collusion Type I. If the same watermark is embedded, collusion will fail and any averaging would fail to produce a marked free copy. Resistance to collusion amounts to statistical invisibility. A video sequence that is watermarked exhibits statistical invisibility if and only if the correlation coefficient between the host frames is equal to that of the corresponding watermarked frames and the host video is highly correlated to the watermark. The aforesaid observation leads to the following design criteria. The first guideline concerns with the original and water-marked video sequences while the second design guideline ensures correlation between the host and the watermark signals.

The first design guideline states that there should be statistical invisibility of the watermark. Statistical invisibility is the consequence of the lack of observable differ-ence between the host and the watermarked sequence [14]. This can be ensured if embedding process is such that the perturbation to the host signal results in a signal that differs from the original signal only by a scaling factor i.e the probability density functions (pdfs) of the signals are similar. Similarity of the pdfs is possible when the variance and other higher moments of the two signals are close to one another. The second design principle enshrined in [36] states that that visually similar regions or frames of the video are marked by consistent watermarks. Finally, the video should be divided into scenes that have the same genre the watermark should be placed in the part of a frame that has low difference between adjacent frames.

4   Proposed Technique

4.1   Basic Idea

As in image watermarking via NPT [37], a watermark is embedded in every frame of the video. The region is chosen on the basis of interframe similarity. The watermark effectively resists collusion based watermark estimation attacks and is impervious to noise and lossy compression to a large extent. This immunity is intrinsic to the process of generation and retrieval of the watermark by NPT and the underlying trans-form employed by NPT. To enhance the resistance to collusion based attacks, a non-overlapping block matching is used to find the region for placing the watermark. The iterative process of watermark reconstruction bestows relative immunity to noise and quantization error in the compression process.

4.2   Watermark Embedding with NPT and Non-overlapping Block Matching

The watermark is embedded and extracted via Naturalness Preserving Transform (NPT), an intermediate spatial-frequency domain representation of a signal. NPT, which is the intermediate transform, is utilized for embedding and extraction of the watermark and representation of the video frames. NPT is a linear combination of two or more transforms in which the individual transforms contribute such that the inverse exists. The NPT of an image P of size $2n{\times}2n$ given as [38]:

$$A = \psi(\alpha)P\,\psi(\alpha)$$

Where α $(0 \leq \alpha_i \leq; \sum \alpha_i = 1)$ is a vector of k components. $\psi(\alpha)$ is the hybrid transform operator and is defined as stochastic weights of the set of orthogonal matrices: $\{M_1, M_2 M_3, \ldots \ldots, M_k\}$.

$$A = \{\alpha I + (1-\alpha)M\}\,P\{\alpha I + (1-\alpha)M\}$$

$$A = \alpha^2 P + (1-\alpha)^2 MPM + \alpha(1-\alpha)\,\{PM + MP\}$$

The three different terms in the right side of the above equation are explained as follows. The first term $\alpha^2 P$ shows the weighted original image and is the cause of visual characteristics of the original image. The second term $(1 - \alpha)^2 MPM$ is the weighted version of orthogonal transform. The third term $\alpha(1 - \alpha)\{PM + MP\}$ is a function of original image and orthogonal transform matrix. This term introduces redundancy and can be employed for reconstruction of the missing portion of an image. In an image watermarked via NPT, the first term preserves the visual imperceptibility of the image while the third term enhances robustness by aiding in the recovery of the watermark in an attacked watermarked image. Each video frame is watermarked separately via the NPTation. Except the first frame, the regions of the subsequent frames to be watermarked are a function of correlation between the frames. Initially, a watermark image that is very small as compared to the frame dimension is inserted after deletion of that portion. The NPT transformed frame is formed by the addition of a large portion of the modified frame and a very small portion from its Discrete Hartley transform (DHT). The transformed image contains both the original spatial characteristics as well as the Hartley domain components of the image and NPT merely redistributes the watermark image into the original video frame. For large values of α, NPT is able to invisibly code an image into the host frame while preserving the original visual appearance of the host. Once the first frame is watermarked as shown in fig. 1, the portion of the subsequent frames where the watermarked is to be insertion prior to NPT is decided on the strength of Interframe correlation. In similar frames, the watermark should be placed in regions that are visually closest. Identical frames should be worked identically. This is achieved by non-overlap block matching to identify the closest regions in the current frame corresponding to the regions in the previous frame where the watermark had been placed. The aim of non-overlap block matching is to allow the placement of the watermark into positions of different frame with similar genre. The procedure is same as conventional block matching with the change that the block is the previous frame that is matches to the block in the current frame does not participate in the block matching process nay further. The current and previous frames divided into 8x8 blocks and the block matched in a limited search region. The watermark broken into 8x8 blocks are placed in the corresponding blocks of the current frame. For example, to determine the portion of the second frame to be replaced by the watermark image, a non-overlapping exhaustive search block matching is performed. The first frame is taken as the reference frame and the second frame as the target frame. Once a block in the reference frame matches with the block in the target frame, the block in the reference frame is excluded from the matching process (non overlapping search). MSE is used as the search criterion to maximize correlation between the blocks. This allows the identification of static scenes in the video that are also highly correlated. After the identification of the portion, watermark is embedded at the intra frame level. This is repeated for subsequent frames till Interframe correlation falls below a certain threshold indicating a new group of frames.

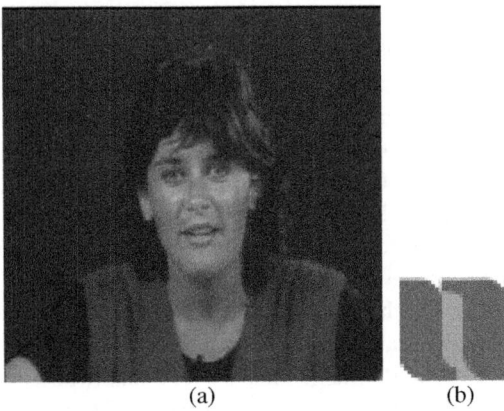

(a)   (b)

Fig. 1. (a) Video Frame (first) to be watermarked; (b) watermark image

The process of embedding the watermark in a video sequence proceeds through the following steps.

1. To embed a watermark image W into a video, the first frame P of the host video V is considered. The size of the watermark W is very small (32×32 in the present case) as compared to the frame size (256×256 in the present case). The watermark image W is substituted at the place where the correlation of W and the part of P is high. The new image P_{append} is transformed via the NPT and generates an image called Q_{append}. The final marked frame is the transformed frame Q_{append} after replacing the watermark image portion by the equivalent parts from the original host frame. To smooth the sharp edges that result because of this substitution, the replacement is made by the NPT of the original frame with very low value of α (0.98 & 0.99).

2. To embed the watermark into the second and subsequent frames, the original first frame and the second frame are divided into 8×8 sub macro blocks and non-overlapping block matching is performed between them using the mean square criterion. In non-overlapping block match, block in the reference frame that has been matched with a block in the second frame does not participate in further matching process. The new P_{append} frame for the second frame is formed after placing the relevant portions of W at the matched places. Step 1 is then repeated to obtain the watermarked second frame. The same procedure is repeated for subsequent frames.

4.3 Watermark Extraction

The Hartley transform component ensures that every point of the final image is a function of the entire image field and contains information about the entire signal. For a range of α below unity, the image displays a high degree of fidelity but there is an imperceptible Hartley component. These two properties allow for a unique approach to reconstruction of an image from an image embedded in noise. If a section is excised, then the total image I can be recovered by positing the contents of the excised area by replacing in the missing region with a suitable candidate region. The reconstruction works by minimization of texture noise. The algorithm adjusts the noisy, or posits the missing pixels so that the variance of the texture measures of the pixel values in the known texture region is minimized by the inverse transform process. The extraction of the watermark requires the original video sequence to supply the candidate region. Once, the region for the placement has been identified by non-overlap block matching, the watermark is embedded independently in each frame.

The reconstruction of the watermark is performed independently for each frame. The iterative process is identical to the procedure illustrated in [37] and proceeds as follows. In the watermarked frame, the portion where the watermark was embedded is set to zero. The corresponding portion of the original frame is also excised. The iterative process starts with the inverse NPT of the watermarked frame. The coefficients generated in deleted portion of the frame as substituted in the original frame and forward NPT is done. The coefficients generated in the excised portion are then used to replace the excised portion of the watermarked image. The process of forward and inverse NPT is repeated till the watermark is regenerated in the deleted portion of the frame. The naturalness-preserving transform is a convex function and the restoration method essentially trades a priori information for incomplete information. A solution, defined as a vector left invariant under a suitable operation, is sought by successive iterations of the suitable operation. The convex nature of the range space ensures convergence.

5   Results and Analysis

To verify the efficacy of the proposed technique, test video sequence of Miss_America video was watermarked via the NPT with non-overlap watermarking. The embedding was performed as follows.

1. The first frame F of Miss_America sequence is considered. A watermark image, IIITM logo, W whose dimensions are small as compared to the frame size is chosen (W=32×32 for F= 256×256 in the present case) and shown in fig. 1. W should be placed in the region of maximum correlation but has been placed arbitrarily for testing in F to form F_1 as shown fig. 2.

2. The NPT F_1 of F_1 is obtained by taking forward NPT with $\alpha=0.93$ as shown in fig. 3.

3. The portion of the watermark W is excised from F_1 and coefficients of that image portion are forced to zero as shown in fig. 4.

The equivalent portion from the original frame replaces the excised image portion. This generates sharp edges. To eschew this, NPT F is taken with α close to unity ($\alpha = 0.98$ in over case) to generate F'. The excised image portion is then replaced by corresponding portion of F' to form the final watermarked frame R as illustrated in fig.4. The extraction of the watermark from the watermarked signal is performed at the frame level.

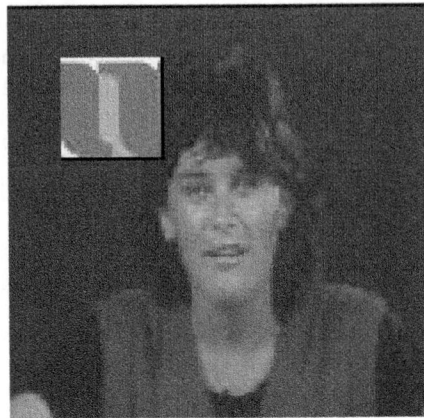

Fig. 2. Watermark Image *S* superimposed over original image after Forward NPT

Fig. 3. Image with watermark portion excised

Fig. 4. Final watermarked Image

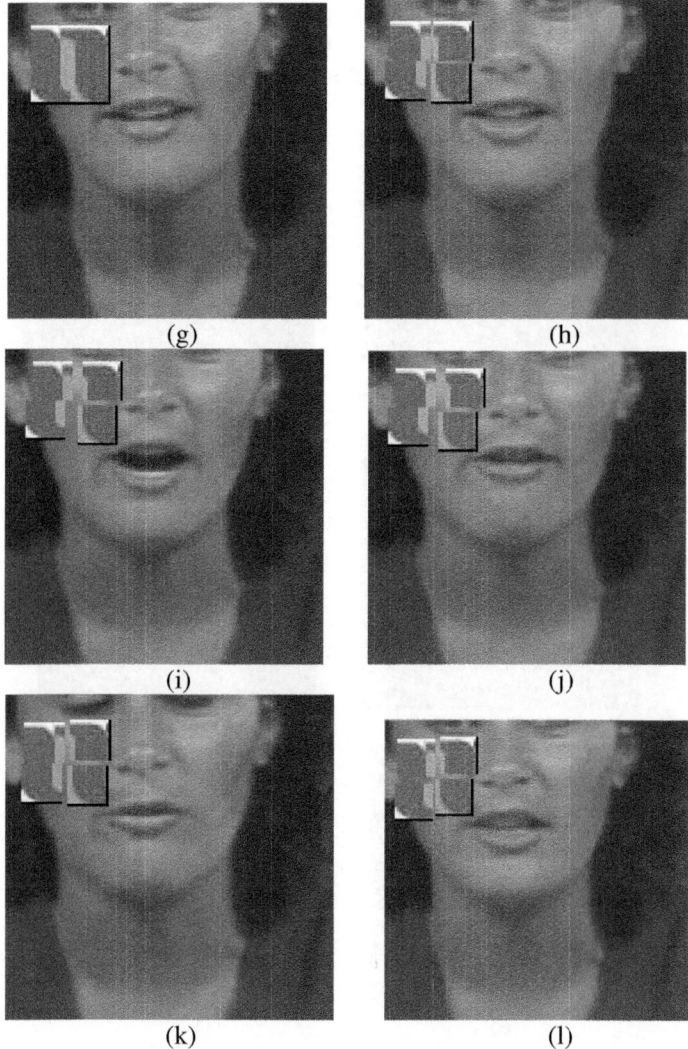

Fig. 5. Optimal Placement of Watermark (size 32×32 pixels) in video frames (Size 256×256 pixel) of video clip of Miss_ America (g) Frame-1; (h)Frame-2; (i) Frame-3; (j) Frame-4; (k) Frame-5; (l) Frame-6;

Different stages of the watermark embedding in various video frames shown in fig. 5. Various stages of extraction process from a video frame are shown in Fig. 6. It is observed that the watermark is restored after a large number (30-40) of iterations. Table1 shows the PSNR values for original and watermarked video frame (first frame of Miss_America) and the Lena image. The watermarked image is subject to noise of different variance, different degree of cropping, collusion I and II attacks by taking average of frames for watermark obliteration and difference of frames for extraction of the watermark, replacement of a frame by replacing it by its average and per frame lossy compression with JPEG. The watermark was then recovered by employing the reverse of the watermarking process. The effect of different attacks on the watermarks was assessed as described in the following sections.

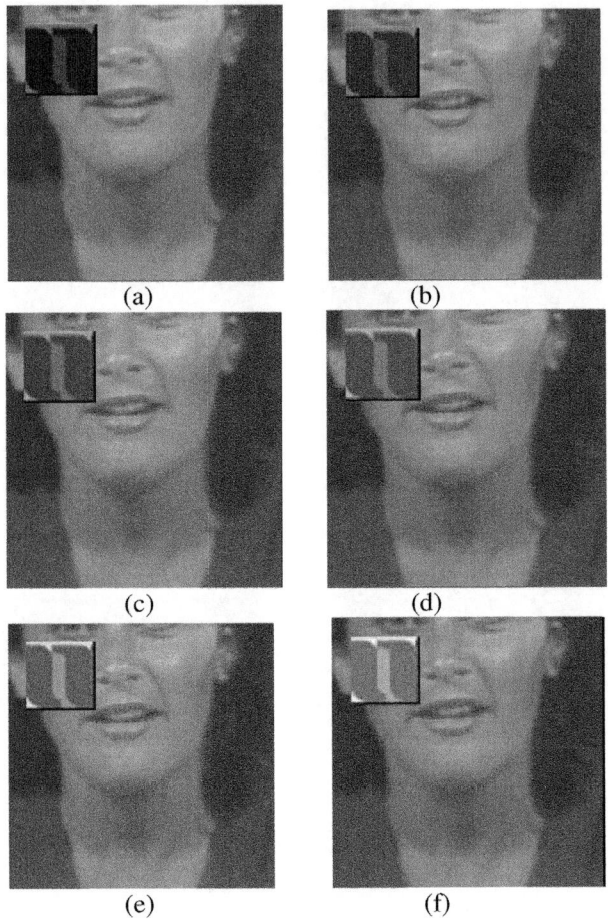

Fig. 6. Recovery of Watermark from Watermarked Video Frame-1with number of iterations as (a) 10; (b) 20; (c) 30; (d) 40; (e) 50; (f) 60

Table 1. PSNR Values of the original and watermarked image without attacks

Original Images	Lena		Miss America	
	$\alpha=0.93$	$\alpha=0.90$	$\alpha=0.93$	$\alpha=0.90$
Original Image & Watermarked Image	41.83	39.75	37.911	36.96
Original Watermark & Extracted Watermark	33.93	36.85	32.38	33.37

5.1 Collision Attacks and Frame Dropping

In this section, an approximate analysis is done that indicates robustness of the NPT based watermarking against estimation attacks like Collusion I & II and to incidental attacks like lossy compression and noise. The analysis assumes that the system under intentional or unintentional attacks can be characterized by second order statistics. The watermark estimation attacks are point estimates that require the first two moments of the process. It is further assumed that the system affected by noise or quantization can be modeled or linear system with additive zero mean Gaussian noise.

The watermarking process essentially perturbs the original signal by adding another signal that is function of the host signal. Let the original video sequence be denoted by $V(k)$, where k is the frame number. The final watermarked signal $X(k)$ that has been watermarked via NPT can be represented as

$$X(k) = \alpha V(k) + (1 - \alpha)W(k)$$

where, $W(k)$ is the perturbation as generated by NPT and post processing. It is known that $W(k)$ can be characterized by a generalized Gaussian Distribution [33] for $\alpha > 0.7$. The value of α in the watermarking process ranges from 0.9 to 0.95 to ensure sufficient frame quality and avoid unnecessary degradation of output video sequence.

Given a watermarked video sequence $X(k)$, $k = 1,2, \dots, i, \dots, n$ the linear summation of the different frames is

$$\sum_{k=1}^{n} X(k) = \alpha \sum_{k=1}^{n} v(k) + (1 - \alpha) \sum_{k=1}^{n} \omega(k)$$

We assume that the perturbation signal has Gaussian distribution with zero mean and a variance of σ^2. For this sequence of these random variables, the variance of the sum, S is

$$Var(S) = \sum_{i=1}^{n} \alpha^2 var(x_i) + \sum_{j=1}^{n}\sum_{j=1}^{n} \alpha_i \alpha_j cov(X_i X_j) \quad i \neq j$$

$Var(S) = \sum_{i=1}^{n} \alpha^2 var(x_i)$ & that of the average is

$$Var\left(\sum_{i=1}^{n} \frac{X_i}{n}\right) = \frac{\alpha^2}{n} + \frac{n-1}{n} \rho\alpha^2$$

$$\text{When } \rho = 0 \; Var\left(\sum_{i=1}^{n} \frac{X_i}{n}\right) = \frac{\alpha^2}{n}$$

Depending on the interdependence between $X(k)$, $V(k)$ and $X(k)$, following cases may arise.

i. $V(.)$ are correlated to one another while $W(.)$ are independent of one another and of the $V(.)$. This case arises when each frame of visually similar sequence is watermarked by a separate watermark that is independent of the frame. (Collusion Type I). The correlation ρ is

$$\rho(W_i, W_{i+j}) \to 0; \rho(V_i, V_{i+j}) \to 1; \rho(W_i, V_j) \to 0$$

ii. $V(.)$ are correlated to one another while $W(.)$ are independent of $V(.)$ but correlated to one another. This case arises when each frame of visually similar sequence is watermarked by a sequence of similar watermarks or a fixed watermark that is independent of the frame.

$$\rho(W_i, W_{i+j}) \to 1; \rho(V_i, V_{i+j}) \to 1; \rho(W_i, V_j) \to 0$$

iii. $V(.)$ and $W(.)$ are neither correlated to one another or among themselves. This happens when a sequence of independent frames are watermark by independent watermarks.

$$\rho(W_i, W_{i+j}) \to 0; \rho(V_i, V_{i+j}) \to 0; \rho(W_i, V_j) \to 0$$

iv. $V(.)$ are independent but the $W(.)$ are correlated. This happens when a group of independent pictures is watermarked by a visually similar watermark sequence. (Collusion Type II)

$$\rho(V_i, V_{i+j}) \to 0; \rho(W_i, W_{i+j}) \to 1; \rho(W_i, V_j) \to 0$$

v. $V(.)$ and $W(.)$ are correlated to one another and each $V(i)$ and $W(i)$ also exhibit pair wise correlation. This is the case of content dependent watermarks.

$$\rho(V_i, V_{i+j}) \sim \rho(W_i, V_j); \rho(V_i, W_j) \to 1$$

In Case (i) frame averaging, frame replacement by average of two similar (contiguous) frames may average out the watermark. Since, the watermark signals are uncorrelated, the variance of the sum reduces by factor n (number of frames). Thus, collusion I attack behaves as, integrates and dump filter and obliterates the watermark defeating the watermarking process. When uncorrelated frames are watermark by a fixed watermark, case (ii), an attack based on difference between the frames is employed to estimate the watermark. The estimated watermark can be used to delete the watermark from the video sequence. When a sequence of independent images are watermarked independently, collusion based attacks are not relevant (case iii). It is a case of watermarking images rather than video. Case (iv) constitutes a Collusion II attack; a difference operation would be similar to a watermark estimation attack, in which the attack shall reveal the watermark to the attacker. The watermark can then be erased from the video sequence. If the watermark is statistically similar to the host signal then attacks based on second order statistics would not estimate or erase the order.

In the proposed watermarking technique, the watermark information is embedded in all the coefficients of the underlying frame through the Hartley transform. Thus, the perturbation in the watermarked signal becomes a function of the host signal. In words, NPT renders the watermark dependent on the contents of the host image apart from the image used for creation of the perturbation. Though the NPT coefficients are characterized approximately by Generalized Gaussian distributions, they tend to other distributions like Gamma distribution. This implies that the first two moments, mean and variance, may not be sufficient to guarantee statistical invisibility to focused statistical estimation attacks. For robustness to attacks, the probability density functions of the host and the watermark must be similar. To embed an effective watermark via NPT, the shape of the transformed signal histogram should be modeled accurately to determine the value of α for which the signal histograms exhibit similarity. The test data generated by the watermarking of images like Lena, Baboon and Vegetable were considered in lieu of Miss_America sequence. It was found that as α varies, the shape of the transformed signal histogram varies. For large values of α, $\alpha > 0.92$, the histogram of the watermarked frame exhibits the characteristics of the original image histogram as shown in fig. 7. It can be seen from the fig.7 a & b that as the value of α decreases below 0.90, the histogram exhibits more Hartley histogram characteristics and becomes more peaked and narrow. The results indicate that the content dependence and statistical similarity of the watermarked signal to the original signal results in a watermark that is tightly coupled with the host signal. This makes it robust to estimation or erasure by collusion.

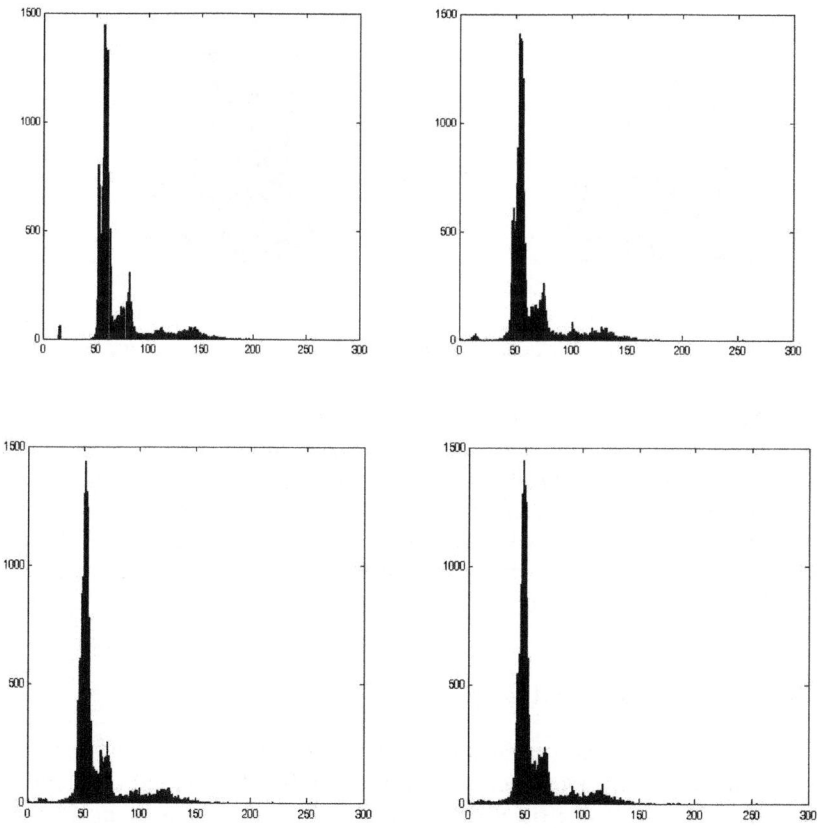

Fig. 7a. Pixel Distribution for Miss_ America - Original and Watermarked images α=0.95, 0.90, 0.85

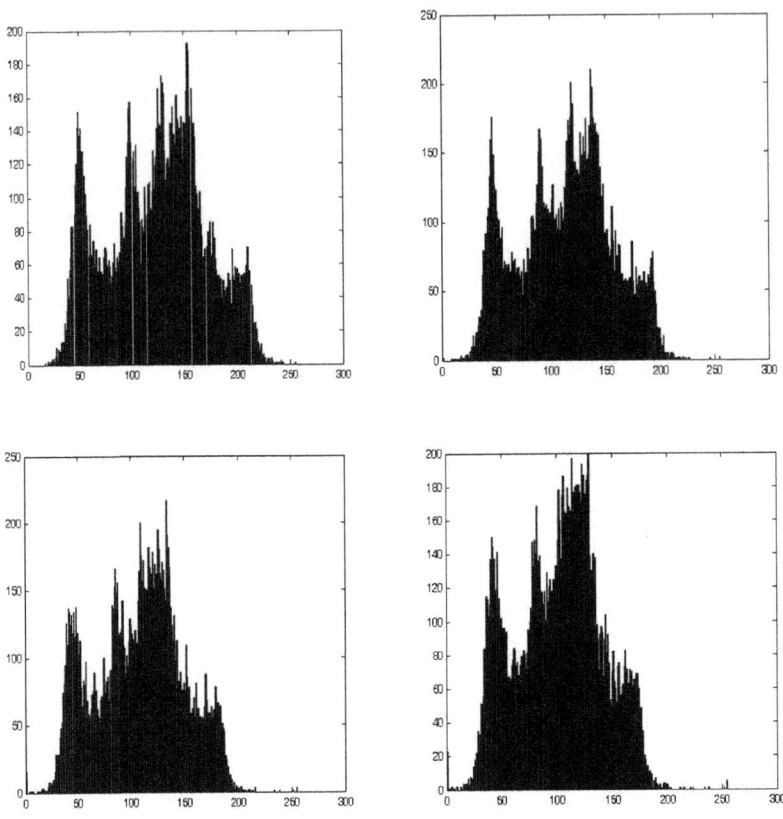

Fig. 7b. Pixel Distribution for Lena - Original and Watermarked images α=0.95, 0.90, 0.85

In the Miss_America test sequence, the following operations were performed to validate the robustness of the technique. First, watermarking was performed on a sequence of 10 frames with an α=0.93. The second, fourth, sixth and ninth frames were dropped and regenerated as an average of the preceding and the succeeding frames. It was observed that the watermark is retained even in the averaged frames. The watermarks extracted from such frames were highly distorted with an average PSNR of around 25. However, the extracted watermarks were distinguishable to the human eye.

5.2 Noise, Compression and Cropping

A video sequence is stored or transmitted in the compressed domain. The process of compression usually involves energy concentration by block based DCT followed by quantization of the DCT coefficients. The Quantization operation incurs a loss and can be characterized as zero mean gaussian noise with a variance σ_q^2. In addition, when a compressed video is transmitted over a channel, it is further corrupted by a zero mean gaussian noise with variance σ_n^2. The Hartley transform is like integration over the entire image. The transform localizes many essential global features of the image in the spectrum. Some of these are the background image intensity, repeating patterns, slow variation in intensity, diffused edges etc. On the contrary, features that are localized in the spatial domain are spread over the spectrum. These are sharp edges, lines and other rapidly varying texture or sharpness of lines and edges are scattered. Any technique that works on the synthesis of the image by working via texture synthesis or restoration shall remove the effect of noise in the process. The inverse NPT transformation process based on Projection on Convex Surfaces is able to achieve this in a precise and extendible fashion. Moreover, the location of the errors is not important in NPT. All the samples are equally protected because all samples are represented in each element of the transformed matrix. The inverse Hartley transform, which is identical to the forward transformation, is integration over the complete frame. It involves pair wise product and addition of the noisy coefficients. The gaussian noise is thus squared and averaged during inverse transformation.

The effect of both noise and compression can be modeled a linear communication system with a square law detector corrupted by zero mean additive white gaussian noise, $N(0, \sigma_T^2)$. Given an observed noisy signal y, hypothesis testing can be formulated for the detection of the signal E as a '0' (null hypothesis H_0) or a '1' (alternate hypothesis H_1). Under H_0, it is assumed that the signal is zero and the contribution is only from noise. Thus, y is the sum of $2M$ independent Gaussian random riables $N(0, K\sigma^2)$ and $y \sim K\sigma^2 \, X_{2M}^2$ where X_{2M}^2 is a (central) chi-squared distribution with $2M$ degrees of freedom [38].

The probability density $W_0(y)$ of y under H_0 is

$$W_0(y) = \frac{1}{K\sigma^2} X_{2M}^2 \left({y}/{K\sigma^2} \right) = \left(\frac{1}{2K\sigma^2} \right)^M \frac{y^{M-1}}{M-1} e^{-{y}/{2K\sigma^2}}$$

Under the alternative hypothesis H_1, the contribution is from the signal plus noise. This generates non-zero mean Gaussian variables. y is the sum of squares of $2M$ independent Gaussian random variables with means $C_{rv}'s$ and $C_{iv}'s$ and therefore is distributed as $y \sim K\sigma^2 \, X_{2M;|E_1|^2}^2$ is a *noncentral* chi-squared distribution with $2M$ degrees of freedom and the noncentrality parameter $|E_1|^2 = \sum_{v=v_1}^{v_2} (c_{rv}^2 + c_{iv}^2)$ [38]. Thus, the probability density of y under H_1 is given by

$$W_1(y) = \frac{1}{K\sigma^2} \sum_{j=0}^{\infty} \frac{|E_1|^2/2 \, (|E_1|^2/2)^j}{j!} \, X_{2M+2j}^2 \left({y}/{K\sigma^2} \right)$$

y is the sum of $2M$ independent random variables (each of them is a squared Gaussian), and therefore, according to the central limit theorem, for large M (> 15), the distribution of y is asymptotically Gaussian. The variance of these gaussian variables are $\sigma_0^2(y)$ and $\sigma_1^2(y)$, indicate that the inverse transformation reduces in the variance by $M^{0.5}$. This makes NPT more impervious to noise & lossy compression as compared to other existing watermarking systems. This is reaffirmed under the test conditions where the watermark could be extracted under mild to severe noise. Gaussian

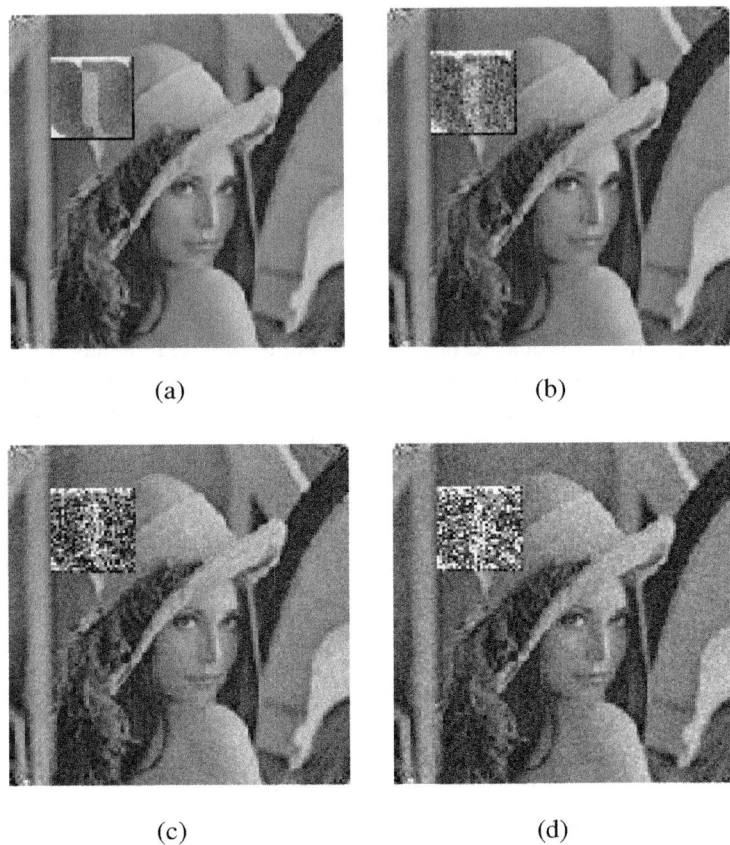

(a)   (b)

(c)   (d)

Fig. 8. Extraction of watermark with addition of Noise in Watermarked Image having value for mean(m) and standard deviation (s): (a) m=0, s=1; (b) m=0, s=10; (c) m=0, s=50; (d) m=0, s=100

noise of different variance (1-100) was added to the watermarked Lena test image as shown in fig.8. The watermark was, then, extracted from the noisy video frames. It was observed that the watermark gets impaired with the increase in the noise. However, the distortion in the extracted watermark is proportional to the distortion in the watermarked image. For large noise, the watermark could not be extracted and was obliterated from the frames. However, the amount of noise required for obliterating the watermark rendered the video useless for consumption because of noise. The watermark was thus found to be robust to additive noise. The progressive impairment is depicted in Fig. 8 for the image under test. The PSNR values in table 2 of the extracted watermarks decrease sharply as the amount of noise in the watermarked image increases. However, there is a corresponding rapid decrease in the Image quality. The watermark is found to be visually discernable even when the image is rendered useless for consumption.

Table 2. PSNR Values for Addition of Gaussian Noise with various standard deviations (SD)

Standard Deviations	SD-1	SD-10	SD-50	SD-100
Original Watermark & Extracted Watermark (α=0.93)	26.39	22.32	15.06	11.86
Original Watermark & Extracted Watermark (α=0.90)	28.31	23.16	16.31	12.01

Similar trends were observed for lossy compression. To test the robustness of the watermark, the watermarked image of Lena was compressed using JPEG compression with different quality (90-50). It was observed that, similar to noise addition, the quality of the extracted watermark declined with the decrease in the quality of JPEG compression as given in table 3. At very low quality of compression as shown in fig.9, the PSNR of the extracted watermark was very low (PSNR 9.22 for 50% quality factor as 50 and $\alpha = 0.93$ and the watermark could not be taken as extracted. However, the decreased quality of compression impaired the image beyond reasonable usage establishing the robustness of the technique against frame level compression. Further, it was found that the watermark is also resistant to cropping. Different degree of cropping from different parts of the watermarked frames was performed. It was observed

Table 3. PSNR Values for various lossy JPEG Compression quality factor index

% Compression	90	80	70	60	50
Original Watermark & Extracted Watermark (α=0.93)	23.09	20.91	15.23	12.90	10.22
Original Watermark & Extracted Watermark (α=0.90)	27.04	25.13	19.34	15.69	11.97

Table 4. PSNR Values for Cropping Attack with various cropped area percentage

Cropped Dimensions (a) Coincide with watermark; (b) No coincide with watermark	(a) 1.25%	(b) 1.25%	(b) 6.25%	(b) 12.25%
Original Watermark & Extracted Watermark (α=0.93)	26.57	26.28	23.30	16.93
Original Watermark & Extracted Watermark (α=0.90)	26.49	25.51	24.85	17.28

Fig. 9. Lossy JPEG Compression Operation with quality factor index are (a) 90; (b) 70; (c) 60

the watermark could be retrieved till about 12% of the frame is cropped out as illustrated in fig.10. The PSNR values of extracted watermark from first video frame for various cropping percentage is given in table 4.

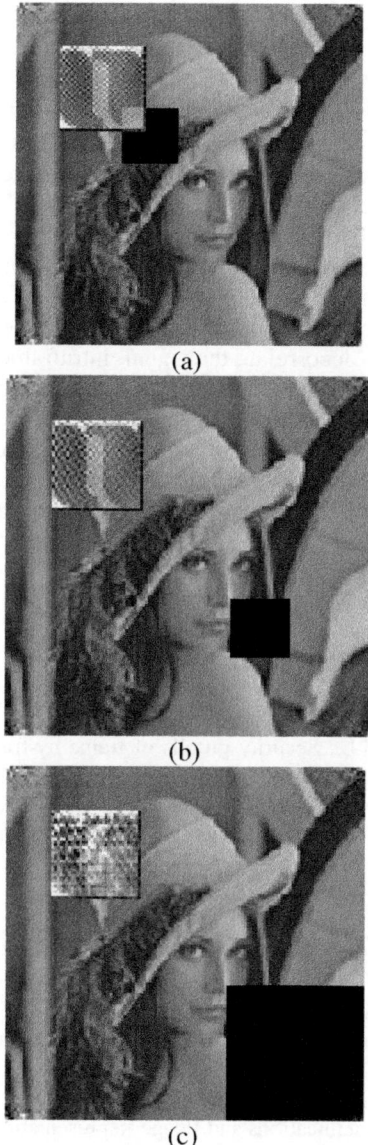

Fig. 10. Cropping Attack (a) Cropped area is 1.25% at the watermarked image coincide with the location of watermark; (b) Cropped area is 1.25% on watermarked image without coinciding with watermark image; (c) Cropped area is 6.25% at the bottom-left corner of the watermarked image

6   Conclusion

The key to design of robust watermarking system is to embed watermarks consistent with the host signal. Watermarking via NPT creates perturbations coherent with the underlying host. The embedded watermark is derived from the host signal and is statistically similar to the host as well as to the watermarked signal. This renders the watermark quite immune to second order statistics based collusion attacks. The vulnerability to compression is a function of strength of the perturbation i.e. α. The value of α governs the recovery of the watermark from watermarked signal distorted by compression, noise etc and also the fidelity of the watermarked signal. Higher values of α ensure recovery at the cost of fidelity and vice versa. The final value is a tradeoff between resistance and fidelity. The texture reduction process of watermark recovery imparts a higher degree of noise immunity as compared to other watermarking process and the watermark is able to sustain substantial amount of noise, lossy compression and distortions like cropping etc. However, since the underlying Hartley transform does not fully decorrelate the signal, intraframe collusion can be devised to estimate or obliterate the watermark. The transform also concentrates the energy of the signal at the corner points distorting the frame at the corners. The watermarking recovery requires the original frame that limits the applicability of the watermarking process. Finally, the robustness of the watermarking technique has not been tested against specialized attacks. The immunity of the watermark to video editing etc. also needs to be investigated.

References

1. Koz, A., Cigla, C., Alatan, A.A.: Watermarking of Free-view Video. IEEE Transactions on Image Processing 19(7), 1785–1797 (2010)
2. Doerr, G., Dugelay, J.L.: Security pitfalls of frame-by-frame approaches to video watermarking. IEEE Transactions on Signal Processing, part 2 52(10), 2955–2964 (2004)
3. Barni, M., Bartolini, F., Caldelli, R., De Rosa, A., Piva, A.: A robust watermarking approach for raw video. In: Proceedings of the 10th International Packet Video Workshop (2000)
4. Hartung, F., Girod, B.: Watermarking of Uncompressed and Compressed Video. Signal Processing 66(3), 283–301 (1998)
5. Langelaar, G., Lagendijk, R.: Optimal differential energy watermarking of DCT encoded images and video. IEEE Transactions on Image Processing 10(1), 148–158 (2001)
6. Lu, C.-S., Chen, J.-R., Liao, H.-Y.M., Fan, K.-C.: Realtime MPEG-2 video watermarking in the VLC domain. In: Proceedings 16th International Conference on Pattern Recognition, vol. 2, pp. 552–555 (2002)
7. Langelaar Gerrit, C., Lagendijk Reginald, L.: Real-time Labeling of MPEG-2 Compressed Video. J. Visual Communications and Image Representation 9(4), 256–270 (1998)
8. Kalker, T., Depovere, G., Haitsma, J., Maes, M.: A video watermarking system for broadcast monitoring. In: Proceeding of SPIE- Security Watermarking Multimedia Contents, vol. 3657, pp. 103–112 (1999)
9. Fridich, J., Goljan, M.: Robust hash functions for digital watermarking. In: Proceeding Conference Information Technology: Coding and Computing, pp. 178–183 (2000)
10. Petitcolas, F., Anderson, R.: Evaluation of Copyright Marking Systems. In: Proceedings IEEE multimedia Systems (ICMCS 1999), Florence, Italy, June 7-11 (1999)
11. Solachidis, V., Pitas, I.: Circularly symmetric watermark embedding in 2-D DFT domain. IEEE Transaction Image Processing 10, 1741–1753 (2001)

12. Duan, F., King, I., Xu, L., Chan, L.: Intra-block algorithm for digital watermarking. In: Proceedings IEEE 14th International Conference on Pattern Recognition (ICPR 1998), vol. 2, pp. 1589–1591 (1998)
13. Reyes, R., Cruz, C., Nakano-Miyatake, M., Pérez-Meana, H.: Digital Video Watermarking in DWT Domain Using Chaotic Mixtures. IEEE Latin America Transactions 8, 304–310 (2010)
14. Pereira, S., Pun, T.: Robust template matching for affine resistant image watermarks. IEEE Transactions on Image Processing 9(6), 1123–1129 (2000)
15. Swanson, M., Zhu, B., Tewfik, A.: Multiresolution Video Watermarking using Perceptual Models and Scene Segmentation. In: Proceedings International Conference on Image Processing (ICIP 1997), Washington, DC, vol. 2, pp. 551–568 (1997)
16. Chan, P.W., Lyu, M.: A DWT-based Digital Video Watermarking Scheme with Error Correcting Code. In: Qing, S., Gollmann, D., Zhou, J. (eds.) ICICS 2003. LNCS, vol. 2836, pp. 202–213. Springer, Heidelberg (2003)
17. Voloshynovskiy, S., Periera, S., Pun, T., Eggers, J., Su, J.: Attacks on digital watermarks: Classification, estimation-based attacks, and benchmarking. IEEE Communication Magazine 39, 118–126 (2001)
18. Pireira, S., Pun, T.: Robust template matching for affine resistant image watermarks. IEEE Transaction on Image Processing 9, 1123–1129 (2000)
19. Holliman, M., Macy, W., Yeung, M.: Robust frame-dependent video watermarking. In: Proceedings of the SPIE Security and Watermarking of Multimedia Contents, San Jose, January 24-26, vol. 3971, pp. 186–197 (2000)
20. Verma, S., Chandra, R., Tomar, G.S., Kim, J.: Video watermark embedding in the compressed domain. In: IET International Conference on Visual Information Engineering, VIE 2006, pp. 53–57 (2006)
21. Kalker, T., Depovere, G., Haitsma, J., Maes, M.: A video watermarking system for broadcast monitoring. In: Proceeding of SPIE, vol. 3657, pp. 103–112 (1999)
22. Mobasseri, B.G.: Exploring CDMA for watermarking of digital video. In: Proceeding of SPIE, vol. 3657, pp. 96–102 (1999)
23. Swanson, M.D., Zhu, B., Tewfik, A.H.: Multiresolution Scene-Based Video Watermarking Using Perceptual Models. IEEE Journal on Selected Areas in Communications 16(4), 540–550 (1998)
24. Darmstaedtev, V., Delaigle, J.F., Nicholson, D., Macq, B.: A block based watermarking technique for MPEG-2 Signals: Optimization and validation on real TV distribution links. In: Proceeding of European Guf Multimedia Applications, Services and Techniques EC MAST, Berlm, Germany (1998)
25. Fridrich, J.: Robust bit extraction from images. In: Proceeding of IEEE International Conference on Multimedia Computing and Systems, vol. 2, pp. 536–540 (1999)
26. Fridrich, J.: Visual hash for oblivious watermarking. In: Proceeding of SPIE, Security and Watermarking of Multimedia Contents III, vol. 4314, pp. 286–294 (2000)
27. Kundur, D., Hatzinakos, D.: Attack characterization for effective watermarking. In: Proceeding IEEE International Conference on Image Processing, vol. 4, pp. 240–244 (1999)
28. Fei, C., Kundur, D., Kwong, R.: Analysis and Design of Secure Watermark-based Authentication Systems. IEEE Transactions on Information Forensics and Security 1(1), 43–55 (2006)
29. Su, K., Kundur, D., Hatzinakos, D.: Spatially Localized Image-Dependent Watermarking for Statistical Invisibility and Collusion Resistance. IEEE Transaction on Multimedia 7(1), 52–66 (2005)

30. Zhang, J., Li, J., Zhang, L.: Video watermark technique in Motion Vector. In: Proceeding of IEEE 14th Brazilian Symposium on Computer Graphics and Image Processing, Brazil, pp. 535–540 (2001)
31. Biswas, S., Das, S.R., Petriu, E.M.: An adaptive compressed. MPEG-2 video watermarking scheme. IEEE Transactions on. Instrumentation and Measurement 54(5), 1853–1861 (2005)
32. Mobasseri, B.G., Berger, R.J.: A foundation for watermarking in compressed domain. IEEE Signal Processing Letters 12(5), 399–402 (2005)
33. Ahmed, A.M., Day, D.D.: Applications of the naturalness preserving transform to image watermarking and data hiding. Digital Signal Processing 14, 531–549 (2004)
34. Kundur Fei, D., Kwong, R.: Analysis and Design of Watermarking Algorithms for Improved Resistance to Compression. IEEE Transactions on Image Processing 13(2), 126–144 (2004)
35. Smith, J.R., Camiskey, B.O.: Modulation and information hiding in images. In: Anderson, R. (ed.) IH 1996. LNCS, vol. 1174, pp. 207–226. Springer, Heidelberg (1996)
36. Su, K., Kundur, D., Hatzinakos, D.: Statistical invisibility for collusion-resistant digital video watermarking. IEEE Transaction on Multimedia 7(1), 52–60 (2005)
37. Ji-Hua, G., Shang, Y., Bei-jing, H.: An NPT Watermarking Algorithm Using Wavelet Image Combination. In: Third International Symposium on Intelligent Information Technology and Security Informatics (IITSI), pp. 378–438 (2010)
38. Biyari, K.H., Lindsey, W.C.: Statistical distribution of hermitian quadratic forms in complex Gaussian variables. IEEE Trans. Inform. Theory 39(3), 1076–1082 (1993)

Author Index